Trends in Plant Tissue Culture and Biotechnology

Editor
L.K. PAREEK
Laboratory of Plant Tissue Culture
and Experimental Morphogenesis
Department of Botany
Rajasthan University, Jaipur 302 001 (India)

Associate Editor
DR. P. L. SWARNKAR
Department of Botany
Rajasthan University
Jaipur 302 001 (India)

AGROBIOS (INDIA)

Published by:
Agrobios (India)
Agro House, Behind Nasrani Cinema
Chopasani Road, Jodhpur 342 002
Phone: 91-0291-2642319, Fax: 2643993
E. Mail: agrobiosindia@gmail.com
www.agrobiosindia.com

AGROBIOS (INDIA)

First Edition 1998
Reprinted 1999
Reprinted 2003
Reprinted 2004
Reprinted 2023

Published by: Dr. Updesh Purohit for Agrobios (India), Jodhpur
Laser typeset at: Yashee Computers, Jodhpur
Cover Design by: Reena
Printed at: Chopra Printing Press. Jodhpur

Preface

Plant Cell and tissue Culture has become major tool in the study of an increasing number of fundamental and applied programmes in the Plant science. Its increasing use to investigate Cell and Developmental Biology, Biochemistry, Physiology, Genetics and Molecular Biology is providing new knowledge about fundamental characteristic of plants. Furthermore, it has become an integral part of *Plant Biotechnology* research as scientists in various universities, research institutions and number of private companies pursue the development of improved plants for agriculture, horticulture and forestry.

The National symposium on recent advances in Plant Tissue Culture and Biotechnology was held in December 1994 at the University of Rajasthan, Jaipur to provide a comprehensive national forum for discussion of the latest developments in this rapidly growing field. More than 300 participants attending the symposium from various universities, research institutes and companies from all over the country provided clear evidence that research in this field has expanded rapidly in the last about two decades. a total of 70 scientific papers and more than 150 posters covering virtually all aspects of current research in the field were presented. The number of developments reported in these presentations and the chapters in this volume clearly indicate that the technology is moving forward on a broad front. O particular significance has been the development of gene transfer technology and the rapidly expanding investigation of the molecular biology of plants in the recent years.

The content of this volume reflects the organization of the symposium which was divided into broad subjects areas. These subjects were discussed at length in concurrent sessions. The chapters in the volume were written by leading experts in the field. Some of the chapters will provide the reader with a broad review of recent developments and progress in the field of tissue culture. It is hoped that the wide range of authoritative reviews and views presented by the authors in this volume will provide a useful and exciting stimulus to researchers and students in the field.

The generous financial support given to the symposium by University Grants Commission of India, Department of Science, and Technology, New Delhi, Department of Biotechnology and Council of Scientific and Industrial Research, New Delhi, ad well as Department of Science and Technology Govt. of Rajasthan, is gratefully acknowledged. This support was especially instrumental in assisting many students and scientists to attend the symposium.

The enthusiastic participation and the devoted and detailed attention to organizing the symposium by many faculty members, post-graduate students and post-doctoral fellows in the Department of Botany, University of Rajasthan, are gratefully acknowledged. Without their generous help the symposium would not have been a success. In the last, I would like to express my gratefulness to Dr. C. R. Bhatia, the then Secretary, Department of Biotechnology, Government of India, New Delhi, Professor P. R. Shewry, Director Long

Ashton Research Station, Bristol, U. K. and Professor S. C. Maheshwari Department of Botany and Plant Molecular Biology, University of Delhi who very kindly agreed to participate in this symposium and contributed their valuable research findings.

I also take this opportunityto express my deep sense of gratitude and profound regards to my teacher Professor Naresh Chandra whose valuable guidance imparted me the basic training of plant tissue culture in early seventies and this inspired me to organize this symposium on the occasion of his 60th birth day, and to publish the present volume.

L. K. Pareek

Contributors

1. **AGNIHOTRI, ABHA,**
Agriculture and Agri-Food
Canada Research Centre,
107 Science Place, Saskatoon,
Saskatchewan S7N OX2,
Canada, & Biotechnology Division,
Tata Energy Research Institute,
Darbari Seth Block, Habitat Place,
Lodi Road,
New Delhi – 110003 (India)
2. **AGRAWAL, ANURADHA**
Department of Botany,
University of Delhi,
Delhi – 110002 (India)
3. **AGRAWAL, VEENA**
Department of Botany,
University of Delhi,
Delhi – 110007 (India)
4. **ANANTANANDTHIRTH, M.A.**
Divya Jyot Ayurvedic Research
Foundation,
Vehelal, Ta. Daskol,
Dist. Ahmedabad - 382 325 (India)
5. **ARORA, F.S.**
Department of Landscaping and
Floriculture, Punjab Agricultural
University,
Ludhianna - 141004, Punjab, India.
6. **ARYA, INDER DEV**
Department of Physiological Botany,
University of Uppsala, Box 540,
751,27, Uppsala, Sweden &
Tropical Forest Research Institute,
Jabalpur - 482 021. (India)
7. **ARYA, SARITA**
Department of Physiological Botany,
University of Uppsala, Box 540,
75L21,, Uppsala, Sweden & Tropical
Forest Research Institute,
Jabalpur - 482 O21(India)
8. **AUGUSTIN, ALICE C.**
Laboratory of Applied Biology,
St. Aloysius College,
Manglore - 575 003 (India)
9. **BALAKRISHNAMURTHY, G.**
Tissue Culture Unit,
Department of Pomology,
Horticultural College and Research
Institute,
Tamil Nadu Agricultural University,
Coimbatore - 641 003 (India)
10. **BAPAT, V. A.**
Biotechnology Division,
Bhabha Atomic Research Centre,
Trombay
Bombay - 400 085 (India)
11. **BARCELO, P.**
ICAR-Rothamsted, Harpenden,
Herts ALS 2]Q.
12. **BARRO, F.**
ICAR-Rothamsted, Harpenden,
Herts ALS 2fQ
13. **BEHERA, PRASANTA, K.**
Plant Physiology Laboratory,
Departments of Botany,
Berhampur University,
Berhampur - 760 O07 (India)
14. **BHATIA, C. R.**
Bhabha Atomic Research Centre,
Bombay-100 085 (India)
15. **BHATLA, S. C.**
Department of Botany,
University of Delhi,
Delhi - 110 007 (India)
16. **BRAGANZA, GRACE**
Dept. of Botany,
Gujarat University,
Ahmedabad – 380009 (India)
17. **CHANDRA, N.**
Department of Botany,
University of Rajasthan,
Jaipur - 302 004 (India)

18. **D'SOUZA, L.**
Laboratory of Applied Biology,
St. Aloysius College
Mangalore - 575 003 (India)

19. **D'SOUZA, L.**
Laboratory of Applied Biology,
St. Aloysius College,
Mangalore - 575 003 (India)

20. **D'SILVA, ICY**
Laboratory of Applied Biology,
St. Aloysius College,
Mangalore 575 003 (India)
Department of Botany,
University of Toronto, Toronto,
Canada

21. **DESWAL, RENU**
School of Life Sciences,
Jawaharlal Nehru University,
New Delhi - 1.10 067 (India)

22. **DOWNEY, R. K.**
Agriculture and Agri-Food
Canada Research Centre,
107 Science Place, Saskatoon,
Saskatchewan S7N OX2, Canada

23. **GANAPATHI, T. R.**
Biotechnology Division,
Bhabha Atomic Research Centre,
Trombay,
Bombay - 400 085 (India)

24. **GIII, R.I.S.**
Forestry Department,
Punjab Agricultural University,
Ludhiana - 1.41 004 (India)

25. **GILL, S. S.**
Forestry Department,
Punjab Agricultural University,
Ludhiana - 1.4't 004 (India

26. **GOSAL, S. S.**
Biotechnology Centre,
Punjab Agricultural University,
Ludhiana - 1,41, OO4 (India)

27. **GREWAL, H. S.**
Executive Engineer (Hort.),
Division Punjab Agricultural
University,
Ludhiana - 141. 004 Punjab (India)

28. **GUPTA, SHRISH C.,**
Department of Botany,
University of Delhi,
Delhi - 110 oo7 (India)

29. **GUPTA, V. K.**
Department of Biochemistry,
Kurukshetra University,
Kurukshetra - 1,32 11.9 (India)

30. **HALFORD, N. G.**
ICAR-Long Ashton Research Station,
Department of Agricultural Sciences,
University of Bristol, Long Ashton,
Bristol BSIBgAF

31. **HEBLE, M. R.**
The Kelkar Education Trust,
Mulund East,
Bombay - 4oO oB7 (India)

32. **HEGDE, SMITHA**
Laboratory of Applied Biology,
St. Aloysius College,
Manglore - 575 003 (India)

33. **FAMBHULKAR, S. J.**
Bhabha Atomic Research Center,
Bombay - 400 OBs (India)

34. **FOSEPH, E. F.**
Department of Agricultural
Biotechnology,
Timiriazev's Agricultural Academy of
Moscow,

35. **TIMIRIAZEVSKAYA ULITSA,**
Moscow - 727 500 (Russia)

36. **KALRA GEETIKA**
Department of Botany,
University of Delhi,
Delhi - 770 007 (India)

37. **KAPOOR, SANGEETA**
Department of Botany,
University of Delhi,
Delhi - 1.1.0 oo7 (India)

38. **KAUL, M. L. H.**
Botany Department,
Kurukshetra University,
Kurukshetra - 732 119 (India)

39. **KAUR, FASWINDER**
Department of Botanical Sciences
Guru Nanak Dev University,
Amritsar - 143 005 (India)

40. **KAVITHA, HEGDE**
Laboratory of Applied Biology,
St. Aloysius College,
Mangalore - 575 003 (India)

41. **KOHLI, U. K.**
Department of Vegetable Crops,
Dr. Y. S. Parmar University of
Horticulture and Forestry,
Solan - 773 23O (India)

42. **KRISHNAMURTHY, K. V.**
Plant Tissue Culture Division,
National Chemical Laboratory,
Pune - 411 008 (India)

43. **KULKARNI, R.**
Laboratory of Applied Biology,
St. Aloysius College,
Mangalore - 575 003 (India)
44. **KUMAR, A. VANDANA**
G.B. Pant University of Agriculture & Technology,
Hill Campus, Ranichauri
Distt. Tehri Garhwal - 249 7gg (India)
45. **KUMAR, ASHOK**
Department of Vegetable Crops,
Dr. Y. S. Parmar University of Horticulture and Forestry,
Solan - 773 230 (India)
46. **KUMAR, ATUL**
G.B. Pant University of Agriculture & Technology,
Hill Campus, Ranichauri,
Distt. Tehri Garhwal - 249 t}g (India)
47. **KUMAR, MEENA**
Botany Department,
University School of Sciences,
Gujarat University,
Ahmedabad - 380 009 (India)
48. **LAZZERI,P,**
ICAR-Rothamsted,
Harpenden, Herts AL5 2]Q, UK
49. **LIU, R. |ANG**
Genetic Engineering Research Institute,
Kist Box 17, Taejon, Korea
50. **MALIK, C. P.**
Department of Botany,
Punjab Agricultural University,
Ludhiana - 147 004 (India)
51. **MATHUR, SHASHI**
Department of Botany,
University of Rajasthan,
Jaipur - 302 004 (India)
52. **MOHAN RAM, H. Y.**
Department of Botany,
University of Delhi,
Delhi - 110002 (India)
53. **NAUTIYAL, M. C.**
G.B. Pant University of Agriculture & Technology,
Hill Campus, Ranichauri,
Distt. Tehri Garhwal - 249 199 (India)
54. **NIGAM, NEENA**
Vincent, Grace Braganza Research Laboratory
for Biological Sciences, LCRD,
St. Xavier's College,
Ahmedabad - 380 009 (India)
55. **NIKIFOROVA, L D.**
Department of Agricultural Biotechnology,
Timiriazev's Agricultural Academy of Moscow
56. **TIMIRIAZEVSKAYA ULITSA,**
Moscow - 727 5OO (Russia)
57. **NIRMALA, C. N.**
Botany Department,
Kurukshetra University,
Kurukshetra - 1.32 1.L9 (India)
58. **PAREEK, L. K.**
Department of Botany,
University of Rajasthan,
Jaipur - 302 OO4 (India)
59. **RANI FANSI**
Tissue Culture Unit,
Department of Pomology,
Horticultural College & Research Institute,
Tamil Nadu Agricultural University,
Coimbatore - 641 003 (India)
60. **RANI, JANSI, G. BALAKRISHNAMURTHY, G.**
Tissue Culture Unit.,
Department of Pomology,
Horticultural College and Research Institute,
Tamil Nadu Agricultural University,
Coimbatore - 641 003 (India)
61. **RAO, MURALIDHARA, M.**
Department of Microbiology & Cell Biology
Indian Institute of Science,
Bangalore - 560 012 (India)
62. **RAO, P. S.**
Biotechnology Division,
Bhabha Atomic Research Centre,
Trombay,
Bombay,- 400 085 (India)
63. **ROIA, G.**
Plant Biotechnology Section,
Bhabha Atomic Research Centre,
Trombay
Bombay - 400 085 (India)
64. **ROUT, GYANARANIAN**
Regional Plant Resource Centre,
Bhubaneswar - 75'1.012 (India)
65. **ROY, SUCHISMITA**
Department of Botany,
Gauhati University
Guwahati - 781. O1.4 Assam (India)

66. **SAGARE, A. P.**
Plant Tissue Culture Division
National Chemical Laboratory,
Pune - 411 008 (India)

67. **SAMANTARAY, PUSPITA**
Pant Physiology Laboratory,
Department of Botany,
Berhampur University,
Berhampur - 760 OO7 (India)

68. **SARMA, C, M,**
Department of Botany,
Gauhati University,
Guwahati - 78101.4 Assam (India)

69. **SAXENA,, O, P.**
Dept. of Botany,
Gujarat University,
Ahmedabad - 380 009 (India)

70. **SAXENA, SHOBHNA,,**
Botany Department,
School of Sciences,
Gujarat University,
Ahmedabad - 380 009 (India)

71. **SEGUIN-SWARTZ, G.**
Agriculture and Agri-Food
Canada Research centre,
107 Science Place, Saskatoon,
Saskatchewan S7N OX2, Canada

72. **SEN, SUMITRA**
Centre of Advanced Study
(Cell and Chromosome Research),
Department of Botany,
University of Calcutta,
Calcutta - 700 019 (India)

73. **SHAH, R. K.**
Divya Jyot Ayurvedic Research
Foundation,
Vehelal, Ta. Daskol, Dist.
Ahmedabad - 3BZ 325 (India)

74. **SHARMA, D. R.**
Department of Biochemistry,
Kurukshetra University,
Kurukshetra - 132 1.1.9 (India)

75. **SHARMA, GOVIND, C.**
Department of Plant & Soil Science,
Alabama A & M University,
Normal, AL 35762, USA

76. **SHARMA, PUSHP**
Department of Botany,
Punjab Agricultural University,
Ludhiana - 1.41 OO4 (India)

77. **SHEWRY, P. R.**
IACR-Long Ashton Research Station,
Department of Agricultural Sciences,
University of Bristol,
Long Ashton, Bristol BSIB9AF

78. **SINGH, BIR**
Department of Botany,
Punjab Agricultural University,
Ludhiana - 1.41. OO4 (India)

79. **SITA, LAKSHMI, G.**
Department of Microbiology and Cell
Biology,
Indian Institute of Science,
Bangalore - 560 012 (India)

80. **SOPORY, K., SUDHIR**
School of Life Sciences,
Jawaharlal Nehru University,
New Delhi - 1.1.0 067 (India)

81. **SRIVASTAVA, D. K.**
Department of Biotechnology,
Dr. Parmar, Y.S. University of
Horticulture and Forestry,
Nauni-Solan - 1.73 230 (India)

82. **SUHASINI, K.**
Plant Tissue Culture Division,
National Chemical Laboratory,
Pune - 411 008 (India)

83. **SUPRASANNA, P.**
Biotechnology Division,
Bhabha Atomic Research Centre,
Trombay,
Bombay - 400 085 (India)

84. **SWARNKAR, P. L.**
Department of Botany,
University of Rajasthan,
Jaipur - 302 OO4 (India)

85. **TAGE, ERIKSSON**
Department of Physiological Botany,
University of Uppsala,
Box 540, 75121, Uppsala, Sweden

86. **TATHAM, A. S.**
IACR-Long Ashton Research Station,
Department of Agricultural Sciences,
University of Bristol, Long Ashton,
Bristol BSIBgAF

87. **TYAGI, VIBHA**
Department of Botany,
University of Rajasthan,
Jaipur - 302 004 (India)

88. **VIRK, G, S.**
Department of Botanical Sciences,
Guru Nanak Dev University,
Amritsar - 143 005 (India

Contents

Recent Developments and Issues in Plant Science

Govind Chandra Sharma

INTRODUCTION

Plant sciences have met some of the most vexing challenges of the twentieth century. As mankind and this science approach a new century, can it meet the challenges of tomorrow with a comparable record of success as it finds socially acceptable solutions to the problems that lay ahead? The present position, the potential promise and the pressing problems that await this field are briefly reviewed here.

The scope of plant science is tremendously vast. The triad of plant genetics, physiology and ecology is buttressed by a myriad of interrelated disciplines-taxonomy, anatomy, morphology, biochemistry, pathology, entomology, and weed science. Then there are the commodity based specializations, all within plant science, and these are diverse indeed. From forestry to agronomy, from range to turf grass management, a very vast canvas has been available to the plant science professionals. Their commission has been quite simple- to meet the ever-growing need for food, fiber and fuel for an expanding world population cost-effectively under the constraints of a dwindling natural resource base and with particular sensitivity to environmental and ecological considerations.

In this review, primary consideration has been given to these issues: 1. What are the major developments in environmental biosphere management that have and will continue to affect the direction of research for plant scientists now and in the future? 2. What are the recent developments in plant science and associated fields that will shape this discipline's direction?

MAN-MADE ENVIRONMENTAL ISSUES

Food and Population Relationship

According to UNESCO estimates, nearly half a billion children suffered from malnutrition and starvation in 1973. It has been suggested that the availability of food per capita worldwide has not increased since 1936, and some contend that it has decreased in the last decade. Although great strides have been made to keep up with the caloric needs, the next challenge for plant scientists is meeting protein needs in human diets. Here also the current situation is quite bleak. In more than half of the world, the protein content of the average diet is about two-thirds of the daily needs or less and the consumption of animal protein is hardly one-fourth that which is required. Some of the scenarios during the Club of Rome Study (Mesarovic and Pestal, 1974) predict that to meet this protein demand in South Asia alone by the year 2025 will require 500 million tons of grain, an amount that is twice that currently produced in North America. As William Saroyan put it, the human race is getting to be too much for itself and too much for the world. Plant scientists must continually work to alleviate malnutrition until population expansion is brought under control.

Loss of Forests

According to English naturalist Norman Myers, the world's forests are being destroyed at the rate of 50 ha/minute corresponding to an annual loss of surface area equivalent to the (U.S.) states of New Hampshire, West Virginia and Maryland combined. If we include the forests that have been converted unsuitably toothier commercial or agricultural uses, the total area would be twice as large, or approximately the size of California. By this reckoning, the tropical forests (both closed cover and open tree formations) in particular, which still occupy some 9 million square kilometers, will last only 40 years (Peccei, 1981). The U.S. National Academy of Science and the Global 2000 Report have both suggested a rate of deforestation as high as 20 million hectares per year. People who support this high estimate predict that except for a few remnant patches, all tropical forests will be gone by the end of the century. Such projections of impending doom are hotly contested, and current projections of the loss of about 7.3 million hectares/year are more commonly accepted. then there are variations among the rate of deforestation among countries largely based on proximity to and the demand for forest products. For example, Costa Rica, Thailand, Ivory Coast and Nigeria are losing forests at a rate of 3% to 6% a year, while Brazil, Cameroon and Zaire are experiencing low deforestation rates of 0.2% to 0.4% a year. Needless to say, deforestation exacerbates soil erosion, siltation of reservoirs, increased flooding and reduced stream flow, loss of rainfall, species extinction, and the greenhouse effect (Chiras, 1985).

Species Extinction

Tropical forests are fundamentally different from temperate forests. First, tropical forests contain far greater numbers of plant and animal species than do temperate forests. One hectare of rich forest in a temperate region may contain 20 to 30 tree species; a single hectare of rich forest in Amazonian Peru can have over 250. Worldwide at least two-thirds of all plant and animal species live in the tropics; only about one-sixth have been named. Second, unlike plant nutrients in a temperate ecosystem, which are available in the soil, nutrients in the tropical ecosystem are recycled in the living matter above the soil, which is essentially a sterile substrata. Thus, most tropical forest land, once cleared, cannot sustain agricultural production (Gentry, 1985).

According to the recent Global 2000 Report to the U.S. President, between 1.5 to 2 million species- 15% to 20% of all species now on earth- are likely to be extinct by the year 2000 mostly due to clear cutting or degradation of tropical forests; most before ever becoming known to science. We know nothing of the potential uses of two-thirds of the world's tropical species. Loss of many species is not only a tragic squandering of the earth's evolutionary heritage but also represents depletion of a significant part of the planet's genetic reservoir, a resource of immense economic value. For example, two drugs used in effectively curing leukemia, Vincristine and Vinblastine, are extracted from the Vinca plant of tropical Madagascar. In coastal Ecuador between 1960 and 1970, a strip of lowland wet primary forest along the base of Andes was deforested after a road was constructed. Only an 0.8 square kilometer now survives as the Rio Palenque Field Station. About 100 plant species new to science have been discovered in this tiny area, many represented by only a few or single individual plants, thus bound for extinction (Gentry, 1985). Although they cover just 7% of the earth's land, tropical moist forests contain more than half of the world's biota (Tangley, 1988). The 1,500 hectare La Selva Field Station located in Costa Rica's Atlantic lowlands contains at least 1,900 vascular plant species, 122 amphibian and reptile species, and 400 bird species. Over 3,000 moth species inhabit La Selva. The loss of these irreplaceable genetic resources with the extinction of tropical forests is one of the most profound losses of diversity of our time (Linden, 1989).

Greenhouse Effect

The earth's climate may be getting warmer in response to the accumulation of atmospheric greenhouse gases, predominantly carbon dioxide (CO_2). Some build-up in atmospheric CO_2 has been documented with one estimate as high as 2.9 billion tons annually. Based on this, some have predicted a doubling of CO_2 by the middle of the next century (Sedjo, 1989). The primary source of CO_2 build-up is the use of fossil fuels (Keeping and Kats, 1988), but large scale destruction of tropical forests is another reason. The controversy associated with the greenhouse effect is not about the elevation of CO_2 concentration but more so about any warming trend associated with it. James Hensen of NASA's Goddard Center believes that the greenhouse effect has been detected and is changing our climate now (Revkin, 1988). Opponents think that in the northern hemisphere temperatures rose about 1 °F during the first forty years of this century, fell for the next forty years and rose again in the 1980s. The magnitude of such change in the past few decades remains somewhat inconclusive (Monastersky, 1989).

If the greenhouse effect does manifest itself in terms of increased temperature, there are several obvious challenges that plant and soil scientists will have to contend with. The effects of global warming will necessitate considerable attention to shifts in areas of food production, judicious use of new or expanded coastal wetlands, and the effects of elevated CO_2 levels and photorespiration on economic yield of agricultural crops. Warmer air will mean more evaporation of ocean water and an overall rise in rain and snow in some areas. Therefore, plant scientists will be called upon to conduct adaptive research on regions newly drier or wetter because of global rainfall shifts.

About 15 years ago, Molina and Rowland (1974) first suggested that chloroflurocarbons (CFC) were destroying the ozone layer in the earth's stratosphere. According to the Rowland-Molina hypothesis, the release of atomic chlorine from solar ultraviolet photolysis of CFC molecules triggers a catalytic chain reaction that causes significant depletion of the ozone layer. A drastic reduction in ozone over the Antarctic caused the so-called ozone hole which was so vividly documented by a remote sensing picture taken by the ozone mapping spectrometer aboard NASA's Nimbus 7 satellite on October 5, 1987. In the hole, the ozone loss had reached nearly 60% The greenhouse effect usually refers to an increase in infrared absorption within the atmosphere because of increasing concentrations of various trace gases. Previously the emphasis had been given to carbon dioxide in this phenomena but was later extended to nitrous oxide. Currently, the focus is also on methane because the incremental greenhouse effect per molecule of methane is about 20 times that for carbon dioxide. More importantly, CFCs are about 10,000 times more efficient in absorbing infrared than CO_2 (Rowland, 1989; Ramanathan, 1975). Plant scientists must remain abreast of the developments created by these major environmental alterations in the biosphere because they will directly affect their own future research.

Highly imaginative remedies have been suggested for cooling the greenhouse which include dumping plane loads of sulphur dioxide in the atmosphere; orbiting parasols equal to 2% of the Earth's surface to block the sunlight; using orbiting reflectors to warm up the polar seas; or supplying phosphates to the seas to increase phytoplankton activity in the ocean. Using lasers to destroy CFCs or designing massive biomass plantations to sequester the additional carbon in the atmosphere are the approaches most intriguing to plant scientists (Revkin, 1989). One estimate suggests that as little as 500 million hectares (an area approximately equivalent to the entire continental U.S. west of the Mississippi) of newly planted, intensively managed, and rapidly growing plantation forests would need to be planted to sequester all of the 5 billion tons or so of carbon (Marland, 1988). Others have estimated 200-400 million hectares of additional rapidly growing forests would be enough to withdraw between 1 to 2 billion of the annual 2.9 billion ton increment of carbon being added to the atmosphere (Sedjo, 1989).

To regulate CFCs, in 1987 fifty-three nations drafted an agreement that would cut CFC production by 50% over the next decade by banning certain chemicals altogether in the future. The regulation of CFCs could be relatively easy since there are only about thirty-eight companies worldwide that produce them (Revkin, 1988). Some in the U.S. have voluntarily announced plans for limiting production of CFCs.

Shrinking Land Base

Another cause of concern for plant scientists is the global trend of per capita decrease of land. Up to 1950, the size of areas under cultivation increased at about the same rate as the world's population. From 1950 to 1975, the rate of increase in agricultural land fell behind the population growth. A few figures illustrate the harsh reality of the difficulties awaiting us. The world area devoted to grain production has dropped from 0.241 ha/capita to 0.184 ha in 1975 and by the end of century will have decreased to 0.120 ha (Brown, 1978).

Expanding Soil Erosion

Even under natural conditions of vegetation, nature takes from 100-400 years or more to generate 10 millimeters of top soil and 3,000 to 12,000 years to generate enough soil to be sufficient for significant food and fiber production (Peccei, 1981). Although soil erosion is considered a natural process, today it has increased to the point where it far exceeds the natural formation of new soil. It has been projected that the amount of top soil per person will decrease by 32% from 1984 to 2000. Many production practices and deforestation activities have a direct effect on soil loss acceleration. For example, in Nigeria a study done on cassava planted on land with varying slopes unprotected by terracing showed that when planted on 1%, 5% and 15% slopes, the soil loss was 3, 87 and 221 mt/ha, respectively. In the U.S., cropping system and soil erosion relationships were brought into focus in a 1936 study in Missouri where corn, wheat and clover rotation lost only 2.7 tons/acre/year while continuous corn planted on the same land lost 19.7 tons/acre/year. The sediment load of the Yellow River in China and the Ganges in India, the world's two muddiest rivers, highlights the magnitude of the soil erosion problems encountered by the world. Draining 1.1 million square kilometers, the Ganges carries an annual sediment load of 1.46 billion tons of soil while Yellow River, draining 0.668 million square kilometers, carries 1.6 billion tons of soil to the ocean each year (Brown and Wolf, 1984; Beasley *et al.*, 1984). Soil conservation is of fundamental importance to plant productivity because once the soil has gone, for all practical purposes it has gone forever, and loss or migration of civilizations from those geographic regions may not be too far behind.

Genetic Erosion In Cultivated Crops

The diversity inherent in seeds and other propagules are plant science's greatest natural heritage. By the 1950s, concern had developed that the natural reservoirs of germplasm resources were rapidly being destroyed, or to use a metaphor from soil science, genetic erosion was taking place at a rapid rate. Curiously enough, such genetic erosion has occurred because plant breeders have been so successful. The new standard cultivars designed to give higher yields are replacing land races which when replaced are being lost forever. The introduction of the Green Revolution, modern farming methods, and destruction of wild habitats have all contributed to this genetic erosion.

To U.S. agriculture, the loss of global germplasm resources is of particular importance since, had it not been for germplasm introduction, the U.S. diet, based on its native resources, would consist merely of sunflower seeds, wild rice, *Jerusalem arthichokes*, concord grapes, craneberries and blueberries (Witt, 1986). To illustrate the power of germplasm introduction consider this: In the U. S. broccoli as a cultivated crop has been prevalent only since 1944; in the Old World red pepper was introduced only after the discovery of the New World. Although our diet primarily relies on just 22 crops. more than 20,000 edible plants exist worldwide (Louder, 1988). Germplasm of each of these edible species and many other economically important fiber, medicinal and commercial crops must be continually preserved. The strategies to do this include exploration, conservation, and gene bank development (Plucknett *et al.*, 1987; Plucknett and Smitton, 1989), evaluation, data storage and retrieval utilization, training and global coordination (Hawkes, 1981). Systematic preservation of gene banks is one of the most essential tasks of plant scientists (Kloppenburg and Kleinman, 1987).

SELECTED ADVANCES IN PLANT SCIENCES

Plant Physiology

Recently it was announced that the Nobel Prize for Chemistry was awarded to Michael, Deisenhofer and Hüber for their work on X-ray crystallographic analysis of photosynthetic reaction centers from bacteria. This deserved award marks the third time in the past two decades that Nobel committees have recognized the importance of plant related research. The other awards honoured Dr. Norman Borlaug's work on the Green Revolution in 1970 and Dr. Barbara McClintock's unique solitary achievement in discerning transposable elements in maize in 1983 (Rabson, 1989).

Chloroplast and its Genome

Similar to these noteworthy achievements in the field of plant physiology, several major initiatives are underway which are shaping the direction of these fields and at the same time tearing down disciplinary walls that have divided plant science or plant biology into isolated subdisciplines. After 18 years of exhaustive work, Chapman and coworkers (1988) produced a three-dimensional map of the 37, 792 atoms in the earth's most abundant protein, ribulose-1-5-bisphosphate carboxylase-oxygenase or RuBisCO. Alterations in RuBisCO structure have already resulted in improved CO_2 uptake and fixation in photosynthetic bacteria. Combine this advancement with an improved understanding of the chloroplast genome (Hoober, 1984; Dean *et al.*, 1989), it is not difficult to imagine that plant science is at the threshold of major breakthroughs in understanding photosynthesis.

Mitochondria

Similar to our expanding understanding of the chloroplast described above, plant mitochondria, powerhouses of the cell, have equally received considerable attention. Many basic features of mitochondrial structure and function are highly conserved between plants and animals despite a billion years of divergent evolution. Such features include morphology, patterns of phospholipids in the membranes, the sequence of electron carriers, etc. The dissimilarities being studied are the nature of electron transport systems, the presence of specific dehydrogenases and anion carriers, and the size and complexities of their DNA. The rate of oxygen consumption is much higher in plant mitochondria and the fatty acid oxidation is low. The plant mitochondrial genome (mt DNA) and its molecular organization and expression differs considerably from that of other organisms. For example, it is much more heterogeneous, from 218 kbp in turnip to about 2,500 kbp in some cucurbits (half the size of the *E. coli* chromoscme). Data available emphasize the great flexibility (recombinatory capacity) of the plant mitochondrial genome. The somatic hybrids of mitochondrial genomes analyzed by RFLP show characteristics of both parents as well as novel non-parental fragments. It is now generally accepted that cytoplasmic male sterility is coded for in the mitochondria and not in the chloroplast (Douce and Neuberger, 1989).

Membrane and Endomembranes

Our understanding of plant function has been greatly advanced by recent progress in the study of plant cell and organelle membranes and endomembranes, the latter term referring to the system largely composed of endoplasmic reticulum, Golgi apparatus, and their transition elements, with the various fascicles, vacuoles and other organelles derived from them (Harris, 1986). For example, we have learned that the primary active transport process in higher plants is the electrogenic transport of H^+ extruded from the cell across the plasma membrane. This important concept is supported in large part by electrical potential and pH differential in vacuole, cytoplasm and the external medium of the apoplast. For, example pH of the external solution and the vacuole contents of an average cell are more acidic than that of the cytoplasm (Sze, 1985; Serrano, 1989). Both cation (e.g., H^+ and K^+ translocation) and anion stimulated ATPases are associated with plasma membranes and other membranes acting as ion pumps or ion carriers. Carrier proteins have a maximum turnover rate of 10^4 - 10^5 transported ions or molecules per second. Ion channels, on the other hand, can pass more than 10^6 ions per second. The number of ion channels discovered in the plasma membrane of the higher plant cells is currently increasing and experimental evidence for K^+, Cl^-, and Ca^{2+} channels has been demonstrated.

Increased understanding of membrane composition, function and structure is aiding in elucidating the greatest reaction of our planet- photosynthesis by obtaining insights into the thylakoid membrane. Most thylakoid proteins are organized into five intrinsic membrane spanning complexes; together with closely associated extrinsic proteins, these perform light harvesting, electron transport, proton translocation and ATP synthesizing functions (Anderson, 1986; Glazer and Melis, 1987). Combined with improved understanding of the RuBisCO and other enzyme systems of C_3 (Woodrow and Berry, 1988) and C_4 pathways and photorespiration (Hoober, 1984; Salisbury and Ross, 1985), the scientific basis of fixation and conservation of carbon is being realized.

Role of Calcium

Although non-hormonal in nature and formerly thought of as just another essential mineral element for plant growth, a revolution is occurring in our thinking about the role of the calcium ion in mediating many different processes including membrane integrity, polarized growth, mitosis and cytokinesis, cytoplasmic streaming, simulating action of red light phytochrome system, gravireception, and its interaction with phytohormones- cytokinin, auxin and gibberellins. The major role of calcium in the realm of stimulus-response coupling has earned the mineral the name the second messenger (Kauss, 1987; Hepler and Wayne, 1985).

Hormonal Regulation

There are two classes of discernible mechanisms that can be employed by multicellular organisms such as plants in their growth and development- (1) systems of chemical messengers and (2) systems of physical or environmental forces (Leopold and Kreidman, 1975). In both areas our understanding has expanded considerably.

Plant growth and development are intriguingly complex processes but the chemical messengers or plant hormones and growth regulators- auxins, gibberellins, cytokinins, inhibitors including abscisic acid and ethylene- are the principal targets by which these natural processes are being regulated. The advent of gas chromatography- mass spectrometry, high pressure liquid chromatography, NMR, immunoassays and other advancements in cell and tissue culture microanalysis and in organic chemistry elucidations have aided in a better understanding of growth and development processes (Wilkins, 1984; Takahashi, 1986). For example, the development of in vitro translation systems for mRNAs and of two-dimensional electrophoresis of translated proteins has made possible the identification of selective m-RNA changes in peas, soybean and corn within 10-20 minutes after exposure to an auxin (Theologis, 1986). In cell elongation such rapid responses seem to manifest by auxin stimulated or induced H^+ secretions. With the use of mutants, we understand that abscisic acid (ABA) may be derived from a carotenoid rather than directly from farnesyl pyrophosphate (FPP) (Creelman, 1989). From Norin 10, the dwarf wheat variety that fueled the green revolution, we now understand that dwarfness is manifested by certain genes that reduce sensitivity to GA rather than an overall reduction in plant GA content (Pharis, 1985). In all three hormone related phenomena it is clear that plant physiologists are utilizing the techniques of plant geneticists. The impetus for such work lies in the fact that expression of genes and their products is a far more reliable and traceable method in following a given growth of developmental process. In effect, therein lies the newly found non-duality in the disciplines of plant physiology and plant genetics- an outcome of advances made in our understanding of plant genomes.

Plant Improvement and Genetic Engineering

The impressive increases in crop productivity achieved over the last several decades have been the result of genetic improvements in crop cultivars and advances in agricultural technology and management practices. Genetic engineering in agriculture in its broadest sense is a post-Mendel plant breeding phenomenon that has led to the development of a plethora of improved cultivars beginning with the application of laws of inheritance and resulting in insect and disease resistance (Sharma and Hall, 1971, 1973), improved harvesting and processing qualities, adaptation to particular environmental conditions, increased yield (through traits such as heterosis and lodging resistance), and enhanced nutritional qualities (e.g. high lysine corn or low erucic acid lines of rape mustard) (Mazur and Falco, 1989).

Plant Cell and Tissue Culture

To accelerate crop breeding processes further, a number of new genetic approaches that require increasing levels of technological sophistication have been developed during the last few decades- seed and pollen mutagenesis, mutagenesis and somaclonal variation in plant cell cultures, regeneration of plants from cultured cells (Sharma et al., 1981), plant protoplast fusion and plant transformation (Mazur and Falco, 1989). With the advent of recombinant DNA technology (Gasser and Fraley, 1989; Glick and Thompson, 1993) transfer of well-characterized genes and traits across the broadest evolutionary boundaries was made possible. Therefore, in the strictest sense, the term genetic engineering denotes isolation, amplification and in vitro manipulation of genes.

Plant cell tissue culture and its development as a specialization (Gautheret, 1985) through the establishment of true tissue culture, demonstration of plant cell totipotency including somatic embryogenesis (Beyl and Sharma, 1985) haploidy and overcoming of sexual barriers by means of protoplast fusion have helped usher in the era of genetic engineering. Most dicotyledonous annuals and foliage plants have been routinely cultured or micropropagated by this technique, and over 1,000 species of plant have been reported to regenerate *in vitro* (Brown and Thorpe, 1986). Considerable success has been achieved in whole plant regeneration from agronomically important monocots such as rice, corn, *Pennisetum, Panicum, Lolium, Festuca* and *Dactylis*. Included among these achievements is the recent report of fertile plants regenerated from the protoplasts of an elite inbred line of maize (Shillito *et al.*, 1989). Woody forest tree species are also increasingly being regenerated in tissue cultures (Hassig, 1989).

Herbicide Resistance in Crops

To illustrate recent progress in the application of genetic engineering to plant agriculture, a review of the development of herbicide resistant crops is interesting. The potential utility of herbicide resistant genes as a dominant selective genetic marker is analogous to antibiotic resistant genes in bacteria. Currently private and public research is focusing on plants
resistant genes in bacteria. Currently private and public research is focusing on plants
resistant to amino acid biosynthesis inhibitors- glyphosate, sulfanyl ureas and imidazolinones and phosphinothricine, and on plants resistant to photosynthesis inhibitors- triazines and bromoxynil (Padgette *et al.*, 1988). The first example of a herbicide-resistant crop, atrazine-resistant canola derived from crossing a resistant weed with a crop, has been released for use. Imidazolinone-resistant corn is in the breeding and evaluation stage. A sulfanyl urea resistant soybean line has been selected from mutagenized seed and is being introduced into soybean cultivars by standard crop breeding methods. Many examples of transgenic herbicide-resistant crops have been reported and are in various stages of evaluation and development. However, attainment of adequate levels of glyphosate resistance has been difficult (Mazur and Falco, 1989). Calgene and Monsanto,

for example, have achieved considerable success in incorporating glyphosate resistance in tomato and tobacco, but incorporation of genetic resistance in soybean and corn is proving to be much more difficult (Eppley, 1989).

Genetically Engineered Biopesticides

Another concerted commercial activity of importance has been the development of plants with genetically engineered biopesticide genes or improved biopesticide product development. Such products are attractive to commercial concerns primarily because pests may not develop resistance to them as rapidly as they do to a chemical. Additional benefits are possible such as reduced potential for groundwater contamination, rapid breakdown in soil, pest-specificity, less toxicity to beneficial species, and relative ease and cost-effectiveness of the registration process. Abbott Laboratory, for example, is engineering several new strains of Bacillus thuringiensis (Bt), and has introduced a new emulsifiable Bt (Gasser and Fraley, 1989). Crop Genetics International is testing a Bt based endophyte, a plant-dwelling bacteria, as an inoculum for corn and soybean seed which multiplies and produces a protein (toxin) systematically in the plant. Ecogen Inc. is working on new Bt strains and has marketed Collego, which uses a live fungus to control northern joint vetch in rice and soybean, and has produced Dagger- the first EPA registered biofungicide using a bacteria. Mycogen Corporation is working on Alternaria cassiae fungus as a bioherbicide, Casst, to control sicklepod and coffee senna weeds. Plant Genetic Systems (PGS), a Belgian company, is advancing a new commercial concept- plant molecular farming. PGS engineered oil rape seeds have become a production system for high value pharmaceutical peptides (Eppley, 1989).

Coat-Protein Mediated Plant Virus Protection

In recent years, significant resistance to tobacco mosaic virus (TMV) infection, termed coat-protein mediated protection, has been achieved by expressing only the coat protein gene of TMV in transgenic plants. Currently engineered plants can tolerate or resist tobacco, alfalfa and cucumber mosaic virus, potato virus x and y, tobacco etch and rattle viruses (Wickelgren, 1989). One public concern has been about the ingestion of various introduced Bt proteins and coat proteins in the produce. Studies in regards to coat protein suggest that the engineered protein will typically comprise 0.01 to 0.5% of the total protein which is well below the coat protein levels found in plants infected with these endemic viruses (Gasser and Fraley, 1989). According to Beachy, for example, a person would have to eat 2,000 to 5,000 transgenic tomatoes to ingest the same amount of protein contained in one naturally virus-infected tomato (Wickelgren, 1989).

Field Testing of Genetically Engineered Crops

After considerable preliminary experimentation, ethical discussion, regulatory approval and legal debates, field tests of genetically engineered microorganisms are underway. The landmark field tests of engineered Ice minus (Ice-) *Pseudomonas syringae* to reduce frost damage to plants have been conducted in California. This genetically engineered (Ice-) bacteria colonizes seedlings and minimizes wild-type natural populations of Ice + bacteria through competition. Early results show one-third less frost damage than in untreated seedlings (Newman, 1989). Transgenic tomatoes carrying the TMV coat protein gene have also been evaluated in the greenhouse and field tests and have been shown to be highly resistant to viral infection. At least 454 field tests of genetically engineered seeds were conducted in the U.S. in 1993 (Anonymous, 1989a). However, not all is well in the recombinant DNA/genetic engineering world of plants. Desirable new phenotypes created by introduction of foreign DNA into plants are frequently unstable following propagation, leading to a loss of newly acquired traits probably by inactivation of transgenes (Finnegan and McElroy, 1994).

Biotechnology- Promise, Potential and Regulatory Aspects

An equitable review of the promise, potential and regulatory aspects of the vast and emerging field of biotechnology is beyond the scope of this paper. Public biotechnology activities (Bloch, 1986), for example, its role in food processing (Harlander and Labuza, 1986), in degradation of toxic chemicals and hazardous waste (Scholze *et al.*, 1988), in processing of municipal wastes for fuels and chemicals (Antonopoulos, 1987) and in environmental management (Omenn and Teich, 1986) are among those that have been reviewed recently. A brief review of regulatory complexities is provided by Gasser and Fraley (1989) and by Muller (1989). The Office of Technology Assessment in a multi-report series has been assessing the new developments in biotechnology (U.S. Congress, OTA, 1988). This report, for example, outlines federal support for biotechnology. In 1987, NIH, Department of Defense and NSF supported greater research than did the USDA; among U.S. States, New York, New Jersey, Michigan and Pennsylvania made substantial contributions for biotechnology R & D, training and facilities; altogether 28 U.S. states have established centers and programs devoted specifically to research in biotechnology areas with most in California. The greatest concentration of biotech companies are in California, Massachusetts, Maryland, New Jersey and New York.

OTHER ISSUES IN PLANT GENETICS

Role of Transposable Elements

Approximately 40 years ago Barbara McClintock proved that the maize transposable element *Ds* (dissociation) moved from one position on chromosome 9 to another position on the same chromosome (McClintock, 1978). She was studying the stability, fusion, and genetic consequences of broken chromosomes. Such chromosomal breakage either liberated or assembled an element, *Ac* (activator). *Ac* as it is understood destablizes or mobilizes *Ds*; Ds is that which responds to *Ac*. Generally transposable elements originate following mechanical chromosome breakage both autonomous elements such as *Ac* and non autonomous elements such as *Ds* move. In some organisms an element may literally change position while in others a copy of the element transposes (Freeling, 1984). The discovery of jumping genes is one of the most profound findings in modern genetics. It is ironic that transposable elements, which for years were pushed aside by mainstream biology as freaks, should now feature in explanations of everything grand (Freeling, 1984), from the somaclonal variation observed in *in vitro* cultures (Lee and Phillips, 1988) and origins of new genes (Jones *et al.*, 1989) to the origin of species (Syvanen, 1984; Vodkin, 1986).

Plant Genome Mapping Initiative and RFLP Research

Mapping the genomes of higher plants and identifying the locations of genes is of critical importance to the advancement of plant sciences. Such maps will permit the manipulation of our plant genetic resources more efficiency and greatly aid in understanding the biochemical pathways, and physiological environmental and developmental bases of variation in crops. This proposed effort by USDA has become possible by restriction fragment length polymorphisms (RFLP) analysis. Numerous laboratories in the U.S. and other countries have demonstrated that they can now track genes from each parent through generations to each progeny line. In collaboration with plant breeders and statisticians, the location and value of genes from parents can be determined (Gelvin *et al.*, 1989). RFLP analysis was initially applied to variations in viruses. The process is complex but technically straightforward for all major crops. DNA is isolated, treated with commercially available restriction endonucleases, subjected to electrophoresis and specific genes are identified by annealing (radioactive) labeled cloned pieces of DNA back to electrophoressed DNA. Excellent RFLP maps have already been developed by scientists for corn, tomatoes, rice, lettuce, broccoli and cauliflower; partial maps have been produced for barley, soybean, and wheat; and projects are beginning for potatoes, cotton, alfalfa, dry beans, lentils , grapes and many other vegetable fruit and forest crops.

As can be observed, much effort is currently being directed to develop biochemical (e.g. isozyme) and molecular markers (e.g. RFLP) for many economically important plants. However, the conventional detection of RFLPs by Southern blot analysis is laborious, costly and sometimes difficult in some species with large and complex genomes. Recently, the swift development of methods for rapid and efficient synthesis of synthetic oligonucleotides with the discovery of polymerase chain reaction (PCR) provides a rapid way to pursue a genome mapping program. PCR has been successfully used to amplify single low copy genes in human genome project and plant genome research is benefiting from such similar strategies (Holzman, 1992 and Tragoonrung *et al.*, 1992).

Soil Science- Research Needs and Issues

Over the past year, the Soil Science Society of America has been assisting the Experimental Station Committee on Organization and Policy (ESCOP) in identifying the nation's major research priorities. Recently the draft of the priorities has been published for review by the membership of the American Society of Agronomy (Anonymous, 1989b). The major research priorities identified include the need to protect and sustain soil productivity and soil ecosystems, the continual enhancement of soil water behaviour and management, protection and enhancement of water quality, incorporation of soil science into systems research, predicting and modeling soil use systems, and global or international issues and concerns.

Such broad areas include several categories of research, for example: characterization, classification and determination of a soil resource base; maintenance or enhancement of structural ability of soils; prevention or management of soil compaction; characterization of tillage systems with respect to surface runoff and groundwater quality; and soil and crop production sustainability. Soil water behaviour and management studies include the need for characterization of micropore and macropore transport, enhancement of water use efficiency, and determination of thermal-induced transport process. In water quality research, needs include development of vulnerability or potential use ratings of soils for production or waste inputs; and prevention of surface water and groundwater pollution through development of improved soil and crop management systems. These include nutrient management, pesticide management, waste and residual management, erosion control and other best management practices. Soils systems research needs should address the influence of soil and management under various cropping systems, and determination of production resource alternatives that are economically sustainable and environmentally compatible. Prediction and modeling soil use systems require research on geographic information systems (GIS) and artificial intelligence systems, determination of crop yield potential vs. resource inputs and their response, and determination of carrying capacities of soil in regions of high population pressures.

Global research needs for soil science include investigation of the greenhouse effect through carbon (organic matter) maintenance; enhancement of soil with tillage options and other management systems; and development of programs and response models for soils when sensitive ecosystems such as tropical rain forests and semi-arid savannahs are disturbed, cleared or altered. The effect of acid rain, ground-level ozone, increased salinity, decreasing soil fertility, desertification and potential climatic changes all affect and shape the future challenges that the soil scientists must address (Anonymous, 1989b).

REFERENCES

Anderson, J.M. 1986. Photo-regulation of the composition, function and structure of thylakoid membrane. Ann. Rev. Plant Physiol. 37: 93-136.

Anonymous. 1989a. Plant field tests flow dot the countryside. Biotechnology News 2(9):1.

Anonymous. 1989b. Research needs and issues in soil science. Agronomy News (September), p. 4-5.

Antonopoulos, A.A. (ed.). 1987. Biotechnological advances in processing municipal waters for fuels and chemicals. Noyes Data Corp., Park Ridge, ND, p. 488.

Beasley, R.P., Gregory, J.M. and McCarty, T.R., 1984. Erosion and sediment pollution control. Iowa State Univ. Press, Ames., p. 354.

Beyl, C.A. and Sharma, G.C., 1983. Picloram induced somatic embryogenesis in *Gasteria* and *haworhia*. Plant Cell Tissue and Organ Culture 2: 123-132.

Bloch, C. 1986. Plant Agriculture. Federal biotechnological activities. Noyes Pub., Park Ridge, NJ, p. 210.

Blowers, J. 1994. Field test permit update. Nat. Biol. Impact Assessment Program (NBIAP) News Rep. August 1993.

Brown, D.C.W. and Thorpe, T.A. 1986. Plant regeneration by organogenes In I.K. Vasil (ed.). Cell culture and somatic cell genetics. Academic Press, New York, Vol. 3: 49-60.

Brown, L.R. 1978. The world wide loss of crop land. World Watch Inst., Washington, DC.

Brown, L.R. and Wolf, S.E. 1984. Soil erosion: Quiet crisis in the world economy. Paper #60. World Watch Inst., Washington, DC, p. 49

Chapman, M.S., Sul, S.W., Curmi, P.M.G., Cascio, D., Smith, W.W. and Eiserberg, D.S., 1988. Tertiary structure of plant RuBisCO: Domians and their contacts. Science 241: 71-73 (July 1).

Chiras, B. 1985. Environmental Science. Benjamin Cummings Pub. Co., Menlo Park, CA., p. 655.

Creelman, R.A. 1989. Abscisic acid and biosynthesis in higher plants. Physiol. Plant 75: 131-136.

Dean, C., Pichersky, E. and Dunsmuir, P., 1989. Structure, evolution and regulation of RbCS genes in higher plants. Ann. Rev. Plant physiol. Mol. Biol. 40: 415-439.

Douce, R. and Neuberger, M. 1989. The uniqueness of plant mitochondria. Ann. Rev. Plant Physiol. 40: 371-414.

Eppley, D. 1989. Biotech's promise- Panacea or pie in the sky? Agrichemical Age (April) p. 14,18

Finnegan, J. and McElroy, D. 1994. Transgenic Inactivation: Plants Fight Back. Bio-Technology 12: 883-888.

Freeling, M. 1984. Plant transposable elements and insertion sequences. Ann rev. Plant Physiol. 35: 277-298.

Gasser, C.S. and Fraley, R.T. 1989. Genetically engineered plants for crop improvement. Science 244: 1293-1299.

Gautheret, R.J. 1985. History of plant tissue and cell culture: A personal account. In I.K. Vasil (ed.). Cell culture and somatic cell genetics of plants. Academic Press, New York, Vol. 2: 1-59.

Gelvin, S.B., Schilperoort, R.A. and Verma, D.P.S. (eds.). 1989. Plant molecular biology manual. Kluwer Academic Pub., Boston.

Gentry, A. 1985. The demise of tropical rain forest. In Environmental Science by D.D. Chiras. The Benjamin Cummings Pub. Co. Inc., Menlo Park, CA., p. 254-255.

Glazer, A.N. and Melis, A. 1987. Photochemical reaction centers: Structure, organization and function. Ann. Rev. Plant Physiol. 38: 11-45.

Glick, B.R. and Thompson, J.E. (eds.). 1993. CRC Methods in Plant Molecular Biology and Biotechnology. CRC Press, Boca Raton, FL. 360 p.

Harlander, S. and Labuza, T.P. (eds.). 1986. Biotechnology in food processing. Noyes Pub., Park Ridge, NJ, p. 323.

Harris, N. 1986. Organization of the endomembrane system. Ann. Rev. Plant Physiol. 37: 73-92.

Hassig, B.E. 89. Status of forest tree vegetative regeneration for biotechnology. Biotech. Lab. 7(1): 48-51.

Hawkes, J.G. 1981. Germplasm collection, preservation and use. In K.J. Frey (eds.). Plant breeding II. Iowa State University Press, p. 57-83.

Hedrich, R. and Schroeder, J. 1989. The physiology of ion channels and electrogenic pumps in higher plants. Ann. Rev. Plant Physiol. Plant Mol. Biol. 40: 539-569.

Hepler, P.K. and Wayne, R.O. 1985. Calcium and plant development. Ann. Rev. Plant Physiol. 36: 397-439.

Holzman, D. 1992. A Tool for all Seasons. Mosaic. 23(2): 2-11.

Hoober, J.K. 1984. Chloroplasts. Plenum Press, New York, p. 147-197.

Jones, J.D.G., Carland, F. and Dooner, H. 1989. Genetic properties of *Ac* in tobacco. J. Cellular Biochem. Suppl. 13D, p. 233.

Kauss, H. 1987. Some aspects of calcium-dependent regulation in plant metabolism. Ann. Rev. Plant Physiol. 38: 47-72.

Keepin, B. and Kats, G. 1988. Global Warming. Science 241: 1027.

Kloppenburg, J. and Kleinman, D.L. 1987. The plant germplasm controversy. BioScience 37: 190-191.

Lee, M. and Phillips, R.L. 1988. The Chromosomal basis of somaclonal variation. Ann. Rev. Plant Physiol. Plant Mol. Biol. 39: 413-437.

Leopold, A.C. and Kreidman, P.E. 1975. Plant growth and development. McGraw Hill Book Co., New York, p. 531.

Linden, E. 1989. Playing with fire. Time, Sept. 18, 1989.

Louder, R. 1988. Meals for the millennium. Discover (Nov.), p. 60-63.

Mackenzie, D.R., Faust, R.M. and Russell, K.W. (eds.). 1988. Plant genomic research conference report. U.S. Dept. Agriculture, S/E. Dec. 1988, p. 29.

Marland, G. 1988. The prospect of solving the CO_2 problem through global reforestation. U.S. Dept. Energy Res. DOE/NBB-0082, p. 66.

Mazur, B.J. and Falco, S.C. 1989. The development of herbicide resistant crops. Ann. Rev. Plant Physiol. Plant Mol. Biol. 40: 441-470.

McClintock, B. 1978. Mechanisms that rapidly reorganize the genome. Stadler Genet. Symp. 10: 2548.

Measarovic, M. and Pestal, E. 1974. Stratosphoric sink for chlorofluoromethane: chlorine atom catalyzed destruction of ozone. Nature 249: 810-812.

Monastersky, R. 1989. Global change: The scientific challenge. Science News, 135: 232-235.

Muller, K. 1989. Biotech's Promise- Pie in the sky? Agrichemical Age (April) p. 15, 19, 21E.

Newman, A.R. 1989. Tracking genetically engineered microorganisms. Anal. Chem. 61: 1001A-1003A.

Omenn, G.S. and Teich, A.M. (eds.). 1986. Biotechnology and environment. Noyes Data Corp., Park Ridge, NJ, p. 169.

Padgette, S.R., Della-Cioppa, G., Shah, D.M., Fraley, R.T. and Kishore, G.M. 1988. Selective herbicide resistance through protein engineering . In J. Schell and I.K. Vasil (eds.). Cell culture and somatic cell genetics of plants. Vol. 6. Academic Press, New York.

Peccei, H. 1981. One hundred pages for the future. New American Library, New York, p. 187.

Pharis, R.P. 1985. Gibbrellins and reproductive development in seed plants. Ann. Rev. Plant Physiol. 36: 517-568.

Plucknett, D.L., Smith, N.J.H., Williams, J.T. and Anishetty, N.M. 1987. Gene banks and the world's food. Princeton Univ. Press, Princeton N.J., p. 247.

Plucknett, D.L. and Smith, N.J.H. 1989. Quarantine and the exchange of crop genetic resources. BioScience 39:16-23.

Postel, S. 1988. Global view of a tropical disaster. American Forests (Nov./Dec.), p. 25-29.

Rabson, R. 1989. It can't happen here. Plant Cell 1: 167-168.

Ramanathan, V. 1975. Greenhouse effect due to chlorofluorocarbons: climate implications. Science 190: 50-52.

Revkin, A.C. 1988. Endless summer: Living with the greenhouse effect. Discover, October, p. 50-61.

Revkin, A.C. 1989. Cooling off the greenhouse. Discover, January, p. 30-32.

Rowland, F.S. 1989. Chlorofluorocarbons and the depletion of stratosphoric ozone. Amer. Scientist 77: 36-46.

Salisbury, F. and Ross, C. 1985. Plant physiology. Wadsworth Pub., Belmont, CA.

Scholze, R.J., Wu, Y.C., Basilico, J.V. (eds.). 1988. Biotechnology for degradation of toxic chemicals in hazardous waste. Noyes Data Corporation, Park Ridge, NJ, p 697.

Sedjo, R.A. 1989. Forests to offset greenhouse effects. J. Forestry (July), p. 12-15.

Serrano, R. 1989. Structure and function of plasma membrane ATPase. Ann. Rev. Plant Physiol. Plant Mol. Biol. 40: 61-94.

Sharma, G.C. and Hall, C.V. 1971. Influence of cucurbitacins, total sugar and fatty acids to spotted cucumber beetle feeding. J. Amer. Soc. Hort. Sci. 96: 675-680.

Sharma, G.C. and Hall, C.V. 1973. Relative attractance of spotted cucumber beetle to fruits of fifteen species of Cucurbitaceae. Environ. Entomol. 2: 154-156.

Sharma, G.C., Bello, L.L., Sapra, V.T. and Peterson, C.M. 1981. Callus initiation and plant regeneration from triticale embryos. Crop Sci. 21: 113-118.

Shillito, R.D., Carswell, G.K., Johnson, C.M., DiMaio, J.J. and Harms, C.T. 1989. Regeneration of fertile plants from protoplasts of elite inbred maize. Biotechnology. 7: 581-587.

Syvanen, M. 1984. The evolutionary implications of mobile genetic elements. Ann. Rev. Genetics. 18: 271-293.

Sze, H. 1985. H^+- translocating ATPases: Advances using membrane fascicles. Ann. Rev. Plant Physiol. 36: 175-208.

Takahashi, N. (ed.). 1986. Chemistry of plant hormones. CRC Press, Inc., Boca Raton, FL, p. 277.

Tangley, L. 1988. Studying and saving the tropics. BioScience.

Theologis, A. 1986. Rapid gene regulation by auxin. Ann. Rev. Plant Physiol. 37: 407-438.

Tragoonrung, S., Kanazin, V., Hayes, P.M. and Blake, T.K. 1992. Sequence-Tagged-Site Facilitated PCR for Barley Genome Mapping. Theor. Appl. Genet. 84: 1002-1008.

Congress, Office of Technology Assessment. 1988. New developments in biotechnology. U.S. Investments in Biotechnology- Special Report, OTA-BA-360. Supdt. Documents, Washington, DC. 20402-9325.

Vodkin, L. 1986. Jumping genes that control plant growth. Yearbook of Agriculture. Supdt. of Documents, Washington, DC, p. 127-130.

Wickelgren, I. 1989. Please pass the genes. Science News. 136: 120-122.

Wilkins, M.B. (ed.). 1984. Advanced plant physiology. Pitman Pub. Ltd., London, p. 514.

Witt, S.C. 1986. Germplasm. The World and I. (No. 7) p. 240-250.

Woodrow, I.E. and Berry, J.A. 1988. Enzymatic regulations of photosynthetic CO2 fixation in C3 plants ann. Rev. Plant Physiol. Plant Mol. Biol. 39:533-594.

Agrobacterium- Mediated Gene Transfer in Plants- A Review

D.K. Srivastava

INTRODUCTION

Plant genetic engineering is a young science. It was only in 1983 that chimeric genes were first expressed in genetically transformed plant tissue (Bevan *et al.*, 1983; Herrera-Estrella *et al.*, 1983). Only after this development, and the concomitant availability of selectable and non selectable marker genes which express in plant tissue, did work on plant genetic transformation systems begin in earnest. Since that time there has been an explosion in the literature of plant genetic transformation.

The *Agrobacterium* system was the first used to transform plant tissue, there have been many developments since then like direct gene transfer of naked DNA to plant cells, a technique that has also been developed to a high degree of sophistication.

The first transgenic plants expressing engineered foreign genes were tobacco plants produced by the use of *Agrobacterium tumefaciens* vectors (Horsch *et al.*, 1984, DeBlock *et al.*, 1984). Transformation was confirmed by the presence of foreign DNA sequences in both primary transformants and their progeny and by an antibiotic resistance phenotype conferred by a chimeric neomycin phosphotransferase gene.

Agrobacterium constitutes an excellent system for introducing genes into plant cells, since i) DNA can be introduced into whole plant tissues, which by pass the need for protoplasts, and ii) the integration of T-DNA is a relatively precise process. The region of DNA to be transferred is defined by the border sequences, occasional rearrangements do occur, but in most cases an intact T-DNA region is inserted into the plant genome. (Spielmann and Simpson, 1986; Jorgensen *et al.*, 1987). This contrasts with free DNA delivery systems in which the plasmids routinely undergo rearrangements and concentration reactions before insertion and can lead to chromosomal rearrangements during insertion in both animal (Mark *et al.*, 1985; Covarrubia *et al.*, 1986) and plant (Potrykus *et al.*, 1985) systems. Sequencing of insertion sites shows that only small duplications or other changes occur in flanking sequences during T-DNA integration (Gheysen *et al.*, 1987). The stability of expression of most genes that are introduced by *Agrobacterium* appears to be excellent. Published work has shown that integrated T-DNAs give consistent genetic maps and appropriate segregation ratio (Horsch *et al.*, 1984; DeBlock *et al.*, 1984; Wallroth *et al.*, 1986; Deroles and Garder, 1988). Introduced traits have been found to be stable over at least five generations during cross breeding and seed increase on in genetically engineered tomato and oilseed rape plants (Gasser and Fraley, 1989). This stability is critical to the commercialization of transgenic plants.

Agrobacterium-Mediated Gene Transfer

Derivatives of the plant pathogen *Agrobacterium tumefaciens* has proved to be efficient highly versatile vehicles for the introduction of genes into plants and plant cells. Most transgenic plants produced to date were created through the use of the *Agrobacterium* system. *Agrobacterium tumefaciens* is the etiological agent of crown gall disease and produces tumorous crown galls on infected species. The utility of this bacterium as a gene transfer system was first recognised when it was demonstrated that the crown galls were actually produced as a result of the transfer and integration of genes from the bacterium into the genome of the plant cells (Chilton *et al.*, 1977). Virulent strains of *Agrobacterium* contain large Ti- (tumour inducing) plasmids, which are responsible for the DNA transfer and subsequent disease symptoms. Genetic and molecular analyses showed that Ti-plasmids contain two sets of sequences necessary for gene transfer to plants, one or more T-DNA (transferred DNA) regions that are transferred to the plant, and the *Vir* (Virulence) genes which are not, themselves, transferred during infection. The T-DNA regions are flanked by 25 base pair right and left border DNA sequences that were shown to be responsible for the T-DNA transfer to infected plant cell (Garfinkel *et al.*, 1981; Zambryski *et al.*, 1983) and that it can be deleted (Phytohormones synthesis interference with plant regeneration and replaced with marker gene or desirable gene.

Early experiments demonstrated that heterologous DNA inserted into the T-DNA genes (Barton *et al.*, 1983). Efficient plant transformation systems were constructed by removing the phytohormones biosynthetic genes from the T-DNA region, thereby eliminating the ability of the bacteria to induce aberrent cell proliferation (Fraley *et al.*, 1985).

Vectors for Agrobacterium Mediated Plant Transformation

Plant transformation vectors based on *Agrobacterium* can generally be divided into two categories: those that cointegrate into a resident Ti-plasmid (cointegrate vectors) and those that replicate autonomously (the binary vectors). Both the vectors have several common features imposed upon them by the requirements of *Agrobacterium*. The T-DNA is delimited by direct 25bp imperfect repeats border sequences. A transformation vector must, then, include at least one border or be capable of cointegrating in such a manner that it becomes flanked by borders. Besides a border, all vectors include a selectable marker for identification of transformed plant cells and a bacterial selectable marker essential for introduction of the vector into *Agrobacterium*.

Cointegrating Vectors

Cointegrating transformation vectors must include a region of homology between the vector plasmid and the Ti-plasmid. This requirement for homology means that the vector is capable of integrating into a limited number of Ti plasmids. The vector is usually designed to cointegrate into one of a few specific Ti-plasmids. two cointegrating system are in use today. The first utilizes the disarmed *Agrobacterium* Ti-plasmid PG 3850 (Zambryski *et al.*, 1983). In this plasmid the phytohormones gene of the C 58 plasmid have been excised and replaced by pBR 322 sequence. Any plasmid containing the pBR 322 sequence homology can be cointegrated into the disarmed Ti-plasmid. The border sequences as well as a nopaline synthase gene are part of the Ti-plasmid, and the cointegration places the new sequences between the T-DNA borders.

A different approach to a cointegrating vector was used by Fraley *et al.*, (1985). In this system, the right border and all the phytohormone genes are removed from the Ti-plasmid. A left border and a small part of the original T-DBA, referred to as the Limited Internal Homology (LIH), remains intact. The vector to be introduced into *Agrobacterium* contains the LIH region for homologous recombination as well as a right border. The cointegrated DNA reconstructs a functional T-DNA with a right border and left border. This system has been used extensively for introduction of many genes into plants. The cointegrate system, while more difficult to use, do offer advantages. Once the cointegrate has been formed, the plasmid is stable in *Agrobacterium* and is virtually impossible to lose. Binary vectors, on the other hand, are not completely stable in *Agrobacterium* in the absence of drug selection. There is also evidence that a cointegrating vector can transform tomato at a higher frequency than a binary vector (McCormick *et al.*, 1986).

Binary Vectors

Binary transformation vectors are some what different from cointegrating vectors. Instead of a region of homology with the Ti-plasmid, they contain origins of replication from a broad host-range plasmid. These replication origins permit autonomous replication of the vector in *E. coli* and *Agrobacterium*. Since the plasmid does not need to form a cointegrate, these plasmids are considerably easier to introduce into *Agrobacterium*. The frequency of introduction into *Agrobacterium* is about 10^{-1} whereas cointegrate formation is usually about 10^{-5}. Since these vectors do not cointegrate, they must contain the T-DNA border sequence (s). Many vectors contain two borders that delimit the portion of the plasmid transferred to the plant. However, a single border is sufficient for transfer (Horsch and Klee, 1986). In this case transfer initiates at the single border, and the entire plasmid becomes the T-DNA.

A major advantage to binary vectors is their lack of dependence on a specific Ti-plasmid. The vector may be introduced into virtually any *Agrobacterium* host containing any Ti or Ri plasmid as long as the *vir* helper functions are provided. This may be important in the transformation of some plant species, since different *Agrobacterium* strains exhibit major differences in their abilities to infect different plant species.

Selectable Markers for Plant Transformation

Several selectable and non selectable marker genes viz. neomycin phospo-transferase (Reiss *et al.*, 1984), luciferase (Ow *et al.*, 1986), streptomycin phosphotransferase (Jones *et al.*, 1987), (Srivastava *et al.*, 1991) chloramphenicol acetyl transferase (Bird *et al.*, 1988), hygromycin phosphotransferase (Waldron *et al.*, 1985), ß-glucuronidase (Jefferson *et al.*, 1987), and dihydrofolate reductase (Eichholtz *et al.*, 1987), are widely available today for plant transformation and have also been transferred in large number of plant species (Table I, II, III & IV). Several requirements must be considered in the development of a truly useful selectable marker system. It is most critical that the selective agent be inhibitory to plant cells. However, not all compounds toxic to plant cells are necessarily useful as selective agents. Cells that are not transformed can be killed in such a manner that they become toxic to adjacent transformed cells. This presumably happens because of leakage of toxic compounds, such as phenols, from the dying cells. If this occurs even high level expression of a resistance gene in the transformed cells is insufficient to rescue these cells. The best selective agents are compounds that arrest growth of non transformed cells or slowly kill them.

TABLE 1. Transgenic field crops

Sr. No.	Plants	Method and Gene Transferred	Reference
1.	Tobacco	At, NPT-II	Horsch *et al.*, 1984
2.	Alfalfa	At, NPT-II	Shahin *et al.*, 1986
3.	Cotton	At, NPT-II	Umbeck *et al.*, 1987
		At, NPT-II	Srivastava *et al.*, 1991
4.	Sunflower	At, NPT-II	Everett *et al.*, 1987
5.	Moth Bean	At, NPT-II	Eapen *et al.*, 1987
6.	Soybean	At, GUS	Hinchee *et al.*, 1988
7.	Safflower	At, GUS	Ying *et al.*, 1992
8.	Green Bean	At, GUS	Franklin *et al.*, 1993
9.	Cowpea	EL* GUS	Akella and Lurquin, 1993
10.	Chickpea	At, NPT-II	Fontana *et al.*, 1993

At: *Agrobacterium tumefaciens*, NPT-II: Neomycin phosphotransferase -II, GUS: ß-glucuronidase, EL*: Electroporation

The most widely used selectable marker has been the neomycin phosphotransferase type II (NPT-II) enzyme, which was originally isolated from the prokaryotic transposon Tn 5 (Garfinkel *et al.*, 1981). This enzyme detoxifies aminoglycoside compounds such as kanamycin and G 418 by phosphorylation. This gene, fused to constitutive plant transcriptional promoters, has been used successfully to transform a large number of plant species (Table, I, II, III & IV), and has been incorporated into numerous plant transformation on vectors.

Transformation of Plants

Two basic approaches have been used to obtain transgenic plants: cocultivation of regenerating protoplasts (Horsch *et al.*, 1984) and the leaf disc procedure used successfully to generate a transgenic plant. The procedure involves (Agrobacterium transformation of regenerating protoplasts/tissue followed by selection and regeneration of plants.

The production of transgenic plants can be divided into four main steps i.e. I. Introduction of foreign genes into modified *Agrobacterium* strain, II. Cocultivation of *Agrobacterium* strains with protoplasts, plant cells or tissues, III. Selection and regeneration of transformants and IV. Analysis and verification of gene expression in transformed plants.

TABLE II. Transgenic Vegetable Crops

Sr. No.	Plants	Method and Gene Transferred	Reference
1.	Carrot	Ar, NPT-II	David *et al.*, 1984
2.	Tomato	At, NPT-II	McCornick *et al.*, 1986
3.	Cucumber	Ar, NPT-II	Trulson *et al.*, 1986
		At, Ar, NPT-II	Srivastava *et al.*, 1990
4.	Potato	At, NPT-II	Ooms *et al.*, 1987
5.	Oil seed rape	At, NPT-II	Fry *et al.*, 1987
6.	Lettuce	At, NPT-II	Michelmore *et al.*, 1987
7.	Cauliflower	Ar, Mannopine	David and Tempe, 1988

Sr. No.	Plants	Method and Gene Transferred	Reference
		At, NPT-II	Srivastava *et al.*, 1988
8.	Celery	At, NPT-II	Catlin *et al.*, 1988
9.	Brinjal	At, NPT-II	Guri and Sink, 1988
10.	Watermelon	At, NPT-II	Srivastava *et al.*, 1991
11.	Sugarbeet	At, NPT-II	Harpster *et al.*, 1988
12.	Muskmelon	At, NPT-II	Fang and Grumet, 1990
13.	Bell pepper	At, GUS	Liu *et al.*, 1990
14.	Peas	At, NPT-II	Dekathen and Jacobsen, 1992
15.	Sweet potato	MB* GUS	Prakash and Varadrajan, 1992
16.	Cabbage	At, GUS	Berthomieu and Jouanin, 1992

At: *Agrobacterium tumefaciens*, Ar: Agrobacterium rhizogenes, NPT-II: Neomycin phosphotransferase-II, GUS: ß glucurn idase, MB*: Micro-projectile bombardment

TABLE III. Transgenic Fruit Crops

Sr. No.	Plants	Method and Gene Transferred	Reference
1.	Walnut	At, NPT-II	McGranahan *et al.*, 1988
2.	Apple	Ar, NPT-II	James *et al.*, 1989
3.	Peach	At, Octopine	Nammerschlag *et al.*, 1989
4.	Grapevine	At, NPT-II	Baribault *et al.*, 1989
5.	Papaya	MB*, NPT-II	Fitch *et al.*, 1989
		At, cp gene	Fitch *et al.*, 1993
6.	Strawberry	At, GUS	Nehra *et al.*, 1990
7.	Citrus spp.	At, NPT-II	Hidaka *et al.*, 1990
8.	Apricot	At, cp gene	Laimer da Camara *et al.*, 1992
9.	Pecan	At, GUS,	McGranahan *et al.*, 1993

At: *Agrobacterium tumefaciens*, cp gene: Coat protein gene, NPT-II: neomycin phosphotransferase-II, GUS: ß-glucuronidase, MB*: Microprojectile bombardment

TABLE IV. Transgenic Forest Plants

Sr. No.	Plants	Method and Gene Transferred	Reference
1.	Populus hybrida (Poplar)	At, NPT-II, `aroA'	Fillatti *et al.*, 1987
2.	Pseudotsuga menziesii (Douglas fir)	At, Octopine, NPT-II	Dandekar *et al.*, 1987
3.	Azadirachta indica (Neem)	At, Octopine, NPT-II	Naina *et al.*, 1989
4.	*Picea mariana** (Black spruce	El*, CAT	Tautorus *et al.*, 1989
5.	*Pinus banksiana** (Jack Pine)	El*, CAT	Tautorus *et al.*, 1989
6.	*Salix* spp. (willows)	At, NPT-II	Vahala *et al.*, 1989
7.	Catharanthus roseus (Madagascar)	Ar, Agropine	Brillanceau *et al.*, 1989
8.	Digitalis purpurea (Foxglove)	At, GUS	Saito *et al.*, 1990
9.	*Picea rubens, P. Glauca* (Red spruce, White spruce)	MB*, GUS, NPT-II	Charest *et al.*, 1993
10.	Larix X eurolepis (Larch)	MB*, GUS, NPT-II	Charest *et al.*, 1993
11.	Liquidambar styraciflua (Sweetgum)	At, NPT-II, Bt gene	Sullivana nd Lagrimini, 1993
12.	*Santalum album* (Sandalwood)	At, NPT-II	Bhattacharya and

Sr. No.	Plants	Method and Gene Transferred	Reference
			Lakshmisita, 1994.

At: *Agrobacterium tumefaciens*, aro A' gene: Glyphosate resistance, Bt gene: -endotoxin gene, NPT-II: neomycin phophotransferase-II, GUS: ß-glucurnonidase, MB*: Microprojectile bombardment, CAT: Chloramphenicol acetytransferase, El: Electroporation

Most transgenic plants produced todate, were created through the use of the *Agrobacterium*- mediated gene transfer system. Other methods that have the potential to influence the production of transgenic plants include, microinjection, electroporation, direct injection into reproductive organs, PEG mediated gene transfer, particle gun and liposome mediated gene transfer.

Herbicide Resistance

Engineering herbicide resistance into crops represents a new alternative for conferring selectivity and enhancing crop safety of herbicide. Research has largely concentrated on those herbicides with properties such as high unit activity, low toxicity, low soil mobility, rapid biodegradation and broad spectrum activity against various weeds. The development of crop plants that are resistant to such herbicide would provide more effective, and less costly weed control. Fillatti *et al.*, (1987), Piruzian *et al.*, (1988) and DeBlock *et al.*, (1989) have successfully transferred the herbicide resistant gene in tomato, poplar, tobacco, oil seed rape and cauliflower. Before commercialization of herbicide resistant crop plants, factors such as potential loss in vigour and yield, herbicide performance crop and chemical registration costs and potential for out crossing to weed species should be considered.

Insect Resistance

The production of insect resistant plant is another application of biotechnology with important implications for crop improvement and for both the seed and agrochemical industries. Progress in engineering insect resistance in transgenic plants have been achieved through the use of the insect control protein genes of *Bacillus thuringiensis* is an entomocidal bacterium that produces an insect control protein which is lethal tolepidopteran larvae, although some strain with toxicity to coleopteran and dipteran larvae have been described. Some of the scientists have successfully introduced the insect resistant gene in tobacco (Vaeck *et al.*, 1987), tomato (Fischhoff *et al.*, 1987), mustard (Basu *et al.*, 1991).

Protease inhibitors accumulate in the seeds of almost every plant, providing them with a generalized defence against insect. The cowpea trypsin inhibitor gene, identified from cowpea (*Vigna unguiculata*) line showed dramatically increased resistance to the bruchid beetle (*Callosobruchus maculatus*) Hilder *et al.*, (1990) reported transformation of tobacco with Agrobacterium carrying cowpea trypsin inhibitor gene and regenerated morphologically normal plants. Survival and growth of the insects and damage to the plants were all significantly reduced, particularly when early instar larvae were tested. Field tests for transgenic plants expressing insect resistance are increasing rapidly. Where there is a grave concern for the clear environment transgenic plants with self protection mechanism would be a healthy development and would obviate the need for environmental pollutants such as toxic insecticides.

Disease Resistance

Disease resistant gene can be categorised under three main heads i.e. virus, bacterial and fungal disease resistance. Of all plant pathogens, viruses are most intimately associated with their hosts, their genomes are relatively small and well characterized. Significant resistance to tobacco mosaic virus (TMV) infection termed coat protein-mediated protection has been achieved by expressing only the coat protein gene of TMV in transgenic plants. This approach produced similar result in transgenic tomato and potato plants against a broad spectrum of plant viruses including alfalfa mosaic virus, cucumber mosaic virus, potato virus X and potato virus Y. Transgenic tomatoes carrying the TMV coat protein gene have been evaluated in green house and field tests and shown to be highly resistant to viral infection (Nelson *et al.,* 1988). Likewise resistance has been created against various groups of viruses (Fitchen and Beachy, 1993).

Recently a new class of bacteriacidal proteins (lysozyme, cecropins and attacins) have been identified in the pupae of giant silk moth *Hyalophora cecropia*. These lytic proteins have also shown a potent *in vitro* antifungal activity against several pathogenic fungi, including *Phytophthora infestans* (*Destifano Beliran et al.,* 1991). Transgenic tobacco plants that express a barley ribosome-inactivating proteins (Stirpe *et al.,* 1992), exhibited heightened protection against agronomically deleterious fungus *Rhizoctonia solani* (Longemann *et al.,* 1992). Similarly, tobacco pathogen-related proteins, namely PR-S and osmotin have been shown to be serologically related to seamatin, an antifungal protein of maize (Vigers *et al.,* 1992) Woloshuk *et al.,* (1991) showed that Osmotin and related proteins from tomato had antifungal activity against *Phytophthora infestans*.

Post Harvest Management

Antisense RNA technology in combination gene transfer technology is being used to increase the shelf life of the plant and plant products. The recent approach to develop this technology aims to mimic mutation by reducing gene expression using RNA that contains the complementary sequences to a given RNA. Transgenic plants have been used to study the effects of antisense RNA on the expression of endogenous genes such as those encoding chalcone synthase. Since chalcone synthase is the key enzyme in flavonoid biosynthesis inactivation of these genes by antisense RNA was scored on phenotypic basis as a change in flower pigmentation in petunia and tobacco (Vander Krol *et al.,* 1988). In transgenic tomato plants expressing antisense RNA of polygalacturonase, endogenous level of this enzyme activity was extremely reduced in ripening fruits (Smith *et al.,* 1988).

CONCLUSION

New emerging gene transfer technologies have enormous potential for plant improvement by introducing foreign genes in plant cells or tissues of both monocot and dicot plants. During the last 10 years, a number of transgenic plants have been produced in many important crops. Besides selectable marker genes, a number of other foreign genes conferring herbicide resistance, insect resistance, viral, bacterial and fungal resistance and genes related to increase shelf life of plants and plant products (via antisense RNA technology), have been transferred to plants from a wide range of plants, bacterial, and viral systems. In the majority of cases, foreign genes show expression in transgenic plants and are stably inherited in the progeny without detrimental effects on the host plant.

Moreover, transgenic plants under field conditions have also maintained increased levels of insect resistance, herbicide resistance and higher levels of virus protection. However, there is a need to have extensive laboratory and field testing of transgenic plants and their progenies before their usefulness can be realized on the commercial scale.

Genetic engineering of plants could ultimately become a reliable tool for crop improvement depending on cost effective production of transgenic plants and their products without compromise on quality, yield and nutrition. So far, no genetically engineered cultivar has been released to farmers and is delayed by 8-10 years. However, numerous, public institutions and private companies are conducting field trials of transgenic plants in different parts of the world. Flavr Savr transgenic tomato (improved shelf life) produced by calgenes Inc. USA, will reach the market very soon. In future, faith in genetic engineering for crop improvement will be dependent on how, well consumer accepts transgenic tomato Flavr Savr which will be the criteria of its commercial success.

ACKNOWLEDGMENT

Author is highly thankful to Prof. D.R. Sharma for critically reading the manuscript.

REFERENCES

Akella, V. and Lurquin, P.F. 1983. Expression in cowpea seedlings of chimeric transgenes after electroporation into seed-derived embryos. Plant Cell Rep. 12: 110-117.

Baribault, T.J., Skene, K.G.M. and Stede-Scott, N. 1989. Genetic transformation of grapevine Cells. Plant Cell Rep. 8: 137-140.

Barton, K.A., Binns, A.N. Matzke, A.J.M. and Chilton, M. 1983. Regeneration of intact tobacco plants containing full length copies of genetically engineered T-DNA and transmission of T-DNA to F1 progeny. Cell. 32: 1033-1048.

Basu, D., Das, S., Bandyopadhyay, D. and Sen, S.K. 1991. Isolation and cloning of *Bacillus thuringiensis* var Kurstaki HD 73 toxin gene and construction of a chimaeric gene for expression in plants. Indian J. Exp. Biol. 29: 1002-1009.

Berthomieu, P. and Jouanin, L. 1992. Transformation of rapid cycling cabbage (*Brassica olerocea* var. Capitata) with *Agrobacterium rhizogenes*. Plant Cell Rep. 11: 334-338.

Bevan, M.W., Flavell, R.B. and Chilton, M.D. 1983. A chimeric antibiotic resistance gene as a selectable marker for plant cell transformation. Nature 304: 184-187.

Bhattacharya, A.a nd G. Lakshmi Sita, 1994. *Agrobacterium* mediated transformation in Sandal wood (*Santalum album* L.). Proceedings of International Union of Biochemistry and Molecular Biology, p. 4-39.

Bird, C.R., Smith, C.J.S., Ray, J.A., Moureau, P., Bevan, M.W., Bird, A.S., Hughes, S., Morris, P.C., Grierson, D. and Schuch, W. 1988. The tomato polygalacturonase gene andripening-specific expression in transgenic plants. Plant Molecular Biology, 11: 651-662.

Brillanceau, M., David, C., Tempe, J. 1989. Genetic transformation of *Catharanthus roseus* G. Don. by *Agrobacterium rhizogenes*. Plant Cell Rep. 8: 63-66.

Catlin, D., Ochoa, O., McCormick, S. and Ohyiros, C.F. 1988. Celery transformation by *Agrobacterium tumefaciens* cytological and genetic analysis of transgenic plants. Plant Cell Rep. 7: 100-103.

Charest, P.J., Calero, N., Lachance, D., Datla, R.S.S., Duchesne, L.C. and Tsang, W.T. 1993. Microprojectile-DNA delivery in conifer species: factors affecting assessment of transient gene expression using the ß-glucuronidase reporter gene. Plant Cell Rep. 12: 189-193.

Chilton, M.D., Drummond, M.H., Merlo, D.J., Saiky, D., Montoya, A.L., Gordon, M.P. and Nester, E.W. 1977. Stable incorporation of plasmid DNA into higher plant cells; the molecular basis of crown gall tumorigenesis. Cell 11: 253-271.

Covarrubias, L., Nishida, Y., and Mintz, B. 1986. Proc. Natl. Acad. Sci.(USA) 83: 6020-6022.

Dandekar, A.M., Gupta, P.K., Durzan, D.J. and Knauf, V. 1987. Transformation and foreign gene expression in micropropagated Douglasfir (*Pseudostuca menziessi*). Bio/Technology, 5: 587-590.

David, C., Chilton, M.D. and Tempe, J. 1984. Conservation of T-DNA in plants regenerated from hairy root cultures. Bio/Technology, 2: 73-76.

David, C. and Tempe, J. 1988. Genetic transformation of cauliflower (*Brassica oleracea* L. Var. Botrytis) by *Agrobacterium rhizogenes*. Plant Cell Rep. 7: 88-91.

DeBlock, M., DeBrouwer, D. and Tenning, P. 1989. Transformation of *Brassica napus* and *Brassica oleracea* using *Agrobacterium tumefaciens* and the expression of the bar and neo genes in the transgenic plants. Plant Physiology, 91: 694-701.

DeBlock, M., Herrera-Estrella, L., Van-Montagu, M., Schell, J. and Zambryski, P. 1984. Expression of foreign genes in regenerated plants and in their progeny. EMBO J. 3: 1681-1689.

DeKathen, A. and Jacobsen, H.J. 1990. *Agrobacterium tumefaciens* mediated transformation of *Pisum sativum* L. using binary and cointegrate vectors. Plant Cell Rep. 9: 276-279.

Deroles, S.C. and Gardner, R.C. 1988. Expression and inheritance of kanamycin resistance in large number of transgenic petunias generated by *Agrobacterium*-medaited transformation. Plant Mol. Biol. 11: 355-364.

Destifano-Beliran, L., Nagpala, P.G., Kim, J., Dodds, J.H. and Jaynes, J.M. 1991. In: Molecular Methods for Potato Improvement. International Potato Centre, Peru, pp. 49-64.

Eapen, S., Kohler, F., Gerdemann, M. and Schieder, O. 1987. Cultivar dependence of transformation rates in moth bean after Cocultivation of protoplasts with *Agrobacterium tumefaciens*. Theor. Appl. Genet. 75: 207-210.

Eichholtz, D.A., Rogers, S.G., Horsch, R.B., Klee, H.J., Hayford, M., Hoffmann, N.L., Bradford, S.B., Fink, G., Flick, J., O'Connell, K.M. and Fraley, R.T. 1987. Expression of mouse dihydrofolate reductase gene confers methotrexate resistance in transgenic petunia plants. Somatic Cell Mol. Genet. 13: 67-76.

Everett, N.P., Robinson, K.E.P. and Mascarenhas, D. 1987. Genetic engineering of sunflower (*Helianthus annuus* L.). Bio/Technology 5: 1201-1204.

Fang, G. and Grumet, R. 1990. *Agrobacterium tumefaciens* mediated transformation and regeneration of muskmelon plants. Plant Cell Rep. 9: 160-164.

Fillatti, J.J., Kiser, J., Rose, R. and Comai, L. 1987. Efficient transfer of glyphosate tolerance gene into tomato using binary *Agrobacterium tumefaciens* vector. Bio/Technology, 5: 726-730.

Fillatti, J.J., Sellmer, J., Mc Cown, B., Hasig, B. and Comai, L. 1987. Agrobacterium mediated transformation and regeneration of populus. Mol. Gen. Genet. 206: 192-199.

Fischoff, D.A., Bowdish, K.S., Perlak, F.J., Marrone, P.G., McCormick, S.M., Niedermeyer, J.G., Dean, D.A., Kusano-Kietzmev, K., Mayer, E.J., Rochester, D.E., Rogers, S.G. and Fraley, R.T. 1987. Insect tolerant transgenic tomato plants. Bio/Technology 5: 807-813.

Fitch, M.M.M., manshardt, R.M., Gonselves, D. and Slightom, J.L. 1993. Transgenic papaya plants from Agrobacterium-mediated transformation of somatic embryos. Plant Cell Rep. 12: 245-249.

Fitch, M.M.M., Manshardt, R.M., Gonsalves, D., Slightom, J.L. and Sanford, J.C. 1990. Stable transformation of papaya via micro-projectile bombardment. Plant Cell Rep. 98: 189-194.

Fitchen, J.H. and Beachy, R.N. 1993. Genetically engineered protection against viruses in transgenic plants. Ann. Rev. Microbiology 47: 739-763.

Fontana, G.S., Santini, L., Caretto, S., Grugis, G. and Mariotti, D. 1993. Genetic transformation in the grain legume *Cicer arietinum* L. (Chickpea). Plant Cell Rep. 12: 194-198.

Fraley, R.T., Rogers, S.G., Horsch, R.B., Eichholtz, D.A., Flick, J.S., Fink, C.L., Hoffman, N.L. and Sanders, P.R. 1985. The SEV system: a new disarmed Ti-plasmid vector system for plant transformation. Bio/Technology, 3: 629-635.

Franklin, C.I., Trieu, T.N., Cassidy, B.G., Dixon, R.A. and Nelson, R.S. 1993. Genetic transformation of green bean callus via *Agrobacterium* mediated DNA transfer. Plant Cell Rep. 12: 74-79.

Fry, J., Barnason, A. and Horsch, R.B. 1987. Transformation of *Brassica napus* with *Agrobacterium tumefaciens* based vectors. Plant Cell Rep. 6: 321-325.

Garfinkel, D.J., Simpson, R.B., Ream, L.W., White, F.F., Gordon, M.P. and Nester, E.W. 1981. Genetic analysis of crown gall: fine structure map by site directed mutagenesis. Cell 27: 143-153.

Gasser, C.S. and Fraley, R.T. 1989. Genetically engineering plants for crop improvement. Science 244: 1293-1299.

Gheysen, G., Von-Montagu, M. and Zambryski, P. 1987. Integration of *Agrobacterium tumefaciens* transfer DNA (T-DNA) involves rearrangements of target plant DNA sequences. Proc. Natl. Acad. Sci. USA 84: 6169-6173.

Guri, A. and Sink, K.C. 1988. *Agrobacterium* transformation of eggplant. J. Plant Physiol. 133: 52-55.

Hammerschlag, F.A., Owens, L.D. and Smigocki, A.C. 1989. Agrobacterium-mediated transformation of peach cells derived from mature plants that were propagated *in vitro*. J. Amer. Soc. Hort. Sci. 114: 508-510.

Harpster, M.H., Townsend, J.A., Jones, J.D.G., Bedbrook, J. and Dunsmuir, P. 1988. Relative strengths of the 35S cauliflower mosaic virus 1', 2' and nopaline synthase promoters in transformed tobacco, sugar beet and oilseed rape callus tissue. Mol. Gen. Genet. 212: 182-190.

Herrera-Estrella, L., Depicker, A., Van-Montagu, M. and Schell, J. 1983. Expression of chimeric genes transferred into plant cell using a Ti plasmid vector. Nature 303: 209-213.

Hidaka, T., Omura, M., Ugaki, M., Tomiyama, M., Kato, A., Ohshima, M. and Motoyoshi, F. 1990. Agrobacterium-mediated transformation and regeneration of *Citrus* spp. from suspension cells. Japan J. Breed. 40: 199-207.

Hilder, V.A., Gatehouse, A.M.R. and Boulter, D. 1990. In: Genetic engineering of crops for insect resistance using the genes of origin Eds. Lycett, G. and Grierson, D. Butterworths, London. pp. 51-56.

Hinchee, M.A.W., Connor-Ward, D.V., Newell, C.A., McDonell, R.E., Sato, S.J., Gasser, C.S., Fischhoff, D.A., Re, D.B., Fraley, R.T. and Horsch, R.B. 1988. Production of transgenic soybean plants using *Agrobacterium* mediated DNA transfer. Bio/Technology 6: 915-921.

Horsch, R.B., Fraley, R.T., Rogers, S.G., Sanders, P.R., Lloyd, A.a nd Hoffman, N. 1984. Inheritance of functional foreign genes in plants. Science 223: 496-498.

Horsch, R.B., Fry, J.E., Hoffman, N.L., Wallroth, M, Eichholtz, D., Rogers, S.G. and Fraley, R.T. 1985. A simple and general method for transferring genes into plants. Science 227: 1229-1231.

Horsch, R.B. and Klee, H.J. 1986. Rapid assay of foreign gene expression in leaf discs transformed by *Agrobacterium tumefaciens*: Role of T-DNA borders in the transfer process. Proc. Natl. Acad. Sci. USA. 83: 4428-4432.

James, D.J., Passey, A.J., Barbara, D.J. and Bevan, M. 1989. Genetic transformation of apple (*Malus pumila* Mill.) using a disarmed Ti binary vector. Plant Cell Rep. 7: 658-661.

Jefferson, R.A., Kavanagh, T.A. and Bevan, M.W. 1987. GUS fusions: ß-glucoronidase as a sensitive and a versatile gene fusion marker in higher plants. The EMBO Jr. 6: 3901-3907.

Jones, J.D.G., Svab, Z., Harper, E.C., Hurwitz, C.D. and Maliga, P. 1987. A dominant nuclear streptomycin resistance marker for plant cell transformation. Mol. Gen. Genet. 210: 86-91.

Jorgensen, R., Synder, C. and Jones, J.D.G. 1987. T-DNA is organized predominantly in inverted repeats structures in plants transformed with *Agrobacterium tumefaciens* C58 derivatives. Mol. Gen. Genet. 207: 471-477.

Laimer da Camara Machado, M., DaCamara Machado, A., Hanzer, V., Weiss, H., Regner, F., Steinkellner, H., Mattanovich, D., Plail, R., Knapp, E., Kalthoff, B. and Katinger, H. 1992. Regeneration of transgenic plants of *Prunus armeniaca* containing the coat protein gene of plum pox virus. Plant Cell Rep. 11: 25-29.

Liu, W., Parrott, W.A., Hildebrand, D.F., Collins, G.B. and Williams, E.G. 1990. Agrobacterium inbduced gall formation in bell pepper (*Capsicum annuum* L.) and formation of shoot like structures expressing introduced genes. Plant Cell Rep. 9: 360-364.

Longemann, J., Jach, G., Tommerup, H., Mundy, J. and Schell, J. 1992. Biotechnology, 10: 305-308.

Mark, W., Signorelli, K. and Lacy, E. 1985. Cold Spring Harbour Symp. Quant. Biol. 50: 453-

McCormick, S., Niedermeyer, J., Fry, J., Barnason, A., Horsch, R. and Fraley, R. 1986. Leaf disc transformation of cultivated tomato (*L. esculentum*) using *Agrobacterium tumefaciens*. Plant Cell Rep. 5: 81-84.

McGranahan, G.H., Leslie, C.A., Dandekar, A.M., Uratsu, S.L. and Yades, I.E. 1993. Transformation of pecan and regeneration of transgenic plants. Plant Cell Rep. 12: 634-638.

McGranahan, G.H., Leslie, C.A., Uratsu, S.L., Martin, L.A. and Dandekar, A.M. 1988. Agrobacterium-mediated transformation of walnut somatic embryos and regeneration of transgenic plants. Bio/Technology 6: 800-804.

Michelmore, R., Marsch, E., Seely, S. and Landry, B. 1987. Transformation of lettuce (*Lactuca sativa*) mediated by *Agrobacterium tumefaciens*. Plant Cell Rep. 6: 439-442.

Naina, N.S., Gupta, P.K. and Mascarenhaus, F. 1989. genetic transformation and regeneration of transgenic Neem plants using *Agrobacterium*. Current Science 58: 184-188.

Nehra, N.S., Chibbar, R.N., Kartha, K.K., Datla, R.S.S., Croshy, W.L. and Stushnoff, C. 1990. Genetic transformation of strawberry by *Agrobacterium tumefaciens* using a leaf disk regeneration system. Plant Cell Rep. 9: 293-298.

Nelson, R.S., McCormick, S.M., Delannay, X., Dube, P., Layton, J., Anderson, E.J., Kaniewska, M., Proksch, R.K., Horsch, R.B., Rogers, S.G., Fraley, R.T. and Beachy, R.N. 1988. Virus tolerance, plant growth and field performance of transgenic tomato plants expressing coat protein from tobacco mosaic virus. Bio/Technology 6: 403-409.

Ooms, G., Burrell, M.M., Karp, A., Bevan, M. and Hille, J. 1987. Genetic transformation in two potato cultivars with T-DNA from disarmed *Agrobacterium*. Theor. Appl. Genet. 73: 744-750.

Ow, D.W., Wood, K.V., Deluca,M., Dewet, J.R., Helinski, D.R. and Howell, S.H. 1986. Transient and stable expression of the firefly luciferase gene in plant cells and transgenic plants. Science 234: 856-859.

Piruzian, E.S., Mett, V.L., Kobets, N.S. and Urmeeva, F.I. 1988. The use of bacterial genes encoding herbicide tolerance in constructing transgenic plants. Microbiological Science 5: 242-248.

Potrykus, I., Shillito, R.D. Saul, M.W. and Paszkowski, J. 1985. Direct gene transfer: State of the art and future potential. Plant Mol. Biol. Rep. 3: 117-128.

Prakash, C.S. and Varadorajan, U. 1992. Genetic transformation of sweet potato by particle bombardment. Plant Cell Rep. 11: 53-57.

Reiss, B., Sprengel, R., Will, H. and Schaller, H. 1984. A new sensitive method for qualitative and quantitative assay of neomycin phosphotransferase in crude cell extract. Gene: 30: 211-218.

Saito, K., Yamazaki, M., Shimomura, K., Yoshimatsu, K. and Murakoshi, I. 1990. Genetic transformation of foxglove (*Digitalis purpurea*) by chimeric foreign genes and production of Cardioactive glycosides. Plant Cell Rep. 9: 121-124.

Shahin, E.A., Spielmann, A., Sukhapind, K., Simpson, R.B., and Yashar, M. 1986. Transformation of cultivated alfalfa (*Medicago sativa*) using disarmed *Agrobacterium tumefaciens*. Crop Sci. 26: 1235-1239.

Smith, C.J.S., Watson, C.F., Ray, J., Bird, C.R., Morris, P.C., Schuch, W. and Grierson, D. 1988. Antisense RNA inhibition of polygalacturonase gene expression in transgenic tomatoes. Nature 334: 724-726.

Spielmann, A. and Simpson, R.B. 1986. T-DNA stuructures in transgenic tobacco plants with multiple independent sites. Mol. Gen. Genet. 205: 34-41.

Srivastava, D.K., Andrianov, V.M. and Piruzian, E.S. 1990. Genetic transformation and foreign gene expression in cucumber tissue (*Cucumis sativa* L. cv. *Kustovoi*). In: Proceedings of VIIth International Congress on Plant tissue and Cell Culture Amsterdam. The Netherlands. Abstract pp. 65.

Srivastava, D.K., Andrianov, V.M. and Piruzian, E.S. 1991. Regeneration and genetic transformation studies in watermelon (*Citrullus vulgaris* L. cv. Melitopolski). In: Horticulture-New Technologies and Applications. J. Prakash and R.L.M. Pierik, Eds. Kluwer Academic Publishers, The Netherlands. pp. 127-130.

Srivastava, D.K., Kolganova, T.V., Mett, V.L. and Piruzian, E.S. 1991. Genetic transformation of *Gossypium hirsutum* L. cv. 108-F. Acta Horticulture. 289: 263-264.

Srivastava, D.K., Ulmasov, T. and Piruzian, E.S. 1991. A chimeric streptomycin resistance gene as a dominant nuclear marker for plant cell transformation. In: Proceedings of IIIrd international Congress on Molecular Biology of Plant Growth and Development. Tucson, USA, Abstract No. 893.

Srivastava, V., Reddy, A.S. and Guha-Mukherjee, S. 1988. Transformation and regeneration of *Brassica oleracea* mediated by an oncogenic *Agrobacterium tumefaciens*. Plant Cell Rep. 7: 504-507.

Stripe, F., Barbieri, L., Battelli, M.G., Soria, M. and Lappi, D.A. 1992. Bio/Technology 10: 405-412.

Sullivan, J. and Lagrimini, M. 1993. Transformation of *Liquidambar styraciflua* using *Agrobacterium tumefaciens*. Plant Cell Rep. 12: 303-306.

Tautorus, T.E., Bekkaoul, F., Pilon, M., Datla, R.S.S., Crosby, W.L., Fowke, L.C. and Dunstan, D.I. 1989. Factors affecting transient gene expression in electroporated black spruce *Picea mariana* and Jackpine (*Pinus banksiana*) protoplasts. Theor. Appl. Genet. 78: 531-536.

Trulson, A.J., Simpson, R.B. and Shahin, E.A. 1986. Transformation of cucumber (Cucumis sativus L.) Plants with *Agrobacterium rhizogenes*. Theor. Appl. Genet. 73: 11-15.

Umbeck, P., Johnson, G., Barton, K. and Swain, W. 1987. Genetically transformed cotton (*Gossypium hirsutum* L.) plants. Bio/Technology 5: 263-266.

Vaeck, M., Reynaerta, A., Hofta, H., Jensens, S., De Beuckeleev, M., Dean, C., Zabeau, M., Van Montogu, M. and Leemans, J. 1987. Transgenic plants protected from insect attack. Nature 328: 33-37.

Vahala, T., Stabel, P. and Erikson, T. 1989. Genetic transformation of willows (*Salix* spp.) by *Agrobacterium tumefaciens*. Plant Cell Rep 8: 55-58.

VanderKrol, A.R., Lenting, P.E., Veenstra, J., Vander Meer, I.M., Gerats, A.G. M., Mol, J.N.M. and Stuiitje, A.R. 1988. An antisence chalcone synthase gene in transgenic plants inhibits flower pigmentation. Nature 333: 866-869.

Vigers, A.J., Wiedemann, S., Roberts, W.K., Legrand, M., Selitrennikoff, C.P. and Fritig, B. Plant Sci. 83: 155-161.

Waldron, C., Murphy, E.B., Roberts, J.L., Gustafson, G.D., Armour, S.L. and Malcolm, S.K. 1985. Resistance to hygromycin B. Plant Mol. Biol. 5: 103-108.

Wallroth, M., Gerats, A.G.M., Roberts, S.G., Fraley, R.T. and Horsch, R.B. 1986. Chromosomal lociliazation of foreign genes in *Petunia hybrida* Mol. Gen. Genet. 202: 6-15.

Woloshuk, C.P., Meulenhoff, J.S., Sela-Burriage, M., Vanden Elzen, P.J.M. and Cornelissen, B.J.C. 1991. Pathogen induced proteins with inhibitory activity towards, *Phytophthora infestens*. Plant Cell 3: 619-628.

Ying, M., Dyer, W.E., and Bergman, J.W. 1992. *Agrobacterium tumefaciens*- mediated transformation of safflower (*Carthamus tinctorius* L. Cv. Centennial. Plant Cell Rep. 11: 581-585.

Zambryski, P., Joos, H., Genetello, C., Leemans, J., Van-Montagu, M. and Schell, J. 1983. Ti plasmid vector for the introduction of DNA into plant cells without alternation of their normal regeneration capacity EMBO J. 2: 2143-2150.

Biotechnology: Miracle or Mirage II. Remarkable Genetic System of *Agrobacterium*

C. Nirmala and M.L.H. Kaul

Agrobacterium tumefaciens, the gram negative, rod shaped aerobic bacterium caused little interest when Smith and Townsend (1907) reported that it is the causative agent of a widespread neoplastic plant disease crown gall, which is characterized by the overproduction of tissues at the infection site. This pathogen adsorbs on the surface of cells exposed at wound sites, most frequently just below the soil surface at the root crown and hence the name crown gall tumour. Later, a key discovery was made in 1974 by Zaenen et al where they found that *A. tumefaciens* contains a large extrachromosomal element (more than 200 kb) harbouring genes involved in crown gall induction. This tumour inducing plasmid was given the name Ti plasmid by van Larebecke *et al* (1975). However, this bacterium received extraordinary attention after the discovery that the bacteria is capable of modifying the genetic material of host cells by transferring part of its Ti plasmid which gets incorporated into the host cell DNA and expresses itself with the normal host DNA. With researchers analysing and unraveling the molecular mechanism underlying crown gall induction, the *Agrobacterium* gene transfer system has now become one of the most significant discoveries triggering a series of existing developments in molecular biology and genetic engineering. The tailoring of the crown gall tumour to fit the needs of the inciting bacterium is a unique phenomenon in plant biology that takes place through the transfer of the bacterial plasmid gene to susceptible plants cells followed by permanent incorporation of these genes in the plant genome. Integration and subsequent expression of these genes lead to the formation of tumours and the production of low molecular weight molecules called opines which serve as specific substrate for food and genetic regulation. Thus, *Agrobacterium* is a clever genetic architect who can precisely design a habit in which it can best survive and ward off any formidable competitors.

Though most researches have been carried out with *Agrobacterium tumefaciens*, now its close ally, *A. rhizogenes* which produces hairy root disease is also being exploited. Both Agrobacteria infect many dicotyledonous plants, gymnosperms and a few monocots. Whereas, *A. tumefaciens* is characterized by the presence of a large Ti plasmid, the T-DNA responsible for transformation in *A. rhizogenes* is located on and Ri (root inducing) plasmid. Though both species transform plant cells in a similar way in terms of T-DNA transfer and the production of opines in the target tissue, transformation with *A. rhizogenes* results into plants that exhibit an abberant phenotype- wrinkled leaves and reduced apical dominance. Also *A. rhizogenes* proliferate at temperatures higher than 30°C and do not grow well in classic *E. coli* media. Hence, it is less used for transformation experiments.

ANALOGY OF T-DNA TRANSFER AND BACTERIAL CONJUGATION

Several evidences support the hypothesis that *Agrobacterium* plant DNA transfer simulates bacterial conjugation, a natural means by which DNA transfer occurs between bacterial cells and are evolutionarily related. This hypothesis was first proposed following the discovery of the T-strand and is based on striking parallels with synthesis and transfer of conjugal DNA. Functional and molecular similarities of T-DNA transfer and conjugation are most completely characterized with respect to the plasmid RP4, a broad host-range plasmid isolated from *Pseudomonas* strain. Points of similarities between these processes are:

1. T-DNA border nicks are analogous to nicks at the origin of conjugal DNA transfer
2. T-stand is analogous to the linear donor single stranded DNA molecules produced during bacterial conjugation,
3. Single stranded DNA is transferred following its displacement from the plasmid occurring directionally from 5' to 3'
4. Protein complexes are formed at the nick sites
5. DNA transfer will occur only if the donor and recipient cell are in intimate contact
6. Similarities exist in the amino acid sequences, gene organization and physical properties of the trans-acting enzymes involved

The most direct support for the analogy 'between T-DNA transfer and bacterial conjugation is the fact that the oriT of the conjugative bacterial plasmid can substitute for T-DNA borders in directing DNA transfer to plant cells.

STEP OPTIMIZATION FOR IN-VITRO TRANSFORMATION

Recent researches have enabled gene manipulators to exploit and modify the natural gene transfer system of *Agrobacterium* to deliver desired genes. However, to use *Agrobacterium* for in-vitro plant transformation, a number of experimental steps must be optimized in both the bacteria and the plant material. These include:

1. identification of an *Agrobacterium* strain which infects the appropriate plant genotype.
2. design and construct a modified T-DNA to allow gene expression in plant cells.
3. transfer to and maintenance of the modified T-DNA in a specified *Agrobacterium* strain.
4. frequency to T-DNA transfer to plant cells should be high enough to be detected.
5. selection and regeneration of transformed plant cells

HOST RANGE

Agrobacterium can induce tumour and hairy root formation in a wide range of plants. Though the hosts are mostly dicotyledonous plants, they also include a few gymnosperms and monocotyledonous plants. The strain is specific for plant genotypes and plant tissue types. The inability of *Agrobacterium* to induce tumours or hairy roots in most monocot species is probably due to

1. reduced activity of T-DNA promoters
2. inhibition of *vir* gene induction
3. abnormal auxin-cytokinin balance in monocotyledonous cells
4. lack of pronounced wound response

5. unavailability of receptor for attachment

However, this does not rule out the possibility that *A. tumefaciens* has the ability to transfer T-DNA into a monocotyledon as there are reports of cereal transformation via *Agrobacterium*

STRAINS USED FOR TRANSFORMATION

Not all *Agrobacterium* strains are tumourigenic. Many strains contain very large cryptic plasmids that do not confer virulence and in some isolates, a cryptic plasmid is present together with a Ti plasmid. While strains containing a wide host range (WHR) Ti plasmid can induce tumours on a large variety of plants, limited host range (LHR) strain are restricted to a few plant species. The plant host range of tumour formation is influenced by genes present on the vir and T-region. Broadly, *Agrobacterium* strains can be classified according to the typical opines present in tumour as octopine, nopaline, leucinopine and succinamopine type strains. Two important strains are A_6 and C_{58}. Whereas strain A_6 utilizes octopine, strain C_{58} utilizes nopaline synthesized in the crown gall.

GENE TRANSFERRING COMPONENTS

Agrobacterium carries three genetic components required for transformation viz.

1. the T-DNA which is the only mobile DNA element which is transferred to the plant cell that ironically does not encode the products that mediate its transfer
2. the Ti plasmid virulence region that provides the trans-acting products for T-DNA transit
3. the *Agrobacterium* chromosome which has 4 regions shown to be directly involved in attachment to plant cells

A plant cell becomes susceptible to *Agrobacterium* when it is wounded. Such wounded cells produce an abundance of low molecular weight phenolic compounds such as acetosyringone (AS) or hydroxyacetosyringone (OH-AS) which act as specific inducers of vir gene expression.

T-DNA structure

The T-DNA also termed as transferred DNA is the only mobile genetic components which is integrated into the plant nuclear genome during invasion of wounded cells by *Agrobacterium*. The T-DNA carries genes that are specifically expressed in the plant and play a direct role in tumorigenesis. These genes catalyze the biosynthesis of growth factors implicated in tumour formation. There are four genes involving tumour morphology and physiology. Two of these genes, the *aux* genes induce shooty tumours and are called *tms* (tumour morphology shoot) or *shi* (shoot inhibition) genes. These two *aux* genes are now known as *iaaM* (*aux1, tms1*) and *iaaH* (*aux2, tms2*). The protein coded by *aux1* is a monooxygenase capable of converting tryptophan into indol acetamide (IAM) whereas *aux2* gene encodes indolacetamide hydrolase whose hydrolase activity converts IAM into IAA. The third gene is the *cyt* gene also called the *tmr* (tumour morphology root) or *roi* (root inhibition) gene. The protein encoded by this gene is an isopentenyl transferase capable of catalyzing the formation of the cytokinin isopentenyl AMP. Hence, the *cyt* gene is now known as *ipt* gene. The fourth gene termed 6b causes tumour formation in certain plant species. This gene is conserved among all T-DNAs examined. An additional T-DNA gene is *ons* which controls octopine and nopaline secretion. Expression of the T-DNA genes lead to production of phytohormones- auxins and cytokinins and the opines. Although the T-DNA genes have their own expressional signals their expression is influenced by the neighbouring chromosomal sequence.

Opines: Opines are condensation products of organic acids with basic amino acids not found in normal tissues. These compounds serve as specific substrates for food consumed specifically by the infecting Agrobacteria and stimulate plasmid conjugative processes as well as virulence gene expression. Together these opines are probably essential for ecological specificity and for fostering competitive advantage to *Agrobacterium* species in nature. The most common ones are octopine and nopaline though agropine and agrocinopine family of opines may be present. Metabolism of opines is a central feature of crown gall disease. Opine synthesis is a property conferred upon the plant cell when it is transformed by *Agrobacterium*. The type of opine produced depends upon the bacterial strain. The bacteria catabolizes the opine through genes for opine catabolism located on Ti plasmid (Fig 1) and uses it as a source and/or nitrogen and energy. Thus bacteria that utilize octopine induce tumours that synthesize nopaline. Certain opines induce not only the opine catabolism genes but also the transfer (tra) genes responsible for conjugative transfer of the Ti plasmid. The inducers for the *tra* gene of octopine, nopaline and leucinopine Ti plasmids are octopine, agrocinopine A and agrocinopine C respectively.

Border sequences: The T-DNA of *Agrobacterium* is flanked by two 25bp imperfect direct repeats or border sequences within which lie the T-DNA ends points. That these repeats are conserved while the surrounding sequence are not suggests that the repeats are important in the transfer and integration mechanism. Any DNA sequence placed between these borders can be transferred into plant cells. Whereas tumour formation is abolished when the right border is removed, removal of left border does not affect it. The border sequences are the only cis-required elements for T-DNA transfer, but the transfer is stimulated by the presence of another 24 bp sequence called overdrive adjacent to right border repeat.

The virulence region

In addition to the T-region, the Ti plasmid contains the virulence or *vir* region which is involved in tumorigenesis. The *vir* region with a size of about 40 kb is located adjacent to the left border repeat of the T-DNA. This region provides most of the products that mediate actual T-DNA movement. Mutation in this region generally lowers the virulence. The T-region and *vir* genes can act in trans to mediate the transfer of the DNA that is present between the border sequences. This forms the basis of the widely used binary vector system for plants.

The vir genes: Besides the oncogenes present in the T-DNA, a large number of other genes involved in tumorigenicity are present in the virulence region. Whereas, the octopine Ti plasmid *vir* region consists of 8 operons- virA to virH, the nopaline Ti plasmid lacks the virF gene buts contains another gene *tzs* which produces a cytokinin transzeatin. While 5 of the operons (virA, virB, virD virE and virG) are absolutely essential for virulence, the remaining 3 (virC, virE, virF and virH) influence plant host range or leads to an attenuation of tumorigenicity. Expression of the virulence genes are positively regulated at the transcriptional level by plant signal molecules. They are phenolic compounds which are exuded from wounds. Also certain monosaccharides synergize the action of the phenolic signals.

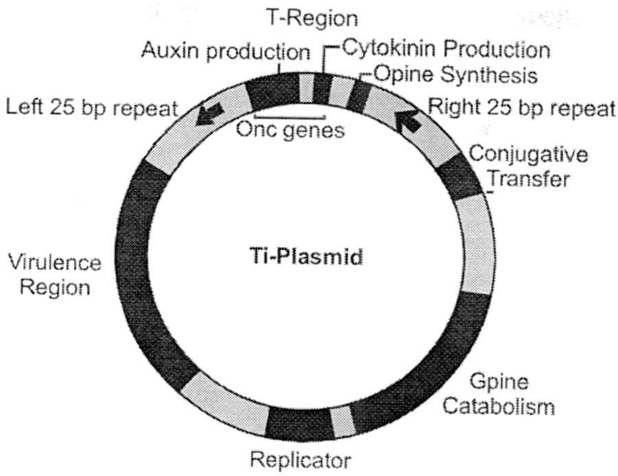

FIG. 1. Genetic map of Ti Plasmid.

The virulence genes virA and virG form a two component regulatory system that induces *vir* gene expression in response to plant phenolic compounds. These genes mediate two distinct type of responses ie (i) chemotaxis at low concentrations of inducer and (ii) induction of vir genes at higher inducer concentrations. Chemotaxis enables the bacterium to migrate towards plant cells whereas vir gene induction results in the production of proteins which are essential for the transfer of T-DNA to plant cell nucleus. The virA protein is an innermembrane protein which acts as a sensor and receptor for plant phenolic and monosaccharide signals. The signal detected by virA is transduced to the VirG protein probably by phosphorylation. The VirG protein is a DNA binding protein and probably acts as a transcriptional activator of the other vir operons. Following vir gene induction, the production of a transfer intermediate begins with the generation of the T-strand, a single stranded copy of the T-DNA. The virD operon is responsible for this process encoding 2 proteins VirD1 and VirD2 which recognize the 25bp border sequence and produces a single endonucleolytic cleavage in the bottom strand of each border. The T-DNA is transferred with the right border leading. This probably accounts for the observation that only the right repeat sequence is essential for transfer. The VirD2 protein attaches to the 5' end of the processed DNA and is thought to act as a pilot protein in the T-DNA transfer process because it has nuclear localization signals and can direct a β-glucuronidase reporter protein to the nucleus of plant cells. Therefore, the VirD protein protects the T-DNA against nucleases, targets the DNA to the plant cell nucleus and helps its integration into the plant genome. The transport of the T-strand across the bacterial membrane is brought about by the virB operon. The virE locus encodes VirE2, an inducible single stranded DNA protein complex that binds tightly and cooperatively completely coating the single stranded T-strand thereby creating a thin and extended protein-DNA complex thus protecting the protein-DNA strand from nucleolytic degradation. Later, binding of VirE2 unfolds and extends single stranded DNA to a narrow diameter of 2nm which may facilitate transfer through membrane channels. The T-strand along with VirD2 and VirE2 are termed the T-complex.

The vir operons virC, virF and virH affect the efficiency of T-DNA transfer only on certain host species. The virC operon codes for two proteins VirC1 and VirC2 both of which appear to be involved in the host dependent attenuated virC mutant phenotype. The VirC1 protein enhances T-DNA processing by interacting with the overdrive sequence but no function has been assigned to VirC2. The VirF protein mediates T-DNA transfer from *Agrobacterium* to the plant cells. The presence of virF gene can be either stimulatory as in *Nicotiana glauca* or inhibitory as in *Zea mays* for T-DNA transfer depending upon the plant species. The virH operon encodes two cytochrome P_{450}- like enzymes which is probably involved in the detoxification of bacteriostatic and bacteriocidal compounds present in plant wound sap.

Chromosomal vir genes

Genetic studies have implicated that genes on the *Agrobacterium* chromosome influence virulence through the following characteristics:

1. causing positive chemotactic movement of the bacterial cells towards wounded sites
2. conferring attachment to exposed plant cell walls
3. promoting growth efficiency in wounded sites, and
4. inducing and regulating *vir* genes on the Ti plasmids

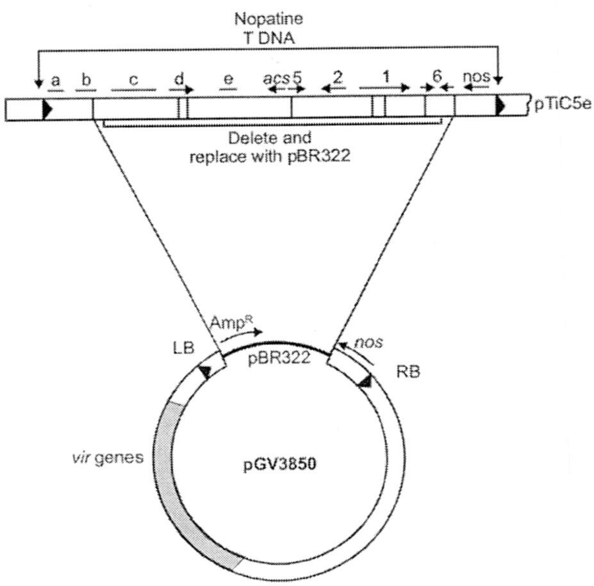

FIG. 2. Disarmed Ti Plasmid (T-DNA replaced).

Chromosomally located loci chvA, chvB and pscA (exoc) are required for virulence since they facilitate attachment of the bacteria to plant cells. These genes affect biosynthesis and secretion of polysaccharides such as beta-1,2 glucan and succinoglycan. Mutations in any of these chromosomal loci result in defective plant cell attachment and consequent lack of virulence. Another, gene *ros* (rough outer surface) independently affects the regulation of vir region. Mutations in the *ros* gene depresses virC and virD operons of the vir region of both octopine and nopaline Ti plasmids and allows for double stranded T-intermediate formation in the absense of induction with inducers such as acetosyringone and virA/virG regulation.

AGROBACTERIUM VECTORS

Exploitation of T-DNA as a transformation vector is based on the fact that it is stably integrated into the plant nuclear genome. Wild type Ti plasmids are not suitable as plant vectors because of their large size and also due to the fact that they cause disorganized growth of the recepient plant cells due to the biosynthetic genes in the T-DNA. This necessiated the development of *disarmed* Ti plasmid in which the phytohormone genes are deleted from the T-DNA or the T-DNA is completely removed (Fig. 2). Zambryski *et al* (1983) disarmed a nopaline Ti plasmid pTiC58 by deleting all the T-DNA except for the left and right border sequences and the *nos* gene and replacing it with a region of pBR322. The resulting plasmid pGV3850, is non-oncogenic and infection of plant tissue with *Agrobacterium* carrying pGV3850 produced callus without any tumour formation. However, disarmed Ti plasmid cannot be used as primary cloning vectors because of their large size and absence of unique restriction sites. Moreover, deletion of the tumour causing genes makes it extremely difficult to identify plant cells that have received the disarmed T-DNA as growth of plant cells become normal. Thus a good selectable marker is needed to identify plant cells that have received the T-DNA which should be located within the T-DNA region of the disarmed Ti plasmid. Genes that are to be transferred are cloned between the border repeats along with a plant selectable marker and a bacterial selectable marker essential for induction of the vector into *Agrobacterium*. All these elements combined together, form a *plant vector expression cassette*.

Plant gene vectors derived from T-DNA are the most commonly used due to the following advantages:

1. accurate transfer of foreign DNA inserted in T-DNA
2. T-DNA encoded oncogenic genes are not required for the transfer of the T-DNA
3. the DNA transfer simulates the natural system as the function of the border sequences is analogous to that the conjugational tansfer origins of bacterial plasmids
4. the T-DNA though lodged in a bacterial plasmid contains plant cell specific expression signals which are expressed only in plant cells

Transformation vectors are either cointegrate or binary,

Cointegrate Vector

A cointegrate vector includes a region of homology between the vector plasmid and the Ti plasmid. The foreign DNA is first inserted by in-vitro recombination into an intermediate vector that cannot replicate in *Agrobacterium* but can recombine with the Ti/Ri plasmid through a single or double crossover event at a homologous site. Formation of a cointegrate vector using disarmed vector pGV3850 is illustrated in Fig. 3. This vector is useful because pBR322 DNA in the T-DNA provides the region of homology with most other cloning vectors. The foreign gene is inserted into a donor vector in *E. coli* eg. vector pLGVneo1103 which carries a plant selectable marker nos-neo gene and a bacterial selectable marker, the *kan* gene. Recombination between the homologous pBR322 DNA in the two plasmids produces the co-integrate which now carries a plant selectable marker and the foreign gene flanked by T-DNA border sequences.

A cointegrate *Agrobacterium* vector carrying a disarmed vector and a donor vector is developed through triparental mating. The three parental bacterial strains are

1. *E. coli* transformed with the donor vector pLGVneo113
2. *E. coli* containing plasmids pGJ28 and R64drd11 lodging *mob* and *tra* genes respectively
3. *tumefaciens* containing disarmed vector pGV3850

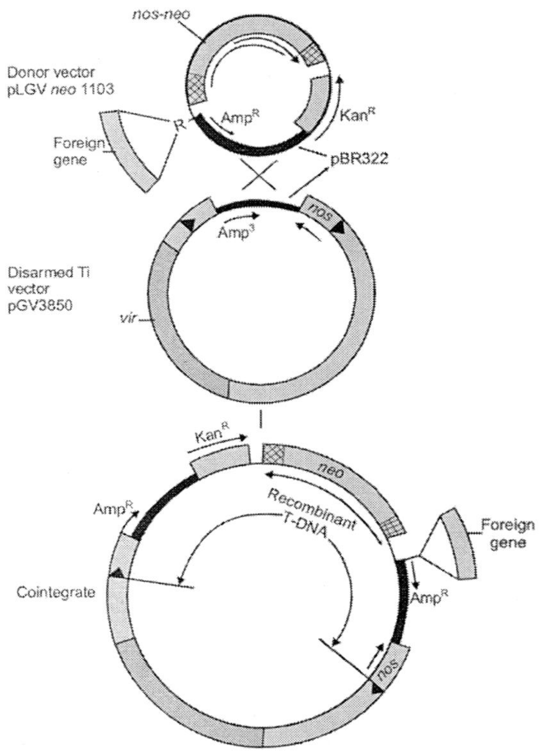

FIG. 3. Production of cointegrate plasmid-vector. The donor vector recombines with the disarmed vector by homologous recombination across the pBR322 sequences to form the cointegrate

When the three bacterial strains are mixed, R64drd11 mobilizes itself and pGJ28 into the *E. coli* strains containing pLGVneo1103. The *mob* proteins encoded by pGJ28 interacts with a specific site *bom* which is retained in pBR322. Conjugal transfer between *E. coli* strains is very efficient and 50% of the recipients contain all three plasmids. Recombination between these plasmids is prevented by a mutation in the recipient recA gene. All three plasmids are then transferred to the *Agrobacterium* recipient, pLGVneo1103 is mobilized by the helper *tra* genes of R64drd11 and the *mob* functions of pGJ28. None of the *E. coli* plasmids can replicate in *Agrobacterium* because they are not broad host range and therefore the only way pLGVneo1103 can be maintained is by recombination with pGV3850 to produce a cointegrate. Strains carrying cointegrates are selected by growth on minimum medium containing kanamycin as they are prototropic and the *kan* gene is provided by the donor vector.

The main advantage of the cointegrate vectors is their high stability in *Agrobacterium*. However, the two disadvantages are the detailed knowledge required of the Ti plasmid before they can be manipulated and the relatively low rates of cointegrate formation.

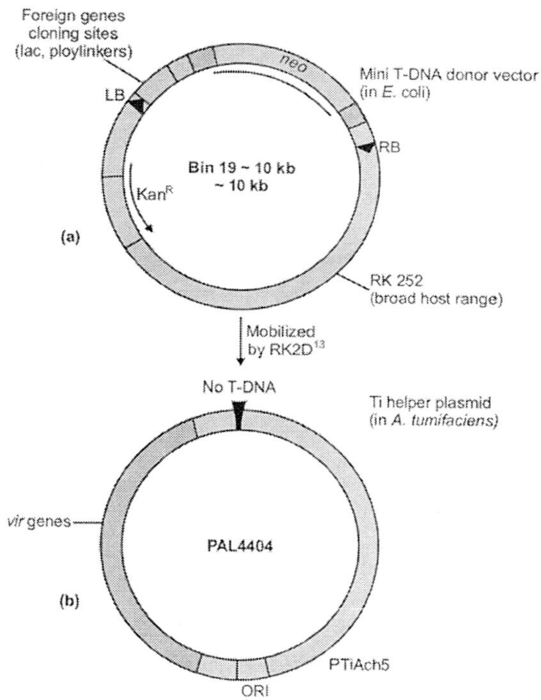

FIG. 4. A binary vector system consisting of (a) a T-DNA donor vector with selectable marker genes and cloning sites flanked by border repeats and (b) a disarmed helper Ti plasmid carrying *vir* genes.

Binary Vector: As in-vivo recombination is required for the production of cointegrates, the precise structure of the recombinant T-DNA is not known compared to vectors constructed in-vitro. In addition, unwanted events can occur in-vivo and the gross structure of the co-integrates must always be checked. These disadvantages can be overcome by using binary vector systems. A binary vector consists of 2 plasmids:

1. a disarmed helper Ti or Ri plasmid carrying the *vir* gene functions
2. a mini T-DNA donor vector containing bacterial and plant selectable marker genes and cloning sites flanked by T-DNA end repeats

Both plasmids must be able to replicate in *Agrobacterium* to avoid the need for cointegrate formation. Most binary vectors are derived from the well characterized wide host range plasmid RK2 that can be efficiently mobilized between *.E. coli* and *Agrobacterium*.

A typical binary vector is shown in Fig. 4. The disarmed helper Ti plasmid is a derivative of the octopine plasmid pTiAch5. The mini T-DNA donor Bin19, contains a truncated T-DNA from the nopaline plasmid pTiT37 comprising the left border and right border sequences and a nos-neo hybrid gene as a selectable marker. A lac polylinker is inserted in the mini T-DNA to provide unique cloning sites and a colour test for insertion. The linker carries the alpha complimentary region of β-galactoside which will provide blue colonies on X-GAL indicator plates and white colonies if it is disrupted by insertion. The T-DNA is inserted into derivative of broad host range plasmid RK2 which has been modified by the addition of kanamycin resistant gene from *Streptococcus*. The plasmid can be mobilized to *Agrobacterium* with a helper plasmid pRK2013 by a mixed plate mating between *E. coli* pRK2013, *E. coli* Bin 19 and *A. tumefaciens* pAL4404 or the mini T-DNA can be directly transformed into *A. tumefaciens*.

An advanced binary system has been developed reffered to as PCV (plant cloning vector) system which is based on a plant vector cassette that carries only the oriR (replication origin) and oriT (origin of conjugational transfer) regions of plasmid RK2 in combination with diverse T-DNAs. The binary system simplifies the transfer of foreign genes to plant cells, the main advantages being their independence of specific Ti or Ri plasmids and the high frequency of introduction into *Agrobacterium*. However, a disadvantage is its instability which requires frequent selection.

T-DNA PROCESSING, TRANSFER AND INTEGRATION

Processing

The first step in the complex process of *Agrobacterium* mediated plant cell transformation is the generation of the T-strand or T-intermediate, specifically excised from the Ti plasmid following *vir* gene induction. In the presence of phenolic compounds, the regulatory proteins VirA and VirG activate the other vir genes, the expression of which results in nicks between the third and fourth base at the bottom strand of the border repeats by an endonuclease encoded by the *virD1* and *virD2* genes. The nicks are used as initiation and termination sites for the displacement of a linear single stranded copy of the bottom strand of the T-DNA region. Single stranded T-DNA copies are generated probably by the replacement synthesis of the bottom strand. T-DNA strand synthesis occurs from right to left border repeat (5' to 3' end). The protein VirD2 remains covalently attached to the 5' end of the T-strand and protects it from exonucleolytic degradation. Another vir protein, the virE2 binds to the single stranded DNA therby creating a thin and extended protein-DNA complex which is subsequently transferred to the plant cell.

Transfer

Export of the T-complex from *Agrobacterium*, passing through the inner membranes as well as the bacterial cell in the transfer process is not well understood. The transfer mechanism seems to involve a number of genes of the vir region. Proteins encoded by virB, virC and virE operons are likely involved in piloting the T-complex for passage through bacterial membrane and culmination in its delivery into the plant cell by a process thought to be analogous to bacterial conjugation and is mediated by further products of the vir genes. All evidence to date supports the hypothesis that virB is involved in the transfer of the T-complex out of the bacterium. Subcellular localization studies confirm the membrane association of seven virB proteins of which two viz VirB4 and 11, provide the energy for translocation of the T-complex. VirB4 has a nucleotide binding site that is essential for virulence and VirB11 is reported to be associated with the cytoplasmic face of inner membrane.

It exhibits both ATPase and protein kinase activity. Sequence data and comparision to counjugative DNA transfer systems indicate that at least some of the virB products form a membrane-associated complex that may be functionally similar to a pilus. The remaining VirB proteins may assist in assembly of the complex or provide ancillary functions during T-complex transfer.

Integration

The final step in the genetic transformation of plant cells is integration of the T-DNA copy, the T-strand into plant cell DNA. The exact T-DNA integration into the plant remains obscure. Integration requires nuclear uptake which is a tightly regulated process that can occur only through the nuclear pore. Nuclear transport of T-complex is most likely mediated by its associated proteins VirD2 or VirE2. Proteins larger that 40kD require a nuclear localization signal (NLS) that mediates their nuclear uptake. Although VirD2 possesses a functional NLS necessary to mediate nuclear uptake of T-complex, the T-complex is an extremely large nucleoprotein complex and a single bipartite NLS would be incapable to mediate this process. Sequence analysis revealed two potential bipartite NLSs VirE2 and experiments with mutants wherein one or the other NLSs were deleted revealed that both NLSs were required for maximum nuclear transport. Thus VirE2 provides NLSs along the entire length of the T-complex.

After T-DNA is transferred into the plant cell, it becomes randomly integrated at various chromosomal sites, invading the target region of the genome by utilizing or creating gaps in the double stranded DNA. This DNA insertion is mediated either by plant host DNA enzymes or the T-DNA encodes its own set of enzymes of efficient integration. In general, T-DNA insertion occur mainly in the transcriptionally actives sites. DNA regions undergoing transcription are presumably more accessible to invading exogenous DNA owing to requisite unravelling of nucleosomes and other chromatin structures. Nucleotide sequences overlapping the T-DNA ends and the target DNA before and after integration was studied. Generally, T-DNA before and after integration was studied. Generally, T-DNA integration does not cause large rearrangements of plant DNA, however, most insertions induce small deletions upto 70 nucleotides at the target sites. Another consistent result is the presence of "filler" DNA from a few to upto 33 nucleotides at the T-DNA plant junctions; such filler DNA is composed of sequences; homologous to plant DNA close to the junctions. Short stretches of homology (5-10 bases) between the T-DNA ends and plant target sites may play a role in integration. Basically, the 5' end of the T-DNA strand invades a nick in the plant DNA. Short stretches of homology in the order of one to a few nucleotides allow partial pairing between both ends of the T-strand and plant DNA. The plant DNA may further unwind to form a gap. Any overhangs to T-DNA are repaired and ligated to the plant DNA ends. Gap repair and DNA synthesis using the T-strand as template result in the final integration product. After its integration, T-DNA adopts eukaryotic features of chromatin organization in terms of nucleosome organisation and DNaseI sensitivity. The integrated T-DNA is genetically stable and segregate in Mendelian fashion. However, from time to time regions within the T-DNA can become methylated and the resultant reduction in gene expression can result in phenotypic instability.

T-DNA AS A GENE TAG

T-DNA can be used as an insertional mutagen and by extension, a gene tag. Gene tagging involves the insertion of a defined piece of DNA in the vicinity of a target gene or DNA sequence. Insertion of the foreign DNA can alter the expression of the neighbouring gene resulting in either a gain or loss of function which produces a screenable phenotype. Once it is established genetically that the gene tag is linked to the DNA sequence in question, hybridization probes based on the sequence of the gene can be used for sequence isolation. A number of vectors have been constructed to use T-DNA as an effective gene tag.

Non-specialized T-DNA transformation vectors were used to generate a population of insertion mutants in *Arabidopsis*. The *Agrobacterium* vector system was used to tag and identify plant genes influencing plant morphology such as plant height, flower and trichome morphology. A simple seed transformation protocol was developed with which large number of independent T-DNA tagged mutants were obtained. Using a T-DNA tagged homeotic mutant, the Arabidopsis gene agamous was identified which was found to encode a transcriptional regulator necessary for flower development. Further isolation and characterization of that T-DNA tagging is an efficient approach to identify genes regulating basic processes such as photosynthesis and differentiation in plants.

A wide variety of special tagging vectors have been constructed with a number of different markers designed specifically for a defined function. Specialized vectors which have a promoterless resistance or marker gene located closely to the border repeat have been developed for plant gene identification. Activation of expression can occur after integration into the transcriptionally active area. By the analysis of gene expression pattern of the reporter construct in the tagged transgenic plant, gene may be identified that have a tissue or organ specific expression pattern. Also, it might be possible to identify genes involved in the regulation of growth and development.

GENE TARGETTING VIA AGROBACTERIUM

Gene targetting refers to targetted integration obtained via homologous recombination between a chromosomal locus and an incoming DNA molecule. Through gene targetting, specific sequences can be modified for genetic engineering purpose or for the functional characterization of newly isolated genes. For example, the coding region of a marker gene can be fused in-vitro to the coding region of the isolated gene of interest. After introducing such a construct into a plant cell, the marker can be targetted to the endogenous gene in the chromosome and after gene replacement, the expressional pattern of the chromosomal gene copy can be studied in-vivo via analysis of marker gene expression. The *Agrobacterium* vector system has been found to be highly suitable and is being exploited for gene targetting. Integration of foreign DNA into the genome of plant and animal cells occurs in random positions via illegitimate recombination. Thus though several methods have been developed for foreign gene introduction into plant cells, a frequent problem encountered in transgenics is that the expression levels of newly introduced genes vary greatly between different individual transformants. Apparently, the site of integration influences the expression of the introduced genes and this phenomena is reffered to as "position effect" which occurs either due to DNA methylation or insertion of multiple T-DNA copies. To avoid position effects, gene targetting was introduced to ensure gene expression.

The plant species *Nicotiana tabaccum* and *Arabidopsis thaliana* have been used for gene targetting via the *Agrobacterium* vector system. There were initial doubts whether the T-DNA transfer intermediate would permit homologous recombination with a chromosomal gene due to the presence of proteins in the T-complex which could interfere with the recognition of sequence homology. But Lee et al (1990) showed that two co-introduced homologous T-DNAs do recombine extrachromosomally. Further, they demonstrated the restoration of a defective kanamycin resistance gene already present in the genome via *Agrobacterium* gene targetting. *N. tabaccum* was transformed with a T-DNA construct that carried a hygromycin resistant HPT gene and a defective kanamycin resistant NPTII gene. Protoplast of a transgenic tobacco line containing 2 copies of this T-DNA construct in an inverted repeat structure were cocultivated in an *Agrobacterium* strain that carried repairT-DNA 101. After selection for kanamycin resistance, PCR analysis and regeneration, a plant line was obtained from a pool of 10^5 transformants in which restoration of the defective gene at the target locus had occured. The shift of the NTPII hybridizing HindIII fragment from 2.5 to 3.7kb was indicative of successful gene targetting event. Two additional HindIII fragments hybridized with the HPT probe in the recombinant line as compared to the target line. This indicated that 2 truncated T-DNA inserts carrying sequences of the hygromycin resistance gene had integrated during the transformation event.

ADVANTAGE OF AGROBACTERIUM GENE TRANSFER SYSTEM

The versatility of T-DNA vectors that exploit the natural gene transfer process is such that T-DNA can be used for a variety of purposes other than simple gene transfer into plants. Hence, *Agrobacterium* vector system is the most extensively used gene tranfer technique. Some of the major advantages of this system are listed below.

1. Foreign DNA inserted in the T-DNA is accurately transferred from *Agrobacterium* to plants. The disadvantages met in direct gene transfer, are the substantial DNA arrangement or integration of multiple DNA copies that may be prone to recombination, rearrangement or silencing. This is probably due to the structure with which the T-complex is delivered into plant cells and the activity of the *vir* proteins in the plant cells.
2. In contrast to most direct DNA uptake methods of plant transformation that require plant cell protoplasts, *Agrobacterium* is capable of infecting intact cells which can be regenerated easily. As a result, tissue culture limitations are much less of a problem in *Agrobacterium* based transformation systems, than in direct DNA uptake.
3. Transformation of plants with *Agrobacterium* has advantages over protoplast uptake of DNA or biolistics as the transformation rate is higher (0.1-5% as compared with 0.01-1%), more efficient and with predictable patterns for integration of the foreign DNA.
4. *Agrobacterium* serves as an ideal insertional mutagenesis vehicle as it introduces one to several copies of the transferred DNA into the plant genome at one or a few loci.

5. This system has also been used to tag and identify plant genes influencing plant morphology such as plant height, flower morphology, trichome formation.

In addition, knowledge about the *Agrobacterium* system has provided basic information fundamental to other areas of biology. The discovery of plant signal molecules that induce *vir* gene expression has led to discoveries of similar types of communication occuring between other soil bacteria and their plant hosts. Also, the identification of nuclear localization signals in VirD2 and VirE2 proteins establishes that such signal sequences are evolutionarily conserved between plant and animal kingdoms. Special T-DNA vectors have been developed for gene identification, complementation and cell ablation experiments.

DISADVANTAGES

Although the introduction of new traits into plant via *Agrobacterium* is a common practice, there are still shortcomings in this system.

1. It is sometimes difficult to transform cells in a tissue that are able to regenerate. It might be that these are in layers too deep to be reached by *Agrobacterium* or simply are not targets for T-DNA transfer.
2. *Agrobacterium* has limited host range and inherent failure to transform monocotyledonous plants which include cereals, the most economically important crops, restrict its widespread use for plant transformation. However, reports of its success in cereals have started to appear.

PERSPECTIVES

The interaction of *Agrobacterium* with plant cells is the only known natural example of interkingdom DNA transfer and has provided many fundamental informations in biology. Development and use of plant vectors based on the *Agrobacterium* T-DNA transfer system and analysis of the molecular of T-DNA transfer has contributed tremendousely to the field of molecular biology and biotechnology by making the genetic modification of plants more precise and sophisticated. Studies in the mechanism of T-DNA transfer have facilitated further improvement of T-DNA vectors and plant transformation systems. T-DNA genes exploited to alter plant morphology and development genetically may also be applied as experimental tools in studies of hormone signal transduction, cell division or organ differentiation. T-DNA induced gene fusion to suicide marker genes that cause cell lethality will help in the isolation of genes that are expressed only in certain cell types or during certain stages of development. Alternatively, T-DNA mediated insertion of strong promoters into the plant genome may be used for dominant activation of life defense genes involved in tolerance to drought, heat, salinity or toxic chemicals. Transformation with T-DNAs carrying segments of the plant genome may result in recombination with homologous chromosomal loci that can be exploited in development of site-specific mutagenesis technique for plants. Furthermore, combination of T-DNAs with telomeric elements of plant chromosomes should facilitate the identification of centromeric and autonomously replicating (ARS) DNA sequences to achieve chromosome engineering in plants. Although monocots, particularly cereals have been considered outside the host range of *Agrobacterium*, information on tumour formation, opine production and presence of T-DNA is accumulating and indicates that monocotyledonous plants can be infected with *Agrobacterium tumefaciens*. With recent advances in understanding the infection process with respect to meristematic target cells, *vir* inducing compounds and wide host range *Agrobacterium* strains and availability of monocotyledon gene promoters and improved selectable markers, monocotyledon transformation systems with *Agrobacterium* have been greatly improved. In due course, its host range may perhaps be extended routinely to include most dicotyledons and monocotyledons.

REFERENCES

Lee KY, Lund P, Lowe K, Dunsmuir P. 1990. Homologous recombination in plant cells after *Agrobacterium* mediated transformation. Pl. Cell 2: 415:425.

Smith EF, Townsend CO. 1907. A plant tumour of bacterial origin. Science 25: 671-673.

van Larebeke N, Genetello C, Schell J, Schilperoort RA, Hermans JP, van Montagu M. 1975. Aquisition of tumour inducing ability by non-oncogenic agrobacteria as a result of plasmid transfer. Nature 255: 742-743.

Zaenen I, van Larebeke N, Teuchy H, van Montagu M, Schell J. 1974. Supercoiled circular DNA in crown gall inducing *Agrobacterium* strains. J. Mol. Biol. 86: 109-127.

Zambryski P, Joos H, Genetello C, Leemans J, van Montagu M, Schell J. 1983. Ti plasmid vector for the introduction of DNA into plant cells without alteration of their normal regeneration capacity. EMBO 2: 2143-2150.

FURTHER READING

Hooykaas PJJ, Schilperoort RA. 1992. *Agrobacterium* and plant genetic engineering. Plant Mol. Biol. 19: 15-38.

Smith RH, Hood EE. 1995. *Agrobacterium tumefaciens* transformation of monocotyledons. Crop Sci. 35: 301-309.

Zambryski P. 1992 Chronicles from the *Agrobacterium* plant cell DNA transfer story. Annu. Rev. Physiol Pl. Mol. Biol. 43: 465-490.

Zupan JR, Zambryski P. 1995. Transfer of T-DNA from *Agrobacterium* to the plant cell. Plant Physiol. 1022: 1041-1047.

Synthetic Seed Technology as a Method for Plant Propagation and Delivery of Tissue Cultured Plants

P.S. Rao, T.R. Ganapathi, P. Suprasanna and V.A. Bapat

Commercial application of plant tissue culture technology has been demonstrated, however constraints of high labour and other expenses like nutrient medium and hardening and delivery of tissue cultured plants remains to be overcome. In this context, synthetic seed technology assumes greater importance because of low cost and high volume propagation system. Many plant systems are known to produce large number of somatic embryos in tissue culture. Such somatic embryos proceed through similar developmental stages like that of zygotic embryos and do possess root and shoot meristem connected by a common vascular system. Like zygotic embryos somatic embryo tolerates severe desiccation. Embryo specific proteins have been located in somatic embryos.

Todate synthetic seeds have been prepared by using a suitable encapsulating matrix in many plants. The main thrust is to prepare a simple inexpensive delivery unit of in vitro propagated plants. Synthetic seed technology offers many useful advantages on a commercial scale. The resultant plant population from the synthetic seed will be uniform and the direct delivery of somatic embryos will save many subcultures to obtain plantlets from regenerated embryos. The encapsulated embryos could also be packed with pesticides, fertilizers, nitrogen fixing bacteria and even microscopic destroying worms (Redenbaugh *et al.*, 1986; Ohishi *et al.*, 1995).

However, successful development of synthetic seeds involves several important features such as (1) the ability of *in vitro* cultures to produce large number of somatic embryos over a long time span, (2) synchronized development of somatic embryos, (3) the capacity of the encapsulated embryos to withstand the dessication shock and to germinate into a plant. The scale-up operation of somatic embryos in a bioreactor followed by mechanization of the process encapsulation is the last step for sowing of synthetic seeds in large numbers.

In addition to using somatic embryos, axillary buds, adventitious buds and shoot tips have also been used (Table 1) in the preparation of synthetic seeds (Redenbaugh, 1993; Bapat and Rao, 1988, 1990; Ganapathi *et al.*, 1992, 1994). The main advantage of using these vegetative propagules is that it can be a useful delivery system for tissue cultured plants by eliminating rooting and hardening phases by directly sowing the encapsulated beads in the soil. Also in a plant species where somatic embryogenesis is not established, these alternative explants could be useful for the production of synthetic seeds.

Redenbaugh *et al.* (1988) identified various stages in the production of somatic embryo and their utilization as artificial seeds.

1. Selection of candidate crops based on both technological and commercial potential.
2. Optimization of somatic embryogenesis system from cultured cells.
3. Optimization of embryo maturation
4. Automation of embryo production
5. Production of mature, synchronized embryos
6. Encapsulation of embryos with necessary adjuvants
7. Coating of encapsulated embryos
8. Optimization of green house and field conditions for the conversion of embryos to plants
9. Delivery system for artificial seed in terms of increasing productivity.

TABLE 1. Crop plants in which artificial seed production and plant conversion has been demonstrated

In vitro propagules for encapsulation	Crop	Reference
Somatic embryos	Alfalfa	Redenbaugh *et al.*, 1984
	Celery	Redenbaugh *et al.*, 1986
	Brinjal	Rao & Singh, 1992
	Carrot	Kittoo & Janick, 1985
	Brassica	Redenbaugh *et al.*, 1987
	Lettuce	Redenbaugh *et al.*, 1987
	Sandalwood	Bapat & Rao, 1988
	Rice	Suprasanna *et al.*, 1994
	Horseradish	Shigeta & Sato, 1994.
Axillary buds/	Mulberry	Bapat *et al.*, 1987
Adventitious buds/		Machi, 1992
	Eucalyptus	Huang & Zhu, 1990
	Vitis	Huang *et al.*, 1990
Shoot tips	Banana	Ganapathi *et al.*, 1992
	Cardamom	Ganapathi *et al.*, 1994
	Carum carvi	Furamanova *et al.*, 1991

The first application of artificial seed technology has been made in alfalfa and celery (Redenbaugh *et al.*, 1986). Somatic embryos were encapsulated as single embryo beads (Ca. 4mm in diameter) to produce individual artificial seeds. The somatic embryos were mixed in sodium alginate and then dropped in a solution of calcium chloride to form calcium-alginate beads. It is now possible to grow alfalfa plants from artificial seeds planted directly in the green house with >50% conversion (Fuji *et al.*, 1989, 1992) and initial field trials have been completed at (Redenbaugh, 1991). The production of seedling quality plants from individual somatic seeds has potential as a novel, universal delivery system.

At the Biotechnology Division of BARC, research on the development of protocols for synthetic seeds using somatic embryos, axillary buds and shoot tips is in progress in five economically important plants, sandalwood, rice, mulberry, banana and cardamom. The following pages describe the results obtained in this direction.

Rice is the world's most important food crop and a primary food source for more than one third of the World's population. This crop has received considerable attention in biotechnological research programmes. Research on artificial seeds in rice is still in infancy and this technology through somatic embryogenesis would offer a great scope for large scale propagation of superior, elite hybrids (Brar and Khush, 1994). We have reported high frequency somatic embryogenesis from indica rice cultivars (Suprasanna *et al.*, 1995) and utilized this embryogenic system for the production of artificial seeds.

Callus cultures were established from mature seed embryos on MS (Murashige and Skoog, 1962) medium supplemented with 2 mg/l 2,4-D. White, compact embryogenic callus developed somatic embryos on medium with low cytokinin, ABA, casein hydrolysate, sucrose and mannitol. Mature somatic embryos (2-3 mm, with scutellum and coleoptile) were used for encapsulation in 3% sodium alginate prepared either in MS or White's (White, 1946) medium. The encapsulated embryos were placed on nutrient media and on different substrates (sterile absorbent cotton, filter paper and macpeat). The encapsulated embryos developed into plantlets on nutrient media and filter paper followed by cotton. Filter paper moistened with MS basal medium gave higher frequency of germination (83.3%) compared to other substrates (25-80%). The encapsulated embryos developed vigorously and rapidly as compared to non-encapsulated embryos.

Redenbaugh (1992) and Brar *et al.* (1994) emphasized the need for research on artificial seeds in rice through somatic embryogenesis and outlined its impact on mass propagation of true breeding hybrids which can lead to commercialization of this technology. Also artificial seeds can be useful in handling, transportation and delivery of embryos in a safer and easier way. Encapsulated embryos can be stored for the preservation of valuable germplasm.

Banana is an economically profitable fruit crop with a large in-country consumption and a considerable export potential. Edible triploid bananas are vegetatively propagated by suckers as viable seeds are generally not produced in these cultivars. New and effective means of propagating bananas would be advantageous over the conventional use of sucker material, for germplasm maintenance, exchange and transportation. Multiple shoot cultures were established from isolated shoot tips of banana cv. Basrai (Cavendish dwarf) grown on MS medium with benzyladenine. Individual shoot tips excised from the multiple shoot cultures were encapsulated in 3% sodium alginate solution containing nutrient medium, antibiotic mixture (rifampicin + cefatoxime + tetracycline) and activated charcoal. Encapsulated shoot tips were placed on different media and substrates such as cotton, filter paper, soilrite, soil (all moistened with 1/4th strength MS salts) for germination. Among the different media tested, use of White's medium gave 100% plantlet development within a week. Among the different substrates, % plantlet formation was comparatively better on moistened cotton followed by filter paper. Encapsulated shoot tips directly sown in petridishes containing autoclaved soil showed the emergence of shoots (10%) within a week, however, these failed to form complete plantlets.

The results suggest that encapsulated shoot tips of banana can be handled like a seed, and could be useful in minimizing the cost of production as 1 ml medium is sufficient for encapsulation of shoot tip compared to 15-20 ml medium required for the conversion of shoot tips into plantlets. Also two stages such as rooting of shoots and hardening of plantlets can be totally eliminated by directly sowing the encapsulated shoot tips in the soil. Furthermore transportation and exchange of banana germplasm through artificial seeds rather than suckers would be easier, relatively inexpensive and safer.

Cardamom, often called as the *queen of spices* is a plantation crop with a considerable export earnings. It is propagated vegetatively as well as through seeds. Since it is a highly cross pollinated crop, plants obtained by seeds show lot of variation. Vegetative propagation is carried out through tiller multiplication, however, the rate of multiplication is slow and only three tillers are produced per year. In recent years, micropropagation of this crop has been demonstrated. We have standardized the procedures for the encapsulation of shoot tips of cardamom akin to the production of synthetic seeds (Ganapathi *et al.*, 1994). Shoot tips of cardamom var. Malabar were isolated from the sprouting buds of green house plants. Multiple shoot formation (Fig. 1A) was induced on MS medium supplemented with BA and NAA. The individual shoot tips excised from multiple shoot cultures were encapsulated in 3% sodium alginate (Fig. 1B) with different gel matrices. The encapsulated shoot tips were cultured on different substrates (soil, macpeat, cotton and filter paper; moistened with 1/4th MS salts). Among these substrates, maximum conversion frequency (83-100%) was recorded on filter paper irrespective of the gel matrix used. Conversion frequency was moderate on cotton and macpeat. Almost 100% conversion of the encapsulated shoot tips into plantlets was achieved on White's medium and the plantlets were grown successfully in soil (Fig. 1C & D). The encapsulated shoot tips can be handled like a seed and has advantages for transportation and delivery. This would minimise the cost of production of tissue cultured cardamom plants.

FIG. 1. Encapsulation of shoot tips of cardamom. A- multiple shoot cultures, B- shoot tips encapsulated in 3% sodium alginate matrix, C- emerging shoot and roots from the encapsulated shoot tips, and D- plantlets derived from encapsulated shoot tips paper cups.

CONCLUSIONS

Although synthetic seed research is being carried out in many laboratories, the principle limitation for commercialization has been the somatic embryogenesis. Despite the fact that somatic embryogenesis has been achieved for many species, during the last decade for many of the important, high value genotypes, somatic embryogenesis has not been reported. Therefore there is a need to shift the focus on somatic embryogenesis in valuable crop plants. Research on encapsulation of axillary buds, shoot tips or any other propagules should be taken up. Equally important is the need to develop new synthetic seed coatings for encapsulating embryos and other vegetative propagules. Synthetic seed technology in years to come would certainly find its application in plant propagation and delivery of tissue cultured plants.

REFERENCES

Bapat, V.A., Mhatre, M., Rao, P.S. 1987. Propagation of *Morus indica* L. (mulberry) by encapsulated shoot buds. *Plant Cell Rep.* 6: 393-395.

Bapat, V.A. and Rao, P.S. 1988. Sandalwood plantlets from synthetic seeds. Plant Cell Rep. 7: 434-436.

Bapat, V.A. and Rao, P.S. 1990. *In vivo* growth of encapsulated axillary buds of mulberry (*Morus indica* L.) *Plant Cell Tissue Org. Cult.* 20: 69-70.

Brar, D.S. and Khush, G.S. 1994. Cell and tissue culture for plant improvement. In: Basra AS (Ed) Mechanisms of plant growth and improved productivity: modern approaches. Mercel Dekker Inc., New York, USA. pp. 229-278.

Brar, D.S., Fujimura, T., McCouch, S. and Zapate, F.J. 1994. Application of biotechnology in hybrid rice, In: Virmani S.S. (Ed.) Hybrid rice technology: new developments and future prospects. International Rice Research Institute, Manila, Philippines, pp. 51-62.

Fuji, J., Slade, D. and Redenbaugh, K. 1989. Maturation and greenhouse planting of alfalfa artificial seed. *In vitro Cell Dev. Biol.* 25: 1179-1182.

Fuji, J., Slade, D., Aguirre-Rascon, J. and Redenbaugh, K. 1992. Field planting of alfalfa artificial seeds. *In vitro Cell Dev. Biol.* 28: 73-80.

Furamanova, M., Sowinaska, D. and Pietrosiuk, A. 1991. *Carum carvi* L. (caraway): *In vitro* culture, embryogenesis and the production of aromatic compounds. In: Bajaj YPS (Ed.) Biotechnology in Agriculture & Forestry Vol. 15. Medicinal and Aromatic Plants III. Springer-Verlag, NewYork, Heidelberg, pp. 176-192.

Ganapathi, T.R., Suprasanna, P., Bapat, V.A. and Rao, P.S. 1992. Propagation of banana through encapsulated shoot tips. *Plant Cell Rep.* 11: 571-575.

Ganapathi, T.R., Bapat, V.A. and Rao, P.S. 1994. *In vitro* development of encapsulated shoot tips of cardamom. *Biotech. Techniques* 8(4): 239-244.

Huang, M.J. and Cheng, Z. 1990. Preliminary studies on artificial seeds of eucalyptus. In: Li X-Q (Ed) Studies on artificial seeds of plants. Peking Univ. Press, Peking China, pp. 139.

Kitto, S. and Janick, J. 1985. Production of synthetic seeds by encapsulating asexual embryos of carrot. *J Amer. Soc. Hort. Sci.* 110: 277.

Machi, H. 1992. *In vitro* growth of encapsulated adventitious buds in mulberry (*Morus indica* L.) *Japan J. Breed.* 42 (3): 553-559.

Murashige, T. and Skoog, F. 1962. A revised medium for rapid growth and bioassays with tobacco tissue cultures. *Physiol. Plant.* 15: 473-497.

Ohishi, N., Sakamoto, Y. and Hirosawa, T. 1994. Synthetic seeds as an application of mass production of somatic embryos. *Plant Cell Tiss. & Org. Cult.* 39: 137-145.

Rao, P.V. and Singh, B. 1991. Plantlet regeneration from encapsulated somatic embryos of hybrid *Solanum melongena* L. *Plant Cell Rep.* 10: 7-11.

Rao, P.S., Ganapathi, T.R., Suprasanna, P. and Bapat, V.A. 1993. Encapsulated shoot tips of banana: a novel propagation and delivery system. *InfoMusa* 3: 4-5.

Redenbaugh, K., Paasch, B.D., Nichol, J.W., Kossler, M.E., Viss, P.R. and Walker, K.A. 1986. Somatic seeds: encapsulation of asexual embryos, *Bio/Tech.* 4: 797-801.

Redenbaugh, K., Fuji, J. and Slade, D. 1988. Encapsulated plant embryos in: Mizrahi, A. (Ed.) Biotechnology in Agriculture, AR Liss, New York, pp. 225-248.

Redenbaugh, K., Nichol, J., Kossler, M.E. and Paasch, B.D. 1984. Encapsulation of somatic embryos for artificial seed production, *In vitro* 20: 256-257.

Redenbaugh, K., Slade, D., Viss, P. and Fujii, J. 1987. Encapsulation of somatic embryos in synthetic seed coats, *Hort. Sci.* 22: 803-809.

Redenbaugh, K. 1991. Application of micropropagation to agronomic crops. In: Debergh, P. and Zimmerman, R. (Ed.) Micropropagation, Kluwer Academic Publ. New York. pp. 285-310.

Redenbaugh, K. 1993. SynSeeds: Applications of Synthetic seeds to Crop Improvement. CRC Press, Boca Raton, USA.

Shigeta, J. and Sato, K. 1994. Plant regeneration and encapsulation of somatic embryos of horseradish. *Plant Sci.* 102: 109-115.

Suprasanna, P., Ganapathi, T.R. and Rao, P.S. 1995. Establishment of embryogenic callus, somatic embryos and plant regeneration in indica rice. Jour. Genet. & Breed. 49: 9-14.

White, P.R. 1943. A Handbook of Plant Tissue Culture, Jaques Catell, England.

Maturation and Desiccation of Somatic Embryos

V.A. Bapat and P.S. Rao

Multiplication of desirable plants *in vitro* through biotechnological methods has emerged in late 1980s as a very valuable approach to increase plant productivity (Gupta *et al.*, 1993). Realising the paramount importance of this technology many laboratories have initiated basic research work to define the parameters for optimising plant production with the ultimate purpose of producing plants on a commercial scale for various benefits. The initial thrust was to identify and to propagate the ornamentals which are in great demand and extensive experiments had been carried out at the laboratory stage which have resulted in the development of effective protocols which were put to test through hundreds of trials in R and D centres as well as in fields. The results of these experiments and trials have proved beyond doubt that plant tissue culture techniques if properly applied is an important and effective method to produce plants and supplement conventional plant propagation programmes. Other related technologies such as processing, packaging, storage life and transportation are also being developed in order to make tissue culture plants more adaptable and easily marketable. It is against this background the idea of encapsulation of somatic embryos in nutrient gel was conceived (Redenbaugh *et al.*, 1991). The encapsulated embryos were termed as synthetic seeds or artificial seeds or somatic seeds. The *in vitro* occurrence of somatic embryogenesis holds promise for the production of synthetic seeds since somatic embryos can be processed to mimic natural seeds with respect to storage and handling, production of disease-free material, true to type clonal propagation and juvenility reversion of perennial species (Janick *et al.*, 1989). The other main advantage of utilizing somatic embryos is the anticipated suitability for rapid and cost effective production of large number of clonal plants in automated culture systems (Aitken-Christie and Connett, 1992).

The steps involved in the process of somatic embryogenesis are analogus to those involved in zygotic embryogenesis (Merkle *et al.*, 1990). Simulation of various physiochemical conditions present *in ovulo* during zygotic embryogenesis has therefore, quantitatively and qualitatively improved somatic embryogenesis (Gray *et al.*, 1993). Zygotic cell is the precursor for zygotic embryogenesis whereas a single cell or a small group of somatic cells are the precursors of the somatic embryos. Zygotic embryos undergo a definite mode of cell divisions which has provided the basis for both classification and laws of embryogeny. Somatic embryogenesis is highly plastic and is influenced by composition of nutrient medium and other factors and laws of embryogeny are not strictly followed (Rangaswamy, 1986). Krochko *et al.* (1992) listed several abnormalities of somatic embryos such as truncated cotyledonary development, precocious germination, recallusing, multiple or fused cotyledons, inability to germinate and stated that the extent to which somatic embryo development represents a faithful recapitulation of zygotic embryo development is still unknown.

Somatic embryogenesis can at best recapitulate events in zygotic embryogenesis with the production of embryos having shoot root apices (Merkle *et al.*, 1990). Somatic embryogenesis has been reported in over 350 species, although on the basis of well recognizable *ab initio* biopolarity and the consequent structural or physiological autonomy of the observed entities the instances of true somatic embryogenesis are 150 (Rangaswamy, 1991). Subcellular studies on somatic embryogenesis are not many. According to Rangaswamy (1991) striking increase in free ribosome, a regular alignment of membrane bound ribosomes, loss of polysomal aggregates, appearance of microtubules and abundance of spherosomes are some of the attributes of a cell after it has become embryogenic. Nearly 60 substances belonging to some 10 classes of organic and inorganic compounds, plant juices, plant extracts and some physical factors have been reported to influence embryogeny *ex ovulo*. But the cause and effect has not been unequivocally established for any except perhaps auxins. Shetty and Mckersie (1993) have postulated that physiological factors activate auxins which in turn induce biochemical events preceding the first cell divisions of somatic embryos.

Induction, development and conversion of somatic embryos to plants takes place generally on a different media combinations. Each step requires a particular concentration of phytohormones. It has been shown in mango that manipulation of culture media was necessary to achieve germination of somatic embryos because of the long maturation period which was required to inhibit precocious germination and other patterns of abnormal development (Dewald *et al.*, 1989). The endogenous level of hormones in somatic embryos is very high which obstruct the normal course of their development to plantlets. Many strategies have been worked out to leach out excess phytohormones from somatic embryos and activated charcoal has been added to the medium to remove as much auxin as possible from somatic embryos (Buchheim *et al.*, 1989). Parrott *et al.* (1989) showed in soybean that only one week's exposure of explants to auxins was necessary to obtain somatic embryogenesis and that increasingly longer exposures resulted in decreased development of shoot meristem in the embryos. Deng and Cornu (1992) showed in walnut that germination percentage increased when somatic embryos were pretreated by storing at 4 °C in the dark for two months.

Although somatic embryogenesis has been reported in a number of plant species (Attree and Fowke, 1993), the corresponding conversion of somatic embryos to plantlets is not high in a number of plants. In addition it has been recorded that the embryogenic cultures are often unstable and average yield do decline with subculture. This blocks the use of somatic embryos on a commercial scale for propagation in many plant species. Recent studies have shown that several factors are known to control this vital conversion. These factors are ammonium ion concentration, organic nitrogen source, hormone concentration, genotype and carbohydrate source (Johnson and Thomas, 1990). According to Parrott *et al.* (1991) ease with which somatic embryos convert to plants differs widely across genotypes, species and culture systems representing the entire continuum from the readily accomplishable to the currently impossible.

In alfalfa (Seitzkris and Bingham, 1988) and soybean (Komatsuda and Ohyama, 1988) it has been reported that conversion of somatic embryos to plants is under genetic control. Effects of various nutrients and carbohydrates formulations on rooting and plantlet regeneration have been communicated in a few species (Lazzeri *et al.*, 1987, Redenbaugh *et al.*, 1987). In recent times, many studies have been conducted to understand basic mechanisms of maturation of somatic embryos since in many cases zygotic embryos undergo a period of dry state and subsequently dormancy prior to germination. During the development process of zygotic embryos into mature seeds, proteins, carbohydrates, lipids, vitamins, minerals and non nutrients are laid down leading to the unique chemical composition of plant seeds. In quiescent stage, the dry zygotic embryo remain viable for many years. Desiccation probably plays an important role in switching seeds from a developmental program to one that is necessary to promote germination and growth. From physiological point of view, desiccation induce many metabolic changes in seeds besides water deficiency. In case of natural seeds which undergo a period of dormancy there could be many causes, chief among them are embryo immaturity, impermeability of the seed coat to water stress or to gases, prevention of embryo development due to mechanical causes, special requirements for temperature or light or the presence of substance inhibiting germination (Mayer and Poljakoff, 1986). It has been identified that during the maturation and drying phase of seed a group of proteins known as Late Embryogenesis Accumulating (*Lea*) induce desiccation tolerance during seed development (Dure *et al.*, 1989). However, in soybean it has been reported by Blackman *et al.* (1992) that the maturation proteins may be an essential part of an early response system that protects against the stresses imposed at the onset of desiccation before carbohydrates reach high levels or they may work in concert with oligosaccharides to function in the development of desiccation tolerance. Hence, induction in somatic embryos of a reversible quiescent phase as occurring in natural seeds will be essential.

According to Gray and Purohit (1991) there are two possible causes of dormancy in somatic embryos: 1) dormancy that is typical for zygotic embryos may also be expressed by somatic embryos of the same species *in vitro* and 2) dormancy may be artificially induced by aspects of the culture environment such as exogeneously supplied growth hormones, further they state that induction of quiescence in rapidly growing somatic embryos by lowering water content is logical since dehydration and rehydration cause arrest and resumption of growth in natural seeds. Parrott *et al.* (1988) found that desiccation in empty petridishes promoted rapid and uniform germination of soybean somatic embryos if they were allowed to mature to a minimum age on basal medium. Inclusion of a desiccation period after embryo maturation resulted in a four fold increase in the percentage of embryo which converted to plants in *Arachis* (Durham and Parrott, 1992). Besides improvement in germination, desiccation of whole somatic embryos is also an alternative method of germplasm storage. Somatic embryos are produced continuously year round and these could therefore be dried and stored until the appropriate planting season or shipped to new locations. Senaratna *et al.* (1991) have stated that the development of desiccation technology would allow long term storage or germplasm preservation of microspore derived embryos of *Brassica napus* L. and if this technology becomes available for early chromosome doubling in culture, dried doubled haploid embryos could be planted directly in soil.

ROLE OF ABSCISIC ACID (ABA) IN MATURATION AND DESICCATION

Abscisic acid has been shown to play a significant role in various plant physiological processes such as senescence and abscission of fruits and leaves, dormancy of buds and tubers, inhibition of stem growth as well as developmental stages in zygotic embryos. In *Arabidopsis thaliana* dormancy sets in during maturation in the wild type but not in the ABA- deficient mutants (Karssen *et al.*, 1983). A review on the use of ABA and GA mutants in elucidating the role of these two hormones by Hilhorst and Karssen (1982) concluded that the action of GA is restricted to the germination process only and ABA is required for desiccation tolerance only. Several reports exist on the changes in ABA levels during seed development in many plant species. ABA level reaches its highest concentration when dry weight in heart stage embryos is increasing storage protein levels in seeds. this inhibits development of abnormal seedlings from immature zygotic embryos. ABA also controls stomatal closures and induces transport of photosynthates towards developing seeds. In embryogenic cultures of caraway, ABA curtailed such aberrations as pluricotyly, secondary regeneration and precocious germination of somatic embryos and promoted normal course of ontogeny and maturation (Ammirato, 1974).

The normalizing effects of ABA on embryo maturation has also been observed in carrot (Kamada and Harada, 1981) and *Pennisetum* (Vasil and Vasil, 1981) and in *Citrus* (Kochba *et al.,* 1978). Nitsch and Nitsch (1969) observed that ABA did not prevent embryo development from *Nicotiana* microspores but did inhibit their germination. Imamura and Harada (1980) demonstrated an increase in response after soaking anthers for three days in ABA prior to explanting. Zygotic embryos of barley matured on ABA containing media (Umbeck and Norstrog, 1979) and ABA was responsible for maintaining normal embryogeny, suppressing precocious germination and in the expression of embryo specific proteins in *Brassica* (Finkelstein *et al.,* 1985) ABA has been shown to be involved in the expression of juvenility and in tissue culture of strawberry (Mohammed *et al.,* 1992). Fujii *et al.,* (1990) have demonstrated in alfalfa that ABA might be promoting somatic embryo storage reserve accumulation. Recently ABA has been shown to influence maturation process of somatic embryos of *Santalum* (Bapat and Rao, 1992).

Numerous *in vitro* experiments have indicated influence of ABA on developing embryos. Soybean embryos showed a high percent germination when the ABA content was less than 4 mg/l fresh weight. The ABA content of embryos *in situ* did not decline to that level until the onset of dehydration. Exogenous ABA prevented germination of immature embryos of several species when cultured *in vitro* and promoted the synthesis of embryo specific proteins and mRNAs (Zeevart and Creelman, 1988; Rhodes, 1990).

ABA is considered as a stress hormone since it enhances adaptation to various stresses. Freezing tolerance could be increased by ABA both in plants and in cell cultures. During cold hardening plants showed a transient increase in ABA content suggesting that ABA may be an endogenous regulator of adaptation to cold stress. ABA also accelerated adaptation of cultured cells to salts at least in part by osmotic adjustment. the synthesis of 26-KD protein has been associated with ABA mediated salt adaptation (Singh *et al.,* 1987). This protein has been shown to be a component in salt adaptation metabolic events. Originally ABA was considered a general inhibitor of RNA and protein synthesis but it is now evident that in certain tissues ABA can induce the formation of its own set of proteins.

Studies have been conducted on mode of action of ABA on somatic embryogenesis. Stimulation of GA_3 synthesis and decrease in ABA level has been reported in germinating seeds of Helianthus (Le Page Degivery, 1990) and maize (Oishi and Bewley, 1990). Senaratna *et al.* (1989) suggested that the expression of desiccation tolerance was dependent on the developmental stage of the embryo. However, the exact mechanism of action of ABA during somatic embryo maturation is not yet known. Recent experimental evidence suggest that ABA may play an important role in regulating some classes of embryo specific genes.

Flinn *et al.* (1991) conducted comparative studies to determine the degree of biochemical similarities between spruce somatic embryos differentiated with ABA and zygotic seed embryos by using storage proteins as biochemical markers. Somatic embryos contained storage proteins which exhibited similar biochemical characteristics to those found in zygotic embryos. The developmental pattern of storage protein accumulation was similar in somatic and zygotic embryos although accumulation kinetics of 22, 24 and 41 kDa storage proteins during the latter stages of maturation differed between zygotic and somatic embryos. According to Tautorus *et al.* (1991) the use of electroporation with somatic embryogenesis using plasmids containing inducible promoters may improve the knowledge of phytohormones receptor molecules and signals. ABA has been reported to promote somatic embryo maturation in conifer tissue cultures and further Tautorus *et al.* (1991) state that such studies using transgenic plants may provide clues as to the tissue specific expression of B-GUS under the control of the ABA inducible promoter. Labelling of proteins with (^{35}S) methionine and analysis by two dimensional gel electrophoresis indicated that the synthesis of several proteins of 20-23 kD molecular weight was enhanced by ABA treatment in carrot (Iida *et al.,* 1992). When added to conifer somatic embryo cultures ABA leads to maturation as well as associated responses such as increase in triglyceride content and strong protein deposition. Although ABA treatment has positive effects in stimulating white spruce somatic embryos (Attree *et al.,* 1991) its use is thought to contribute to subsequent inhibition of germination and growth. It has been proposed that high levels of ABA lead to the accumulation of protective proteins and hence prepare the embryo for desiccation. A gradual water loss allows sufficient time for the protective changes to occur in cells and hence increases the embryo resistance to dehydration. Once the embryos are at the cotyledonary stage of development, heat shock, cold stress, osmotic stress and nutrient deprivation can also be employed to induce similar responses to act by stimulating the synthesis of ABA within the embryo itself. Subsequently somatic embryos can be slowly dried to less than 15% moisture and stored for prolonged periods of time, similar to true seeds. Kvaalen and Arnold (1991) studied effects of various partial pressures of O_2 and CO_2 on initiation, proliferation and maturation of somatic embryos of *Picea abies* and concluded that gaseous environment influences maturation of somatic embryos upto certain extent and results are more complex than expected, however, none of the artificial atmosphere used could substitute for ABA to initiate maturation. It has been suggested that epigenetic control of embryogeny *ex ovulo* could be responsible since the aquisition of totipotency to the development of mature somatic embryos shifts in nutritional requirements, a situation similar to embryogeny *in ovulo*. Carman and Campbell (1990) studied the effects of desiccation of wheat somatic embryos on post imbibition germination and concluded that the genes that control dormancy are more readily influenced by ABA during differentiation of somatic embryos rather than during their maturation because somatic embryos rather than during their maturation because somatic embryos had been exposed to high levels of ABA during maturation but dormancy increased only in those that had also been exposed to ABA during differentiation. It is apparent that tissue culture has potential to be a valuable tool in the study of the biosynthesis of ABA. As this process is clarified it will be possible to consider and evaluate techniques to suppress or enhance ABA synthesis in somatic embryos. Such an insight would greatly aid any efforts to use ABA in an attempt to improve germination percentage of somatic embryos.

ROLE OF SUCROSE AND OTHER FACTORS IN MATURATION AND DESICCATION

During the course of seed development, the accumulation of soluble sugar is always increased and sugar is known to be responsible as adaptive agent for desiccation tolerance during seed development and germination (Koster and Leopold, 1988). In addition, soluble enzymes in seeds are protected by sucrose from salt-induced damage (Schwab and Gaff, 1990).

Plant tissue culture medium contains carbohydrates generally supplied as sucrose or glucose which serve as osmotic agents and as carbon and energy sources. Carbohydrates other than sucrose and glucose, also, have been shown to support growth of cultures. Maltose has been added instead of sucrose to increase efficiency of embryo to plant conversion in *Trifolium* (Weissinger and Parrott, 1993). According to Levi and Sink (1990), successful somatic embryogenesis in asparagus may require high sucrose levels during embryo formation and maturation followed by transfer of embryos to a lower sucrose level for germination. The concentration of sucrose in the somatic embryo maturation medium has a profound effect on the development of the embryo, particularly with respect to its acquisition of desiccation tolerance and seedling vigour (Anandarajah and Mckersie, 1990). Based on the work on *Medicago sativa* (Anandarajah and Mckersie, 1990), it has been concluded that increased sucrose in the maturation medium enhanced embryo dry weight and at the same time blocked precocious germination. High levels of sucrose in the medium resulting into high osmolarity may be one of the factors that promoted maturation and inhibited precocious germination. This has also been demonstrated in soybean (Komatsuda *et al.*, 1988). Alfalfa embryos matured on the medium with high sucrose levels lost chlorophyll and turned yellow. The synthesis of chlorophyll and lamellar development in tissue cultures is inhibited by high sucrose although the exact metabolic events leading to this degreening are not known. The increased osmolarity at high sucrose concentration has been associated with the control of embryo maturation and elevation of endogenous ABA. In Cacao (Pence, 1992), ABA and sucrose are both needed for the initiation of events associated with maturation of somatic embryos *in vitro*. Charcoal has been added into the maturation medium of soybean somatic embryos to adsorb ABA which leads to high osmolarity created by sucrose may have evoked endogenous ABA synthesis thus mimicking the *in vitro* osmotic environment of developing zygotic embryos.

Plasmolysing osmotica such as sucrose, mannitol or salts are absorbed by the symplast of plant cells. Roberts (1991) observed that a one week pulse treatment of 2-6% mannitol included in medium with 3-4% sucrose and 40-80 mM ABA doubled the frequency of globular stage interior spruce somatic embryos compared to ABA alone. Such absorbtion facilitates adjustment of tissue osmotic potential without lowering the tissue water content. Further more direct or indirect metabolic effects on specific plant metabolism can occur due to utilization of the solute or its toxic effects. Osmoregulation apparently maintains the osmotic potential of actively growing embryos at a level lower than surrounding tissues and can prolong embryo hydration (Carman, 1989). One of the hypothesis is that the accumulation of sugars maintains the stability of membranes in the dry state by replacing the water molecules at the charged exterior surface of the membranes (Anandarajah and Mckersie, 1990).

According to Attree *et al.* (1991) alternatives to simple sugars and salts are high molecular weight compounds such as polyethylene glycol (PEG). PEG stimulates a non osmotic moisture stress at the cellular levels because the large molecular size of the solute excludes its passage through plant cell walls preventing entry into cells and plasmolysis. Consequently the non plasmolysing effect of PEG is similar to the effects of water stress observed in cells of plants subjected to drought conditions. In tomato (Finch-Savage and McQuistan, 1991) it has been reported that a period of controlled seed hydration in PEG or salt solutions significantly enhanced the germination and subsequent seedling emergence. These osmotic priming were demonstrated after seeds had been dried back to their original moisture content, a prerequisite for seed sowing. During treatment the osmoticum prevents the progress and germination beyond the point of desiccation tolerance (Flinn *et al.*, 1991). Attree *et al.* (1991) have demonstrated that plants regenerated from PEG treated desiccated somatic embryos were of better quality than control which they attribute to precocious germination of embryos in absence of PEG in control. This has direct correlation to water content since PEG treated embryos has less moisture content.

Heat shock and cold acclimation are known to induce *de novo* synthesis of proteins which may be involved in adaptation to stress and this may be related to its induction of desiccation tolerance in somatic embryos to varying extent. According to Anandarajah *et al.* (1990) sublethal temperature stress pretreatment induce the expression of desiccation tolerance and subsequent regeneration of plants from desiccated embryos. The clear mechanisms by which heat shock treatments induce desiccation tolerance is still not known, however there seems to be a link between involvement of proteins in adaptations to stress. Chen *et al.* (1983) have demonstrated a correlation between endogenous level of ABA and heat/cold treatment. Cold treatment has been shown to reduce the level of endogenous ABA in grape (Rajasekaran *et al.,* 1992) and in walnut somatic embryos (Tulecke *et al.,* 1992).

The accumulation of cytoplasmic proline during water deficit is a common metabolic response to drought or salt stress (Kim and Janick, 1991). Exogenous proline benefits the adaptation of cultured cells by adjusting osmotic potential in response to changing external water potential. It has been assumed that proline acts as an active solute and as an enzyme protectant (Kim and Janick, 1991). Proline also acts as a regulator of fatty acid composition in both somatic and zygotic embryos as has been shown in soybean (Shoemaker and Hammond, 1988), *Brassica* (Avjioglu and Knox, 1989) and jojoba (Wang and Creelman, 1986). In fact free proline accumulation is frequently observed in plants grown under such conditions. The proline biosynthetic pathway in higher plants is not well characterized and the mechanism of proline accumulation under stress is still unclear (Mori *et al.,* 1989). Another amino acid glutamine has enhanced embryo dry weight and stimulated synthesis of 11S storage proteins (Mckersie *et al.,* 1992). Lai and Mckersie (1993) concluded that their work on alfalfa somatic embryos showed that glutamone seems to have a regulatory role as well nutritive role in storage protein synthesis and optimal embryo quality would be achieved with a low plating densities and frequent transfer to large volumes of fresh nutrient media and supplemented with glutamine during maturation. The increased ratio of S-2:S.1 proteins with 50 mM glutamine supplements indicate that glutamine stimulated S-2 (Medicagin) storage protein synthesis (Lai *et al.,* 1992). This would enable the deposition of substantial quantities of storage proteins and acquisition of desiccation tolerance (Skriver and Mundy, 1990).

CONCLUSION

From the results which are mainly based on concrete experimental data it can be concluded that the conversion of somatic embryos to plantlets is a very complex process. Although somatic embryogenesis is often presented as a promising technology for micropropagation, the application of this technology remains limited due to the lack of proper knowledge of the very process. Another problem has been to precisely identify the factors influencing conversion of somatic embryos to plants. Future basic research is needed to gain insight into the biochemical pathways that regulate zygotic embryogenesis so that similar parameters can be investigated in somatic embryogenesis. For example, recently LEA like proteins have been shown to be present in embryos of many plant species, however, acquisition of desiccation tolerance and presence of these proteins remains to be clarified. To determine the differences between zygotic embryos and somatic embryos a wide range of influencing factors have to be evaluated. A factor of great importance is to induce synchrony in the development of somatic embryos and to develop *in vitro* techniques that prevent precocious germination of somatic embryos and yield a large number of uniform somatic embryos. Maturation and desiccation are necessary for somatic embryos so that the embryos can withstand the harsh environmental conditions encountered when directly seeded in field.

In-depth investigations on somatic embryogenesis and equivalent process in zygotic embryogenesis at morphological, biochemical, physiological and molecular level would yield valuable information on the differences between these two types of embryos. This would help to overcome the barriers in several economically important species for obtaining somatic embryogenesis.

REFERENCES

Ammirato, P.V. 1974. The effects of abscisic acid on the development of somatic embryogenesis from cells of caraway (*Carum carvi* L.). Bot Gaz. 135: 328-337.

Aitken-Christie, Connet and J. 1992. Micropropagation of forest trees In: Kurata, K. and Kozai, T. (Eds.) Transplant production system, Kluwer Academic, Netherlands, pp. 163-194.

Anandarajah, K. and Mckersie, B.D. 1990. Enhanced vigor of dry somatic embryos of *Medicago sativa* L. with increased sucrose. Plant Sci. 71: 261-266.

Anandarajah, K., Kot, L., Beversdorf, W.D. and Mckersie, B.D. 1991. Induction of desiccation tolerance in microspore derived embryos of *Brassica napus* L. by thermal stress. Plant Sci. 77: 119-123.

Attree, S.M., Moore, D., Sawhney, V.K. and Fowle, L.C. 1991. Enhanced maturation and desiccation tolerance in white spruce. (*Picea glauca* (Moench) Voss) somatic embryogenesis: Effects of a non-plasmolusing water stress and abscisic acid. Ann. Bot. 68,: 519-525.

Attree, S.M. and Fowle, L.C. 1993. Embryogeny of gymnosperms: advances in synthetic seed technology of conifers. Plant cell Tissue and Org. Cul. 35: 1-35.

Avjioglu, A. and Knox, R.D. 1989. Storage lipids accumulation of zygotic and somatic embryos in culture. Ann. Bot. 63: 409-420.

Bapat, V.A. and Rao, P.S. 1992. Plantlet regeneration from encapsulated and non encapsulated desiccated somatic embryos of a forest tree: Sandalwood (*Santalum album* L.). J.Pl. Biochem. and Biotech. 1: 109-113.

Blackman, S.A., Obendorf, R.L. and Leopold, A.C. 1992. Maturation proteins and sugars in desiccation tolerance of developing soybean seeds. Plant Physiol, 100: 225-230.

Buchheim, J.A., Colburn, S.M. and Ranch, J.P. 1989. Maturation of soybean somatic embryos and transition to plantlet growth. Pl. Physiol. 89: 768-775.

Carman, J.G. 1989. The *in ovulo* environment and its relevance of cloning wheat via somatic embryogenesis. *In Vitro* Cell Dev. Biol. 25: 1155-1162.

Carman, J.G.a nd Campbell, W.F. 1990. Factors affecting somatic embryogenesis in whet. Y.P.S. Bajaj (Ed.) In: Biotechnology in Agri. and Forestry. Springer Verlag, Berlin. pp. 68-87.

Chen, H.H., Li, P.H. and Brenner, M.L. 1983. Involvement of abscisic acid in potato cold acclimation. Pl Physiol. 71: 362-365.

Deng, M.D., and Cornu, D. 1992. Maturation and germination of walnut somatic embryos. Plant Cell, Tissue & Org. Cul. 28: 195-202.

DeWald, S.G., Litz, R.E. and Moore, G.A. 1989. Maturation and germination of mango somatic embryos. J. Amer. Soc. Hort Sci. 114: 837-841.

Dure, L., Crouch, M., Harada, J., Ho, T.D., Mundy, J., Quatrano, R., Thomas, T., and Sung, Z.R. 1989. Common amino acid sequence domains among the *Lea* proteins of higher plants. Plant Mol. Biol. 12: 475-486.

Durham, R.E., and Parrott, W.A. 1992. Repetitive somatic embryogenesis from peanut cultures in liquid medium Plant Cell Rep. 11: 122-125.

Finch-Savage, W.E. and McQuistan, C.I. j1991. Abscisic acid: an agent to advance and synchronise germination for tomato (*Lycopersicon esculentum* Mill) seeds. Seed Sci. and Technol. 19: 537-544.

Finkelstein, R.R., Tenbarge, K.M., Shumway, E.a nd Crouch, M.L. 1985. Role of ABA in maturation of rapeseed embryos. Pl. Physiol. 78, 630-636.

Flinn, B.S., Roberts, D.R., and Taylor, I.E.P. 1991. Evaluation of somatic embryos of interior spruce. Characterization and developmental regulation of storate proteins. Physiol. Plant. 82: 624-632.

Fujii, J.A., Slade, D., Olsen, R., Ruzin, S.E. and Redenbaugh, K. 1990. Alfalfa somatic embryo maturation and conversion to plants. Pl. Sci. 72: 93-100.

Gray, D.J. and Purohit, A. 1991. Somatic embryogenesis and development of synthetic seeds technology. Critical Rev. in Pl. Sci. 10: 33-61.

Gray, D.J., Trigiano, R.N. and Conger, B.V. 1993. Liquid suspension culture production of orchardgrass somatic embryos and their potential for the breeding of improved varieties. In: Redenbaugh, K. (Ed.). Synseeds. Application of synthetic seeds to crop improvement 1993, CRC Press, USA, pp. 351-366.

Gupta, P.K., Pullman, G., Timmis, R., Kretinger, M., Carlson, W., Grob, J., and Willy, E. 1993. Forestry in the 21st century. Biotechnology 11:454-459.

Hilhorst, H.W.M. and Karssen, C.M. 1982. Seed dormancy and germination: The role of abscisic acid and giberllins and the importance of hormone mutants Plant Growth Reg. 11: 225-238.

Iida, Y., Kamada, U., and Harada, H. 1992. Effects of abscisic acid on the induction of desiccation tolerance in carrot somatic embryos. In: Oono, K., Hirabayashi, T., Kikuchi, S., Handa, H., Kajiwara, (Eds.). Plant tissue culture and genetic manipualtion for breeding and formation of phytochemicals, 1992. NIAR, Japan, pp. 117-132.

Imamura, J. and Harada, H. 1980. Effects of abscisic acid and water stress on embryo and plantlet formation in anther cultures of *Nicotiana tabacum* cv samsum. Z. Pflanzenphysiol. 100: 285-

Janick, J., Kitto, S.L. and Kim, Y.H. 1989. Production of synthetic seeds by desiccation and encapsulation. *In Vitro* cell Dev. Biol. 25: 1167-1172.

Johnson, L.B., and Thomas, M.R. 1990. Somaclonal variations in alfalfa. In: Bajaj, Y.P.S. (Ed.). Biotechnology in Agri. and Forestry 11, Springer Verlag, Berlin, 654-678.

Kamada, H. and Harada, H. 1981. Changes in endogenous levels and effects of abscisic acid during somatic embryogenesis of *Daucus carota* L. Plant Cell Physiol. 22: 1423-1429.

Kim, Y.H., and Janick, J. 1991. Abscisic acid and proline improve desiccation tolerance and increase fatty acid content of celery somatic embryogenesis. Plant Cell, Tissue & Org. Cul. 24: 83-89.

Kochba, J., Spiegel-Roy, P., Neumann, H., and Saad, S. 1978. Stimulation of embryogenesis in *Citrus* ovular callus by ABA, ethapon, CCC, and alar and its suppression by GA. Z. Pflanzenphysiol. 89: 427-432.

Koster, K.L. and Lepold, A.C. 1988. Sugars and desiccation tolerance in seeds. Plant Physiol. 88: 829-832.

Komatsuda, T., and Ohyama, K. 1988. Genotypes of high competence for somatic embryos and plantlet regeneration in soybean (*Glycine max*). Theor. App. Genet. 75: 695-700.

Krochko, J.E., Pramanik, S.K. and Bewley, J.D. 1992. Contrasting storage protein synthesis and messenger RNA accumulation during development of zygotic and somatic embryos of alfalfa (*Medicago sativa*, L.). Plant Physiol, 99: 46-53.

Kvaalen, H. and Arnold, S. 1991. Effects of various partial pressures of oxygen and carbon di oxide on different stages of somatic embryos in *Picea abies*. Plant Cell Tissue & Org. Cul. 27: 49-57.

Lai, F., Senaratna, T. and Mckersie, B.D. 1992. Glutamine enhances storage protein synthesis in *Medicago sativa* L. somatic embryo Plant Sci. 87: 69-77.

Lai, F., and Mckersie, B.D. 1993. Effect of nutrition on maturation of alfalfa (*Medicgo sativa* L.) somatic embryos. Plant Sci. 91: 87-95.

Lazzeri, P.A., Hilderbrand, D.F. and Collins, G.B. 1987. Soybean somatic embryogenesis: Effects of nutritional, physical and chemical factors. Plant Cell, Tissue & Org. Cul. 10: 209-220.

Le Page-Degivry, M.T., Barthe, P. and Garello, G. 1990. Involvement of endogenous abscisic acid in onset and release of *Helianthus annus* embryo dormancy. Pl. Physiol. 92:1164-1168.

Levi, A. and Sink, K.C. 1990. Differential effects of sucrose, glucose and fructose during somatic embryogenesis in *Asparagus*. J.Pl. Physiol. 137: 184-189.

Mayer, A.M. and Poljakoff-Mayber, A. 1982. The germination of seeds, Pergamon Press, Oxford.

Mckersie, B.D. and Bowley, S.R. 1993. Synthetic seeds of Alfalfa In: Redenbaugh, K. (ed.) Synseeds Application of synthetic seeds to crop improvement, CRC press USA, pp. 231-255.

Merkle, S.A., Parrott, W.A. and Williams, E.G. 1990. Application of somatic embryos and embryo cloning In: Bhojwani, S. (ed.), Elsiever Pub., pp. 67-101.

Michler, C.H. and Lineberger, R.D. 1987. Effects of light on somatic embryo, embryo development and abscisic acid levels in carrot suspensions cells. Plant Cell Tissue & Org. Cul. 11:189-207.

Mohamed, F., Swartz, H.J. and Buta, J.G. 1991. The role of abscisic acid and plant growth regulators in tissue cultures induced rejuvenation of strawberry *ex vitro* Plant Cell Tissue & Org. Cul. 25: 75-84.

Mori, S., Hasegawa, H., Che, R., Nakanishi, H. and Murakami, M. 1989. Free proline contents in two different groups of rice mutants resistant to hydroxy-L proline. Theor. Apple. Genet. 77: 44-48.

Nitsch, J.P. and Nitsch, C. 1969. Haploid plants from pollen grains. Science, 163: 85-87.

Oishi, M.Y. and Bewley, J.D. 1990. Distinction between the responses of maize kernels to fluridine and desiccation in relation to germinability, amylase activity and abscisic acid content. Pl. Physiol. 94: 592-598.

Parrott, W.A., Dryden, G., Voght, S., Hilderbrand, D.F., Collins, G.B., and williams, E.G. 1988. Optimization of somatic embryo germination in soybean *In vitro* Cell Dev. Biol. 24: 817-820.

Parrott, W.A., Williams, E.G., Hilderbrand, D.F. and Collins, G.B. 1989. Effect of genotype on somatic embryogenesis from immture cotyledons of soybean. Plant Cell, Tisue & Org. Cul. 16: 15-21.

Parrott, W.A., Merkle, S.A. and Williams, E.G. 1991. Somatic embryogenesis: potential for use in propagation and gene transfer systems. In: Murray, D.R. (ed.) Advanced methods in plant breeding and Biotechnology, CAB Internat. pp. 158-200.

Pence, V.C. 1992. Abscisic acid and the maturation of Cacao embryos *in vitro*. Pl. Physiol. 98: 1391-1395.

Rajasekaran, K., Vine, J., and Mullins, M.G. 1982. Dormancy in somati embryos and seeds of *Vitis*. Changes in endogenous abscisic acid during embryogeny and germination. Planta, 154: 139-144.

Rangaswamy, N.S. 1986. Somatic embryogenesis in angiosperm cell tissue and organ culture. Proc. Ind. Acad. Sci. (Plant Sci) 96: 247-271.

Rangaswamy, N.S. 1991. Seed embryos, somatic embryos and pollen embryos. In: Development and differentiation in plants. Cellular and molecular aspects. Ministry of Sci. and Technol Pub. Govt. of India. New Delhi. pp. 58-59.

Redenbaugh, K., Viss, P., Slade, D. and Fujii, J.A. 1987. Scale up: artificial seeds. In: C. Green, D. Somers, W. Hackett and B. Biesboers (Eds.) Plant Tissue and Cell Cul. Alan Liss. NY. pp. 473-493.

Redenbaugh, K. 1990. Application of micropropagation to tropical crops. Hort. Sci. 25: 251-255.

Redenbaugh, K., Fujii, J., Slade, D., Viss, P. and Kossler, M. 1991. Artificial seeds-encapsulated somatic embryos. In: Biotechnology in Agriculture and Forestry, Y.P.S. Bajaj (Ed.). Springer Verlag, Berlin, pp. 2-46.

Rhodes, D. 1987. Metabolic responses to stress. In: The Biochemistry of plants, Vol. 12, Academic Press Inc. pp. 201-241.

Roberts, D.R. 1991. ABA and mannitol promote early development, maturation and storage protein accumulation in somatic embryos of interior spruce Physiol Plant, 83:247-254.

Schwab, K.B. and Gaff, D.F. 1990. Influence of compatible solutes on soluble enzymes from desiccation tolerant *Sporobolus stapfianus* and desiccation sensitive *Sporobolus pyramidalis* Plant Physiol. 137: 208-215.

Seitz Kris, M.H. and Bingham, E.T. 1988. Interactions of highly regenerative genotypes of alfalfa (*Medicgo sativa*) and tissue culture protocols. In Vitro Cell and Dev. Biol. 24: 1047-1052.

Senaratna, T., Mekersie, B.D. and Bowley, S.R. 1989. Artificial seeds of alfalfa (*Medicago sativa* L.). Induction of desiccation tolerance in somatic embryos. In Vitro Cell Biol 26: 90-95.

Senaratna, T., Kotte, L., Beversdorf, W.D. and Mckersie, B.D. 1991. Desiccation of microspore derived embryos of oilseed rape (Brassica napus L.). Pl Sci. Rep. 10: 342-344.

Shetty, K. and Mckersie, B.D. 1993. Proline, thioproline and potassium mediated stimulation of somatic embryos in alfalfa (*Medicago sativa* L.) Plant Sci, 88: 185-193.

Shoemaker, R.C. and Hammond, E.G. 1988. Fatty acid composition in soybean (*Glycine max* L.) somatic embryogenesis. In Vitro, 24: 829-832.

Singh, N.K., LaRossa, P.C., Handa, A.K., Hasegawa, P.M. and Bressan, R.A. 1987. Hormonal regulation of protein synthesis associated with salt tolerance in plant cells Proc. Natl Acda. Sci, USA 84: 739-743.

Skriver, K. and Mundy, J. 1990. Gene expression in response to ABA and osmotic stress. The Plant Cell, 2: 503-512.

Tautorus, T.E., Fowle, L.C. and Dunstan, D.I. 1991. Somatic embryogenesis in conifers. Can. J. Bot. 69: 1873-1899.

Tulecke, W., McGrananhan, and Ahmadi, H. 1988. Regeneration of somatic embryogenesis and triploid plants from endosperm of walnut *Juglan regia* L. cv. Mangerian. Pl, Cell Rep. 7: 301-304.

Umbeck, P.F. and Norstog, K. 1979. Effects of abscisic acid and ammonium ions on morphogenesis of cultured barley embryos. Bull Torrey Bot Club. 106: 110-116.

Vasil, V. and Vasil, I.K. 1981. Somatic embryogenesis and plantlet regeneration from tissue cultures of *Pennisetum americanum* and *P. americanum* X *P. purpureum* hybrid. Amer. J. Bot. 68: 864-872.

Wang, Y.C. and Janick, J. 1986. *In vitro* production of jojoba liquid wax by zygotic and somatic embryos. J. Amer. Soc. Hortic. 111: 798-806.

Weissinger, A.K. and Parrott, W.A. 1993. Repetitive somatic embryogenesis in white clover Plant Cell Rep. 12: 125-128.

Zeevart, J.A.D. and Creelman, R.A. 1988. Metabolism and physiology of abscisic acid. Ann. Rev. Pl. Physiol. 39: 439-473.

Commercial Applications of Medicinal Plant Biotechnology

M. R. Heble and G. Roja

Medicinal plants constitute nearly 25% of the prescribed drugs. In recent years screening of such plants for biological activities has resulted in the development of therapeutics used in the treatment of Cancer, AIDS and others. A large number of medicinal plants are exploited from the natural flora for the commercial production of the drugs (Paul *et al.*, 1992). Most of the drugs in the tree species accumulate either in the root or the stem bark of the tree after several years of growth. As a result, for the isolation of medicinals the trees have been indiscriminately exploited from the natural habitat. In order to conserve the natural flora and meet the increasing demand for plant based drugs certain alternate methods have become more important. The biotechnological approach such as tissue cultures initiated from medicinal plants is a viable method for the production of therapeutic compounds. Medicinal plant biotechnology has applications towards micropropagation of elite trees containing important phytochemicals. It is also possible to reduce the entire tree to small cell masses and use them for the synthesis of high value constituents. The present review concerns the production of certain plant components from the whole plant and from tissue cultures.

ANTI-CANCER AGENTS

Taxol: Numerous anticancer agents have been obtained from medicinal plants but only a few of them have proved effective in clinical trials. Amongst the dimeric alkaloids Vinblastine and Vincristine from the plant *Catharanthus roseus* and more recently the diterpenoid Taxol (Fig 1A) (Hamburger and Hostettmann, 1991) from the bark of the pacific yew tree *Taxus brevifolia* have proved to be suitable candidates for tissue culture production (Wani *et al.*, 1971). The levels of taxanes in the plants are generally low and are difficult to produce economically. The stem bark contains 0.01% D.W. of taxol. The trees are slow growing and the accumulation of the compounds are susceptible to geographic and environmental conditions.

Taxol is a promising drug for the treatment of advanced ovarian cancer (Rowinsky *et al.*, 1990). It has a unique mechanism of action. Taxol promotes tubulin polymerisation and stabilises microtubules against depolymerisation (Manfredi *et al.*, 1984). Currently the drug obtained from *T. brevifolia* is approved by Food and Drug Administration of US and is marketed by Bristol-Myers Sqibb. The projected demand for Taxol for treating patients with ovarian cancer is around 25 kg/year. About one gram of Taxol is required for treating one cancer patient and is obtained by cutting 3-6 trees of *T. brevifolia*. Therefore, in order to obtain adequate supply of Taxol alternate approaches were proposed by various groups of scientists all over the world. These methods are the identification of other species of *Taxus* such as *Taxus baccata, T. cuspidata* and *T. canadensis* (Wani *et al.*, 1971; Witherup, 1990), isolation and identification of precursors from the needles and twigs of yews common in Europe and Asia for chemical synthesis preferably the total synthesis of Taxol.

Related species of *Taxus* have been identified as a source material and some of these are *T. baccata, T. cuspidata* and *T. wallichiana*. The needles and stems of these species were found to contain the precursors Baccatin III and 10-deacetyl baccatin III (Fig IB). These precursors can be converted into Taxol in high yield. The Taxol analogue Taxotere (Fig IC) is synthesised from 10-deacetyl baccatin III. The availability of these precursors from the needles of *Taxus* species is therefore advantageous for the production of Taxol and related compounds without damaging the tree and thus represents a renewable source of the compound.

The chemical synthesis of Taxol molecule is difficult to achieve due to its complex structure and is unlikely to become a commercial source for Taxol. To date, two methods of synthesis have been achieved. The first involves 28 steps and is based on the coupling of precursors to rings A (C1, C 11-15) and C (C3-8) which also have to be synthesised (Nicolaou *et al.*, 1994). The second synthesis (Holton, 1994) involves 40 steps to arrive at Taxol via 10 deacetyl baccatin III. A short route synthesis using pinene, a natural product obtained from pine trees and a major component of turpentine and other solvents is likely to be a successful source (Blume, 1992). Pinene has the same three dimensional structures as the taxane ring and contains 10 of the 20 carbons that form Taxol skeleton.

Plant tissue culture has been suggested as an alternative method to provide large amounts of Taxol and related compounds. Production of Taxol from cell cultures has been patented (Christen *et al.*, 1991). Fetto-Neto *et al.* (1992) reported cell cultures of *Taxus* species *T. cuspidata* and *T. canadensis*. The callus cultures of *T. cuspidata* produced Taxol in amounts equivalent to the stem of the intact plant (0.02% D.W.) after two months of culture. Immobilised cell cultures of *T. cuspidata* produced Taxol equivalent to those found in the bark of the tree (0.012% D.W.) after six months of culture. These results indicated the possibility of using tissue cultures for the production of taxanes. The USDA has signed an agreement with Phyton Catalytic of USA to develop a process for commercial production. Another Biotechnology company ESCA genetics Corporation has also announced success in getting significant amounts of the compound from plant cell cultures. This company has collaborated with Genencor International to scale up the production of Taxol.

The major problems that delayed Taxol development as an anticancer drug was the short supply of the material and secondly the solubility of the drug. The problems of limited supply of Taxol was overcome to some extent by using related *Taxus* species. *Taxus brevifolia* is propagated by cuttings (C. Hetz-Phillips, Personal communication) and tissue cultures may be useful for screening and propagation of high yielding varieties. The solubility problem was solved by development of a formulation using Cremophor EL, a polyethoxylated castor oil and absolute ethanol (Joel, 1994). The entire mixture was diluted with 5% dextrose in water or normal saline before use.

Presently Taxol is being tried for various other types of Cancers such as cervix, colon, skin and prostate. Mcguire and others (1989) in NCI cancer treatment programme started to mix Taxol with other better known treatments and are beginning trials with cancer patients in the early stages of their disease. The demand for Taxol and other derivatives is ever increasing. The existing methods cannot satisfy the requirement. Plant tissue culture method, therefore, have scope for the production of these choice drugs.

Camptothecin: Camptothecin (Fig 1D) is a quinoline alkaloid reported earlier from the Chinese tree *Camptotheca acuminata* (Wall *et al.*, 1966) and later discovered in the indigenous tree *Nothapodytes foetida* (Govindachari *et al.,* 1972). Clinical trials of Camptothecin were conducted during the 1970's but due to the severe toxicity of the drug leading to gastrointestinal cysts, it was discontinued. In the peoples Republic of China, clinical studies were carried out with a fine suspension of Camptothecin rather than the sodium salt, which gave encouraging results in gastrointestinal cancers, head and neck tumours and bladder carcinoma (Gottlieb *et al.,* 1970). Future research efforts to develop Camptothecin in analogues with greater solubility and lower toxicity resulted in the development of two promising compounds viz. CPT 11 and Topotecan (Fig 1D). CPT 11 has been shown to be effective against refractory leukemia and lymphoma in a phase II trial (Ohno *et al.,* 1990) and Topotecan has shown promising results in Phase I trials in USA and Europe (Kingsbury *et al.,* 1991). Both Camptothecin and Topotecan are potent inhibitors of DNA synthesis by a rare mechanism involving inhibition of the enzyme topoisomerase I. The novel mechanism of action of both CPT-11 and Topotecan along with their clinical efficacy indicated excellent prospects for the development of anti-cancer agents.

C. acuminata is a tree species having low levels of Camptothecin (0.005% D.W.). The growth rate of the tree is also slow. To improve the production of Camptothecin for clinical studies, Plant tissue culture methods were used. Sakato *et al.* (1974) reported 0.0003% D.W. of Camptothecin in cell cultures of *C. acuminata*. Govindachari *et al.* (1972) reported the occurrence of Camptothecin and related compounds in an indigenous plant *Nothapodytes foetida*. Our studies using tissue cultures of *N. foetida* have shown the presence of Camptothecin in callus as well as in differentiated cultures. Callus and plantlet cultures of *N. foetida* were established from the immature embryos of the tree on Murashige and Skoog's medium supplemented with different plant growth hormones such as BA, Kn, NAA, IAA and 2,4-D singly or in combination. Callus cultures from BA + 2,4-D medium synthesised low levels of Camptothecin compared to the differentiated cultures from BA + NAA, Kn + NAA and Kn + IAA media. The level of 9- methoxy camptothecin in the differentiated cultures were more than callus cultures (Roja and Heble, 1994). The development of tissue culture methods for the production of *Camptotheca* alkaloids would therefore be of considerable interest.

Other anti cancer agents: Some of the other well known medicinal plants containing anticancer compounds are *Cephalotaxus harringtonia, Podophyllum peltatum, Tryptergium wilfordii* and *Putterlickia* (Misawa and Nakanishi, 1988). Tissue cultures of these plants have been established and the cultures contained the compounds of the parent plant. However, differences were observed in the yield of the compounds. In certain cases such as *T. wilfordii* the cultures synthesised significantly higher quantities of tripdiolides than the intact plant. (Kupchan, 1972). The production of podophyllotoxins from *P. peltatum* were also higher (7.1 X 10^{-1} % D.W.) than the field plant (6.4 X 10^{-1} % D.W.) (Kadkade, 1982). The cultures of *C. harringtonia* contained cephalotaxine in the differentiated tissues in concentrations higher than the suspension cultures. These illustrations indicated that the anticancer compounds can be produced by tissue cultures but commercial exploitation is possible only if the process can be scaled up in large bioreactors and condition can be optimised for active production.

Anti-Malarials: Artemisinin obtained from the Chinese plant *Artemisia annua* is an effective antimalarial drug against Chloriquine- resistant and Chloroquinine sensitive strains of *Plasmodium falciparum*. The drug is also effective in the treatment of Cerebral malaria (Klayman, 1985). Artemisinin (Fig 1E) is a sesquiterpenoid lactone and has a endoperoxide moiety. It is highly liphophilic in nature and certain problems are encountered in the administration of the drug. The significant biological activity, novel chemical structure and low yield from natural sources have let to studies directed towards its chemical synthesis. The total synthesis of artemisinin was achieved but the yields are low and economically unattractive (Schmidt and Hofheinz, 1983; Avery *et al.*, 1992). A series of semisynthetic derivatives such as artemether, artoether, artelinic acid and sodium artesunate (Fig 1F) have been prepared and extensive clinical trials have been conducted in China and in Burma. These derivatives have proved to be effective in severe *falciparum* malaria (WHO, 1986). sodium artesunate is water soluble and has been tested for the intravenous treatment of *P. falciparum* infections.

A. annua plants contain Artemisinin mainly in the leaves of the plant and the concentration varies from 0.01 to 0.05% D.W. in the Chinese varieties. The plants collected in US where it is not native but was introduced from Asia, the maximum concentration was about 0.06% D.W. The plants collected from different places exhibited wide variability in the contents of Artemisinin. This problem may be overcome to a certain extent by the use of clonal propagation through multiple shoot cultures. Establishment of shoot cultures of *A. annua* have been reported by several scientists (Fulzele *et al.*, 1991; He *et al.*, 1983). These cultures in general contained lower levels of Artemisinin than the parent plant but the growth of the cultures and Artemisinin contents remained stable. Fulzele *et al.* (1991) reported 0.012% D.W. of Artemisinin in shoot cultures and He *et al.*, 1983 reported 0.008% D.W. of the compound. Root cultures of *A. annua* synthesised 0.05-0.192 mg% of Artemisinin (Nair *et al.*, 1986). The callus and cell cultures of A. annua did not synthesise any Artemisinin. Artemisinin synthesise requires a certain degree of organ differentiation. Recently Woerdenberg *et al.* (1983) reported an yield of 0.16% D.W. of Artemisinin in shoot cultures growing on MS 1/2 + BA (0.2) + NAA (0.05) medium containing 1% sucrose. The Artemisinin content increased with 0.5% sucrose. Supplementing the medium with GA (10 mg/l), CH (0.5 mg/l) and 10 or 20 mg/l of naftifine enhanced the production of Artemisinin in shoot cultures of A. annua. These observations suggested that the shoot cultures could be a viable alternative for Artemisinin production.

Certain other constituents other than anticancer agents are also important candidates for tissue culture methods. The best examples are Forskolin (Fig 1G) and Ginkgolides (Fig 1H). *Coleus forskohlii* is an Indian medicinal plant and the active constituent Forskolin is a diterpenoid isolated from the roots of the plant. Forskolin in possesses the property of lowering the blood pressure and its anti spasmodic activity is also reported (Bhat *et al.*, 1977). Forskolin is a potent stimulator of adenylate cyclase activity and has been used extensively to investigate the physiological consequences of increased intracellular cAMP (De souza, 1991). The total synthesis of Forskolin has been reported (Hashimoto *et al.*, 1988; Corey *et al.*, 1988) but a commercially viable method has not yet been achieved.

Coleus plants collected from different places exhibit variations in Forskolin content. Cell cultures of *Coleus forskohlii* synthesised Forskolin in the production medium but the cultures lost the biosynthetic capacity after 3-4 years of subculturing (Mersinger *et al.*, 1988). Using differentiated cultures of *Coleus* the problem of stability could be overcome to some extent. The *in vitro* propagation of *C. forskohlii* by shoot cultures obtained from shoot tips has been reported by Neelam *et al.* (1991). The Forskolin content in the tissue culture plants compared with that of the wild plant (0.1% D.W.). Recently, Sen *et al.* (1993) reported the presence of Forskolin in untransformed root cultures of *C. forskohlii*. The roots were grown in different basal media and different cultural conditions. The media used for the growth of root cultures comprised of MS 25%, 1% sucrose and either IAA, IBA or 2 chlorinated derivatives of IAA viz. 4- Cl- IAA and 5,6, C12 IAA. The cultures grew well in all the dosages of chlorinated IAA's and also produced more Forskolin than in cultures grown in IAA or IBA singly. In medium containing 50 ug/l of 4 Cl-IAA the Forskolin content was 0.09% D.W. which was comparable to Forskolin content in the wild plant (0.1% D.W.). At a concentration of 500 ug/l of IAA or IBA only small quantities of Forskolin was produced.

Forskolin derivatives have been successfully employed for the treatment of glaucoma and for the treatment of bronchial asthma. The unique biological activity and the least toxicity of Forskolin makes it a potential bioactive ingredient for future applications. In this context, the report on hairy root cultures of *C. forskohlii* cultivated in shake flasks and 20 litre capacity glass jar bioreactors is of significance (Krembholz *et al.*, 1992).

Ginkgo biloba, chinese medicinal plant, contains Ginkgolides A, B, C, and M (Fig. 1H). The Ginkgolides are a group of terpenoids and are extracted from the leaves of the slow growing tree. They antagonise platelet aggregations and are presently undergoing clinical trials in Europe for the treatment of Asthma and allergy (Braquet, 1987). The therapeutic activity of Ginkgolides is attributed to the cage molecule structure with 6 five membered rings and a t-butyl group. The synthesis of Ginkgolides is complex and is difficult to achieve economically. (Corey *et al.*, 1980). *In Vitro* studies related to *Ginkgo* has been limited. Initial efforts to grow *Ginkgo* cultures for elucidation of biosynthetic pathways were attempted by Nakanishi and Habaguchi (1971). However, they were unable to detect Ginkgolides in their cultures due to the low concentrations.

Recently, there are reports pertaining to the cultivation of *G. biloba* immobilised cells in 2 and 6 litre reactors (Carrier *et al.*, 1991). The technique was found to be simpler than suspension cultures. The cultures produced Gingkolides but at lower concentrations than the plant. The leaves of the plant contained 2.5 mg/g D.W. of Ginkgolide A and 0.8 mg/g D.W. of Ginkgolide B. (Okabe *et al.*, 1967). The shake flask cultures contained 7 ng/g D.W. of Ginkgolide A during a period of 18 days and 17 ng/g D.W. of the compound during a 27 day period. The cultures grown in 6 litre bioreactors for 31 days contained 19 ng/g D.W. of Ginkgolide A and cultures grown in 20 litre bioreactors for 27 days produced 7 ng/g D.W. of the compound. The cultures of *G.biloba* grown in shake flasks and in bioreactors did not undergo any differentiation such as the formation of root or leaf primordia. The observations indicated that there is scope for improving the production levels by selection of high yielding cell lines and other standard methods.

CONCLUSIONS

The global turnover of medicinal plants and their extractives is around US $ 4 billion. This could reach US $ 40 billion if compounds effective against AIDS are procured from them. In addition to the anticancer agents in the pipeline there is significant scope for obtaining anti AIDS products from the plant kingdom. Some complex molecules belonging to this group are Hypericin from *Hypericum*, trichosanthin from *Trichosanthum* and Castanospermine from *Castanospermum* (Heble, 1993).

The plants continue to remain as major source for novel phytochemicals and the developments in Medicinal Plant Biotechnology have opened up avenues both for the propagation of elite medicinal plants and mass cultivation of cells in bioreactors for novel products.

The discussion so far, was restricted to Taxanes and Artemisinin which are already in the market as choice drugs and Camptothecin, Forskolin and ginkgolides which are undergoing clinical trials.

REFERENCES

Avery, M.A., Chong, W.K.M. and Jenninger White, C. 1992. J. Am. Chem. Soc. 114: 974.

Bhat, S.V., Bajwa, B.S., Dornauer, H. and De Souza, N.J. 1977. Tetrahedron Letters, 19: 1669.

Blume, E. 1992. J. Natl. Cancer Inst. 84: 674.
Braquet, P. 1987. Drugs Future 12: 643.
Carrier, D.J., Chaureti, N., Manuin, M., Coulombe, P., Neufeld, R., Weber, M. and Archambault, J. 1991. Plant Cell Rep. 10, 256.
Christen, A.A., Gibson, D.M. and Blandt, J. 1991. United States patent Number, 5: 019, 504.
Corey, E.J., Jardine, P.D. and Rohloff, J.C. 1988. J. Am. Chem. Soc. 110: 649.
Corey, E.J., Jardine, P.D. and Rohloff, J.C. 1988. J. Am. Chem. Soc. 110: 3672.
De Souza, N.J. 1991. Coleus forskohlii Briq: The Indian plant source for Forskolin. In S.P. Choudhuri (ed.) Recent Advances in Medicinal, Aromatic and Spice crops. Today & Tomorrow's Printers and Publishers, New Delhi, India. Vol 1, pp. 83-91.
Fett-Neto, A.G., DiCosmo, F., Reynolds, W.F. and Sakata, K. 1992. Bio/Technol. 10: 1572.
Fulzele, D.P., Sipahimalani, A.T. and Heble, M.R. 1991. Phytotherapy Research, 5: 149.
Gottlieb, J.A., Guarino, A.M., Call, J.B., Olivero, V.T. and Block, J.B. 1970. Cancer Chemother. Rep. 54: 461.
Govindachari, T.R. and Viswanathan, N. 1972. Phytochemistry 11: 3529.
Hamburger, M. and Hostettmann, K. 1991. Bioactivity in Plants: The link between Phytochemistry and Medicine. In Thirty years of Phytochemistry, 1961-1991. Phytochemistry, 30(12), pp. 3849-3874.
Hashimoto, S., Sakata, S., Souegana, M. and Ikegami, S. 1988. J. Am. Chem. Soc. 110: 3670.
He, X.C., Xheng, M.Y., Li, G.F. and Liang, Z. 1983. Acta. Bot. Sin. 25: 87.
Heble, M.R. 1993. High value chemicals by tissue culture. Chemical Industry Digest (Sept.), pp. 113-118.
Holton, R.A. 1994. J. Am. Chem. Soc. (in Press).
Joel, S.P. 1994. Chemistry & Industry, No. 5(March) 172.
Kadkade, P.G. 1982. Plant Sci. Lett. 25: 107.
Kingsbury, W.D., Boehm, J.C., Jakas, D.R., Holden, K.G., Hecht, S.M., gallagher, G., Caranfa, J.J., McCabe, F.L., Faucette, L.F., Johnson, R.K. and Hertzberg, R.P. (1991). J. Med. Chem. 34: 98.
Klayman, D.L. 1985. Science 228: 1049.
Krembholz, R., Meringer, R., Kreis, W. and Reinhard, E. 1992. Planta med. 58: 328.
Kupchan, S.M., Court, W.A., Dailey, R.G., Jr. Gilmore, C.J. and Bryan, R.F. 1972. J. Am. Chem. Soc. 94: 7194.
Manfredi, J.J. and Horwitz, S.B. 1984. Pharmacol. Ther. 25:83.
McGuire, W.P., Rowinsky, E.K., Rossenhein, N.B., Grumbine, F.C., Ettinger, D.S., Armstrong, D.K. and Donehower, R.C. 1989. Ann. Intern. Med. 111: 273.
Mersinger, R., Donauer, H.a nd Reinhard, E. 1988. Planta Med. 3, 200.
Misawa, M. and Nakanishi, T.M. 1988. Antitumour compounds production by plant cell cultures In, Bajaj, Y.P.S. (ed.) Biotechnology in Agriculture & Forestry 15, Medicinal & aromatic plants II. Springer Verlag, Berlin, Heidelberg, New York, Tokyo, pp. 192-207.
Nair, M.S.R., Acton, N., Klayman, D.L., Endrick, K., Basile, D.V. and Mante, S. 1986. J. Nat. Prod. 49: 504.
Nakanishi, K. and Habaguchi, K. 1971. J. Am. Chem. Soc. 93: 3546.
Neelam Sharma, Chandel, K.P.S. and Srivastava, A.K. 1991. Plant Cell. Rep. 10: 67.
Nicolaou, K.C., Yang, Z., Liu, J.J., Ueno, H., Nantermet, P.G., Guy, R.K., Claiborne, C.F., Renaud, J., Couladouras, E.A., Paulvannan, K. and Sorenson, E.J. 1994. Nature 367: 630.
Ohno, R., Okada, K. Masaoka, T., Kuramoto, A., Arima, T., Yoshida, Y., Ariyashi, H., Ichimaru, M., Sakai, Y., Oguro, M., Ito, Y., Morishima, Y., Yokomaku, S. and Oka, K. 1990. J. Clin. Oncol. 8: 1907.
Okabe, K., Yamada, K., Yamamura, S. and Takada, S. 1967. J. Am. Chem. Soc. (C), 2201.
Paul, L. Huang, Philip, L. Huang, Peter Huang, Henry, I. Huang and Sylvia Lee Huang 1992. Developing Drugs from Traditional Medicinal Plants. Chemistry & Industry, No. 8. pp. 290-293.
Roja, G. and Heble, M.R. 1994. Phytochemistry 36, 65.
Rowinsky, E.R., Cazenave, L.A. and Donehower, R.C. 1990. J. Natl. Cancer Inst. 82: 1247.
Sakato, K., Tanaka, H., Mukai, N. and Misawa, M. 1974. Agr. Biol. Chem. 38: 217.
Schmidt, G. and Hofheinz, W. 1983. J. Am. Chem. Soc. 105: 624.
Sen, J., Sharma, A.K., Sahu, N.P. and Mahato, S.B. 1993. Phytochemistry, 34: 1309.

Wall, M.E., Wani, M.C., Cook, C.E., Palmer, K.H., McPhail, A.T. and Sim, G.A. 1966. J. Am. Chem. Soc. 88: 3888.
Wani, M.C., Taylor, H.L., Wall, M.E., Coggon, P. and McPhail, A.T. 1971. J. Am. Chem. Soc. 93: 2325.
Witherup, K.M., Look, S.A., Stasko, K.W., ghiorzi, T.J. and Muschik, G.M. 1990. J. Nat. Prod. 53: 1249.
Woerdenbag, H.J., Luers, J.F.J., Uden, W.V., Pras, N., Malingre, T.M. and Alfermann, A.W. 1993. Plant Cell Tissue and Organ Culture, 32: 247.
World Health Organisation. 1986. Report on the chemotherapy of Malaria. WHO, Geneva.

Genetic Variation in *in vitro* Cultured Plant Cells and in the Progenies of Regenerated Plants

S.J. Jambhulkar and C.R. Bhatia

Nucleotide changes at the DNA level, visible spontaneous mutations and chromosomal anomalies are ubiquitous in all living systems. Mutations arise due to errors during DNA replication that escape proof reading. Various stresses, as experienced during *in-vitro* culture, can further enhance replication errors and chromosomal anomalies. Meiotic sieves, competition between gametes, gametic and zygotic lethality eliminate a large proportion of chromosomal anomalies and deleterious mutations in sexually reproducing plants. *In vitro* culture of cells does not provide similar mechanism to reduce the genetic load as it permits normal or near normal propagation of variant cells. The spontaneous mutation rates vary from 10^{-4} to 10^{-7} per locus. Certain genes are more mutable than others. Therefore heritable changes in cultured plant cells and tissues are to be expected. Such heritable changes are a limitation when genetic fidelity of the tissue culture raised plants is crucial as in the propagation of elite genotypes. At the same time, tissue culture induced variation provides a new tool for enhancing genetic variability. Variation in cultured plant cells and plants regenerated from them is referred as somaclonal variation (Larkin and Scowcroft, 1981; see Scowcroft *et al.*, 1983; Evans *et al.*, 1984; D'Amato, 1985 see Evans, 1986). Plant cells in culture undergo various visible, morphological and genetic changes, at ploidy level, chromosomal aberrations (D'Amato, 1985; Buyser *et al.*, 1988; Ye and Yu, 1989; Stelly *et al.*, 1989; Detrez *et al.*, 1989 and mutations (see Bhatia *et al.*, 1985). Besides, changes at DNA, protein and enzyme level (Landsman and Uhrig, 1985; see Ball and Seilleur, 1986; Berlyn, 1982; Cullis, 1983; Brown *et al.*, 1990; Brettell *et al.*, 1986) also occur. An overview of the type of somaclonal variation, factor responsible, its significance and possible usefulness for crop improvement is presented in this article.

SOMACLONAL VARIATION- THE GENETIC BASE

The transient variation or the epigenetic changes induced by culture conditions are not expressed in the progeny of the regenerated plants. The heritable variation obtained through tissue culture has been attributed to the following (Scowcroft, 1984; Larkin, 1987; see Bajaj, 1990). Genetic implication of somaclonal variation has been described by Peschke and Phillips (1992).

- Gross chromosomal changes
- Structural rearrangements

- DNA amplification
- Gene mutations
- Transposable elements
- Altered expression of multigene families
- Mitotic recombination
- Genetic mosaics and chimeras

Gross chromosomal changes

Karyotypic instability has been described among the regenerants (Alicchio *et al.*, 1984). The tendency of chromosome number to increase in plant cell cultures is expressed often as aneuploidy and mixoploidy (Skirvin, 1978; Bayliss, 1980; D'Amato, 1985; see Sree Ramulu, 1986; Dahleen and Eizenga, 1990; Koma and Griesbach, 1993). Polyploidy has been reported in barley (Pickering, 1989), tomato (Bulk *et al.*, 1990) and *Aloe barbadensis* (Cavellini *et al.*, 1993).

Structural rearrangements

Cytologically detectable structural chromosomal changes have been observed in cultured cells and regenerated somaclones (D'Amato, 1985). The observed changes include deletion, duplication, inversion and translocations (Whelan, 1990). Larkin *et al.* (1989) obtained evidence of duplicated segment on the alpha arm of chromosome 4A in a wheat cultivar cv. Banks somaclone (BaB32*). The proximally located reduced height gene (Rht1) was duplicated as evidenced by the short heterozygous plants and very short homozygotes. In addition, an increased number of publications report on sequence alteration of DNA in nuclei, plastids and mitochondria (Cullis, 1983; Rode *et al.*, 1987; Breiman *et al.*, 1987b; Depaepe *et al.*, 1982; Muller *et al.*, 1990) are available.

DNA amplification

The DNA content shows extensive variation in plant cells ranging from less than 1 to over 200 pg (Bennett and Smith, 1976). The haploid genome size ranges from 10^9-10^{10} nucleotide pairs. About 60% of the plant genome is composed of repeated DNA (see Mitra and Bhatia, 1986). Most of the repeated DNA is low or moderately repetitive with reiteration frequency of less than 10^3 per sequence. Some highly repetitive DNA (10^5-10^6 copies) also occurs in most plants and can comprise up to 20% of the total DNA as in maize (Sorenson, 1984). The copy number of specific sequences in the genome may vary after cell culture. Yung and Wernsman (1990) suggested that variation in quantitative trait may arise due to DNA amplification. De Paepe *et al.* (1982) identified both heritable quantitative and qualitative changes in the nuclear DNA content in regenerated plants. Amplification of enolpyruvyl shikimic acid phosphate (EPSP synthase gene in carrot, conferring glyphosate resistance (Nafziger *et al.*, 1984) and for glutamine synthase gene conferring resistance to herbicide phosphorothricin (Donn *et al.*, 1984) have been reported. Kidwell *et al.* (1993) observed specific repeated nuclear DNA sequences have changed copy number in plants regenerated from tissue culture. The T-DNA in transformed cells has been reported to amplify after prolonged cell culture (Peerbolte *et al.*, 1987). Some examples of deamplification of repetitive sequences in potato (Landsman and Uhrig, 1985), wheat (Breiman *et al.*, 1987a) and *Hordeum spontaneum* (Breiman *et al.*, 1987b) and

DNA polymorphism (Endo *et al.*, 1988; Roth *et al.*, 1989) are known. Besides, gross alterations in methylation due to tissue culture resulting in altered phenotypes (Brown and Lorz, 1986; see Ball, 1990) are reported.

Gene mutations

Heritable changes occur spontaneously and are related to genome size (see Bhatia *et al.*, 1985). They could be due to addition or substitution of one or several bases (Drake, 1969). The point mutations could be dominant, co-dominant or recessive. In maize single base changes for alcohol dehydrogenase 1 locus, resulted in an altered electrophoretic mobility in one variant and null mutation in the other (Brettell *et al.* 1986; Dennis *et al.*, 1987). The rate of deleted nucleotide substitution in this gene was 2 in 218 cultured embryos producing 1382 somaclones examined. A chlorosulfuran resistant tobacco mutant isolated from culture was due to a resistant tobacco mutant isolated from culture was due to a resistant acetolactate synthase and therefore mutation in coding region of the corresponding gene (Chaleff and Mauvais, 1984).

Somaclones with distinct phenotypes have been reported by many investigators (Evans and Sharp, 1983; Lorz and Scowcroft, 1983; Mathews *et al.*, 1985). Evans *et al.*, (1984) obtained 13 mutants in 230 regenerated plants (5.6%) while Mathews *et al.* (1985) reported segregation of 10 out of 70 (14%) progenies of cotyledonary plants in mungbean.

Transposable elements

Involvement of transposable element in somaclonal variation has been demonstrated by Peschke *et al.* (1987). A number of somaclonal variants behave as if they were caused by altered location and activity of transposable elements. Williams *et al.* (1991) showed that there was no movement of the *Ac* related sequences to cause the tissue culture induced mutation. Differences in methylation of the genome can also generate the transpositional burst, leading in turn to more permanent changes (Schwartz and Dennis, 1986).

Altered expression of multigene families

Plant genes found in clusters at one or more locations in the genome are referred as multigene families. Alteration of their products in tissue culture raised plants have been reported, such as, gliadin storage protein of wheat (Larkin *et al.*, 1984; Maddock *et al.*, 1985; Cooper *et al.*, 1986), the hordeins of wild barley (Breiman *et al.*, 1987b), and esterases of potato (Alicchio *et al.*, 1987). Often these mutant express proteins not present in the mother plants which is attributed to the expression of previously repressed gene following tissue culture. Such rearrangements that alter gene families have the potential to give access to the hidden gene(s). Alteration in gene copy number of maize calli was observed by Brown *et al.* (1991).

Mitotic recombination

The mechanism of mitotic recombination have been investigated in both prokaryotic and eukaryotic systems. D loop formation between double stranded circular DNA and single stranded fragment is considered as initiation step of recombination (Atsushi-Higashitani *et al.*, 1990). In *in vitro* recombination activity assay high D-loop forming, and recombinatory activity was observed in rice callus. Such mitotic recombination can occasionally lead to the expression of a recessive mutant in the first generation itself (Evans *et al.*, 1984; Sibi *et al.*, 1984).

Genetic mosaics and chimeras

Organisms or tissues having genotypically different cell populations are described as genetic mosaics or chimeras. The variation could be either in nuclear or organelle genomes. Variegation seen in many ornamental plants is a common expression of genetic mosaicism. However, all variegated plants need not be genetic mosaics (Marcotrigiano and Stewart, 1984).

Many of the vegetatively propagated plants are chimeric and such cells can easily dissociate *in vitro* and the recovered plants would be considered as somaclonal variants. Such somaclonal variants can also arise due to the existing somatic mutations and somatic crossing over (Evans and Paddock, 1976).

In general, it is difficult to determine whether the genetic variability pre-existed in the explant or occurred after *In vitro* growth (Seeni and Gnanam, 1981). Segregation of chimeras in tissue culture could be useful in horticultural or ornamental plants. Marcotrigiano *et al.* (1987) excised strawberry runner's tips from a green-white-green periclinal chimeras and cultured on the media which was not conducive to shoot multiplication., The integrity of histogen was not altered in the developed plants implying that if the shoot apex is undamaged development remains normal. Similar results were obtained with Geranium chimeras (see Cassells *et al.*, 1980). The micropropagation of periclinal chimeras generally leads to the dissociation, while a certain percentage of shoots identical to the original explant are also recovered. Johnson (1980) reported isolation of new chimeras, identical chimeras and non chimeral plants of *Dianthus caryophyllus*. Some non chimeral plants were also isolated in *Dracaena marginata* (Chau *et al.*, 1981), *Episcia cupreata* (Chin, 1980), *Euphorbia pulcherrima* (Preil and Engelhardt, 1982). Shoot organogenesis *in vivo* is common in species where roots and leaves are used for vegetative propagation. Shoot may be derived from one cell (Broertjes and Keen, 1980) or more than one cell (Marcotrigiano, 1986). Therefore, depending upon the pathway of organogenesis there is a possibility of chimeral expression.

Chimeras can also appear during anther culture. When diploid tissues surrounding the pollen are included in the explant and the regeneration is via organogenesis, it is possible to get cytochimeras. Chromosome doubling can also occur in some cells which after organogenesis lead to cytochimeras.

In protoplast fusion experiments nuclei and cytoplasmic organelles are combined in a single fusion product. In some instances nuclei do not fuse and the cell becomes binucleate. Subsequent cell division can lead to chimeral cell cluster which, depending on mode of regeneration (organogenesis or embryogenesis) could result in the formation of chimeral plants. Plastids or mitochondrial chimeras may result following plant regeneration from fused protoplasts. Although protoplast fusion causes cytoplasmic mixing it is not possible to predict which plastids or mitochondrial genome will eventually dominate (Scowcroft and Larkin, 1981; Kumar and Cocking, 1982).

INHERITANCE OF SOMACLONAL VARIANTS

The genetic basis of the somaclones can be determined by testing the sexual progeny of the regenerated plants. For this, the regenerated plants should be raised to maturity, selfed and the harvested seeds planted in progeny rows. Genetic changes resembling single gene mutations have been found in many crops. Recessive single gene mutations do not appear in the first (Ro or R1) generation. In the selfed progeny the recessive, variant morphological trait segregates in 3 (normal) : 1 (variant) ratio. The mutant should breed true in the following generation. Genetic analysis to confirm single gene traits was reported in tomato (Evans and Sharp, 1983), maize (Edallo *et al.*, 1981), rice (Fukui, 1983; Sun *et al.*, 1983), wheat (Larkin *et al.*, 1984) and mungbean (Bhatia and Mathews, 1988).

FACTORS RESPONSIBLE FOR SOMCLONAL VARIATION

From the above, it is clear that the observed heritable variation in the plants raised through *in-vitro* culture could be due to various reasons. Considering this, the genotype of the explant, type of explant and culture conditions can influence the frequency of the variants observed.

Genotype of the explant

Genotype of the explant used for *in vitro* culture has a significant effect on the extent of variation generated during culture (Rietveld *et al.*, 1993, Bebeli *et al.*, 1993) significantly high number of chromosomal aberrations were observed in one culture of oat than the other (McCoy *et al.*, 1982). Roest *et al.* (1981) found that in one variety of Begonia hiemalis, 43% of the regenerants showed variation whereas in the other variety only 7% were variants. This could be due to somatic mutations, chimeras and genetic heterozygosity of the mother plant.

Explant type

The explant can have a significant effect on the extent of variation generated. Barbier and Dulieu (1981) found somaclonal variation not only among the plants regenerated from callus and cultured protoplast of tobacco but also among the plants derived from adventitious shoots. Substantial variation for stem, leaf, flower and skin variants was found among the plants regenerated by rapid adventitious shoot technique from rachis petiole and leaf disc explants (van Harten *et al.*, 1981). Among the plants obtained from rachis and petioles explants, 50% were variant. The rate of somaclonal variation was low in the explant of single large plantlet and high in basal small plantlet in banana (Reuveni *et al.*, 1993).

Culture condition

Media composition and the duration in culture are two important factors contributing to somaclonal variation. The tissue culture media, *per se*, are generally not mutagenic but prolonged period in culture are known to result in an increased frequency of gross chromosomal aberration (Bayliss, 1980; Meins, 1983; Reuveni *et al.*, 1983). Excess of $CaCl_2$ in the nutrient medium increased the frequency of gross chromosomal aberration (see Sree Ramulu, 1986). Further the type and relative concentration of auxins and cytokinins in the culture medium are reported to influence the ploidy level and the phenotype of potato plantlets regenerated from callus or protoplasts (Shepard, 1981; see Sree Ramulu *et al.*, 1983). Increased frequency of chromosomal abnormalities was associated with higher concentration of EDTA (see Sree Ramulu, 1986) and atrazin concentration (Wang *et al.*, 1991).

SOMACLONAL VARIATION IN TREE SPECIES

Unlike the crop plants there are few reports on somaclonal variation in tree species. This is possibly due to comparatively less efforts in tissue culture of tree species as compared to the annual crop plants and long generation time to screen the second generation. Various levels of polyploidy and aneuploids in callus cultures of tree species have been observed (Torok de 1968; Winton, 1971) Berlyn *et al.* (1987) observed nine times more DNA content in roots of regenerated plants of caribbean pines than the haploid amount, however, buds regenerated from dormant embryo explants were genetically stable. Renfroe and Berlyn (1985) reported variation in nuclear DNA content of tissue culture raised plants of *Pinus taeda* L. DNA content of nuclei in regenerant of poplar tended to be tetraploid (Antonetti and Pinon, 1993). Variation for morphological traits and isozyme pattern were observed in spruce somaclones (Eastman *et al.*, 1991). Somaclonal variation for more height and stem girth and early scented heartwood formation in sandalwood was observed by Sita (1991). Differences were noticed in mean specific gravity of wood in tissue culture plantlet and seedlings of loblolly pine (see Frampton and Jett, 1989). Somaclonal variants, resistant to toxin produced by Xanthomonas campestris in peach (Hammerschlag and Ognjanov, 1990) and Hypoxylon mammatum in poplar (Antonetti and Pinon, 1993) are reported.

SOMACLONAL VARIATION FOR CROP IMPROVEMENT

The potential value of the variation among the plants regenerated from tissue culture for crop improvement was first recognized in the mid 1970's by the sugarcane workers (Heinz *et al.*, 1977). Skirvin *et al.* (1993) described the proven usefulness of somaclonal variation. Haines (1993) reviewed the use of *In-vitro* methods to select for traits, such as resistance to disease, salinity, metal, heat and moisture tolerance. Somaclonal variants showing resistance to Fiji virus disease, downy mildew, eye spot and smut were reported (Krishnamurthi, 1981). Significant variation for growth habit, tuber colour, tuber production and resistance to early and late blight was reported in plants regenerated from potato protoplasts (Shepard, 1981). Both of these are vegetatively propagated crops. Subsequently somaclonal variants useful for improvement programs have been reported in a number of crop plants. These are summarized in Tables 1 and 2. Many of the reports included are based on early generation data and precise information on the stability of the traits in the following generation or their genetic basis have not been ascertained. Many tissue culture laboratories also do not have easy access to field facilities necessary for proper agronomic evaluation. It is likely that some of the claims made may not be true. The ultimate test of superior somaclones is to establish their yield or other agronomic advantage in multilocation trials and final acceptance as a new, improved cultivar. Development of such varieties through somaclones have been reported in Tomato (see Evans, 1986) and *Brassica napus* (Swanson *et al.*, 1989).

TABLE 1. Somaclonal variats for traits contributing to yield and enduse quality.

Crop and Character	References
Cereals	
Wheat	
Plant height, spike length, grain/spike and 100-grains weight increase in spike density and biomass	Mohmand and Nabors (1990)

Crop and Character	References
	Craver and johnson (1989)
Plant weight, tiller/plant	Hashim *et al.* (1990)
Spike length, and length and seeds/spikes	Cravera nd johnson (1990)
Maize	
Morphology and agronomic character	Cowen *et al.* (1990)
Oat	
Agronomic traits	Dahleen *et al.* (1991)
Sorghum	
Plant height, tiller number and grain weigh	Smith and bhaskaran 1988)
Rye	
Secalina nd storage protein	Bebeli *et al.* (1990)
Pearl Millet	
Quantitative characters	Morrish *et al.* (1990)
Triticale	
Yield and its components	Kaltsikes and Bebeli (1993)
Rice	
Grain number and weight/plant	Vajranabhaiah and Reddy (1993)
Oilseeds	
Brassica Juncea	
Early growth and flowering, increased branching and high yield	Katiyar and Chopra (1990)
Early flowering	Eapen *et al.* (1989)
Morphology and yield character	Jain *et al.* (1989)
Yellow seed coat colour	George and Rao (1989)
Soybean	
Agronomic traits	Stephens *et al.* (1991)
Growth habit (determinate and indeterminate)	Fretag *et al.* (1989)
Sunflower	
Head diameter, 100 seed weight oil content and fatty acid composition	Encheva *et al.* (1993)
Tuber crops	
Potato	
Desirable traits	Rietveld *et al.* (1991)
Fruits	
Banana	
High yield	Drew and Smith (1990)
Dwarf	Smith and Drew (1990)
Black Berry	
Growth habit, Vigour and flower number	McPheeters and Skirvin (1989)
Thorneless	Nelson *et al.* (1989)
Strawberry	
Yield and its components	Camerson *et al.* (1989)
Ornamentals	
Hemerocallis	

Crop and Character	References
Dwarf clone of commercial importance	Griesbach (1989)

Based on close familiarity with radiation and chemical induced mutations in self pollinated, sexually reproducing crops, limited personal experience with somaclonal variants and published reports in the same crops we find that in general, the somaclonal variants obtained are phenotypically similar to the radiation induced mutations. Spectrum of somaclonal variants reported in wheat mungbean and were not different from those obtained in induced mutation experiments. A green cotyledon mutant of mungbean was allelic to a similar somaclone (Bhatia and Mathews, 1988). Somaclonal variation would have a distinct advantage when *in-vitro* selection of the desired trait is feasible at the cellular level. However, there should be a good correlation between the *in vitro* cellular response and adult plant reaction in the field.

TABLE 2. Somaclonal variants for resistance to biotic and abiotic stresses

Stress	Crop	References
Biotic Stresses		
Diseases		
Helminthosporium sativum	Wheat and barley	Chawla and Wenzel (1987)
Rhychosporium secalis	Barley	Pickering (1989)
Albuqo candida	Brassica	Katiyar and Chopra (1990)
Alternaria brassicae		
Puccinia melanocephala	Sugarcane	Jalaja Sreenivasan *et al.* (1989)
Phytopthora cactorum	Apple	Rosati *et al.* (1990)
Xanthomonas campestris and Pseudomonas syringe	Peach	Hammerschlag and Ognjanov (1990)
Fusarium oxysporium	Cellary	Toth and Lacy (1991) Heath-Pagliuso *et al.* (1989).
Alternaria dauci	Carrot	Dugdole *et al.* (1993)
Blank shank and Bacterial blight	Tobacco	Daub and Jenns (1989)
Bacterial canker	Tomato	Bulk and Dons (1993)
Potato Y potyvirus (PVY)	Tobacco	Witherspoon *et al.* (1991)
Potato virus X (PVX)	Potato	Wenzel and Uhrig (1981)
Potato virus Y (PVY)		
Potato leafroll virus		
Insect Pests		
Leaf hopper	Potato	Lentini *et al.* (1990)
Sunflower beetle	Sunflower	Craig *et al.* (1991)
Nematodes		
Potato cyst nematode	Potato	Wenzel and Uhrig (1981)
Root-knot nematod	Egg plant	Fassuliotis (1992).
Abiotic Stresses		
Salt	Sorghum	Waskom *et al.* (1990)
	Brassica	Jain *et al.* (1991)
	Tomato	Sala *et al.* (1990)
	Flax	McHughen (1987)

Stress	Crop	References
Drought	Tomato	Galiba and Sutka (1989)
Frost	Wheat	Galiba and Sutka (1989)
Acid	Sorghum	Waskom *et al.* (1990)
Heat shock	Carrot	Hwang and Zimmerman (1989).
Ultraviolet rays	Sugarbeet	Levall and Bornman (1983).

Serious crop improvement programmes are always sharply focused with clear objectives which should be reached within a time frame and with minimal cost. Often, the same goal can be reached following different approaches. The breeders therefore use the most appropriate approach to reach the goal. Somaclonal variation is one of the tools available which may have distinct advantages over the other methods in reaching some breeding objectives.

REFERENCES

Alicchio, R., Antonioli, C., Graziani, L., Roncarati, R., and Vannini, C. 1987. Isozyme variation in leaf-Callus regenerated plants of Solanum tuberosum. Plant Sci. 53: 81-86.

Alicchio, R., Antonioli, C. and Palenzona, D.L. 1984. Karyotypic variability in plants of *Solanum melongena* regenerated from callus grown in presence of culture filtrate of *Verticillum dahliae* Theor. Appl. Genet. 67: 267-271.

Antonetti, P.L.E. and Pinon, J. 1993. Somaclonal variation within poplar. Plant Cell Tissue and Organ Culture. 35: 99-106.

Atsushi-Higashitani, Satoshi, T., Kiyohara, O. and Hotta Yasuo 1990. D-loop forming activity and general recombinatory activity in extracts from shoot, flower of callus of variety of rice. Japanese J. Genet. 65: 53-64.

Bajaj, Y.P.S. 1990. Somaclonal variation- Origin, induction, cryopreservation and implication in plant breeding. In: Y.P.S. Bajaj (ed.) Biotechnology in Agriculture and Forestry II- Somaclonal Variation in Crop Improvement I. Springer-Verlag Berlin, Heidelberg, New York, Tokyo pp. 1-48.

Ball, S.G. 1990. Molecular basis of somaclonal variation. In: Y.P.S. Bajaj (ed.) Biotechnology in Agriculture and Forestry II. Somaclonal Variation in Crop Improvement I. Springer-Verlag, Berlin, pp. 134-152.

Ball, S.B. and Seilleur, P. 1986. Characterization of somaclonal variations in potato: a biochemical approach. In: Semal, J. (ed.) Somaclonal Variation and Crop Improvement. Nijhoff, Dordrecht. pp. 229-235.

Barbier, M. and Dulieu, H. 1981. Early occurrence of genetic variants in protoplast cultures. Plant Sci. Lett. 29: 201-206.

Bayliss, M.W. 1980. Chromosomal variation in plant tissue culture. Int. Rev. Cytol. 11A: 113-143.

Bebeli, P.J., Kaltsikes, P.J. and Karp, A. 1993. Field evaluation of somaclonal variation in triticale lines differing in telomeric heterochromatin. J. Genet. Breed. 47: 249-257.

Bebeli, P.J., Karp, A. and Kaltsikes, P.J. 1990. Somaclonal variation from cultured immature embryos of sister lines of rye differing in heterochromatic content. Genome 33: 177-183.

Bennett, M.D. and Smith, J.B. 1976. Nuclear DNA amounts in antiosperms. Phil. Trans. Royal Soc. London. P. Biological jSciences 274: 227-274.

Berlyn, M.B. 1982. Variation in nuclear DNA content of isonicotinic acid hydrazide-resistant cell lines and mutant plants of Nicotiana tabacum. Theor. Appl. Genet. 63: 57-63.

Berlyn, G.P., Anoruo, A.O., Beck, R.C. and Cheng, J. 1987. DNA content ploymorphism and tissue culture regeneration in caribbean pine. Can. J. Bot. 65: 954-961.

Bhatia, C.R., Joshua, D.C. and Mathews, Helena, 1985. Somaclonal variation: A genetic interpretation based on the rates of spontaneous chromosomal aberrations and mutations. In: C.M. Govil and V. Kumar (eds.). Trends Pl. Res. pp. 317-326.

Bhatia, C.R.a nd Mathews, H. 1988. Inheritance of two somaclonal variants in mungbean (*Vigna radiata* (L.) Wilczek) J. Heridity 79: 122-124.

Breiman, A., Felsenburg, T. and Galun, E. 1987a. Nor loci analysis in progenies of plants regenerated from the scutellar callus of bread wheat. Theor. Appl. Genet. 73: 827-831.

Breiman, A., Rotem-Abarbanell, D. Karp, A. and Shaskin, H. 1987b. Heritable somaclonal variation in wild barley (*Hordeum spontaneum*) Theor. Appl. Genet. 74: 104-112.

Brettell, R.I.S., Dennis, E.S., Scowcroft, W.R. and Peacock, J.W. 1986. Molecular analysis of somaclonal mutant of maize alcohol dehydrogenase. Mol. Gen. Genet. 202: 235-239.

Broertjes, C. and Keen, A. 1980. Adventitious shoots: Do they develop from one cell? Euphytica 29: 73-87.

Brown, P.T.H., Gobel, E. and Lorz, H. 1991. RFLP analysis of *Zea mays* callus cultures and their regenerated plants. Theor. Appl. Genet. 81: 227-232.

Brown, P.T.H., Kyozuka, J., Sukekiyo, Y., Kimura, Y., Shimamoto, K. and Lorz, H. 1990. Molecular Changes in protoplast-derived rice plants. Mol. Gen. Genet. 223: 324-328.

Brown, P.T.H. and Lorz, H. 1986. Molecular changes and possible origin of somaclonal variation and crop improvement. Nijhof, Dordrecht. pp. 148-156.

Bulk, R.W. van den and Dones, J.J.M. 1993. Somaclonal variation as tool for breeding tomato for resistance to bacterial canker. Acta Horticulturae. No. 336: 347-355.

Bulk, R.W. van den, Loffler, H.H.M., Lindhout, W.H. and Moornneef, M. 1990. Somaclonal variation in tomato: Effect of explant source and comparison with chemical mutagenesis. Theor. Appl. Genet. 80: 817-825.

Buyser, J. De, Hartmann, C., Henry, Y. and Rode, A. 1988. Variation in long term wheat somatic tissue culture. Can. J. Bot. 66: 1891-1895.

Camerson, J.S., Hancock, J.F. and Flore, J.A. 1989. the influence of micropropagation on yield components, dry matter partitioning and gas exchange characteristics of strawberry. Scientia Horticulturae 38: 61-67.

Cassells, A.C., Minas, G. and Long, R. 1980. Cultue of Pelargonium hybrids from meristems and explants: chimeral and beneficially-infected varieties. In: D.S. Ingram and J.P. Helgeson (eds.). Tissue culture methods for plant pathologists. Blackwell, Oxford. pp. 125-130.

Cavallini, A., Natali, L., Cionini, G., Sassoli, O. and Catorena-Sanchez, I. 1993. In-vitro culture of *Aloe barbadesis* Mill.: Quantitative DNA variation in regenerated plants. Plant Sci. (Limerik) 91: 223-229.

Chaleff, R. and Mauvais, C. 1984. Acetolactate synthase in the site of action of two sulfonylurea herbicides in higher plants. Science. 224: 1443-1445.

Chau, B.U., Kunisaki, J.T. and Sagawa, Y. 1981. *In vitro* propagation of Dracaena marginata Tricolour Hort. Sci. 16: 494.

Chawla, H.S. and Wenzel, G. 1987. *In vitro* selection of barley and wheat for resistance against *Helminthosporium sativum* Theor. appl. Genet. 74: 841-845.

Chin, C.K. 1980. Growth behaviour of green and albino plants of *Episcia cupreata* Pink Brocade *in vitro*. *In vitro* 16: 847-850.

Cooper, D., Sears, R., Lockhart, G. and Jones, B. 1986. Heritable somaclonal variation in gliadin proteins of wheat plants derived from embryo callus culture. Theor. Appl. Genet. 71: 784-790.

Cowen, N.M., Thompson, S.A. and Wilkinson, T.C. 1990. Culture associated variation in maize inbreds. Plant Breeding 104: 134-143.

Craig, R., Roseland, A.E.a nd Teryl, J.G. 1991. Somaclonal variants of sunflower with modified coumarin expression under stress. Euphytica 54: 183-190.

Craver, B.F. and Johnson, B.B. 1989. Partitioning of variation derived from tissue culture of winter wheat. Theor. Appl. Genet. 78: 405-410.

Cullis, C.A. 1983. Environmentally induced DNA changes in plants. CRC Critical Rev. Plant Sci. 1: 117-129.

D'Amato, F. 1985. cytogenetics of plant cell and tissue culture and their regenerates. CRC Critical Rev. Plant ci. 3: 73-112.

Dahleen, L.S. and Eizenga, G.C. 1990. Meiotic and isozymic characterization of plants regenerated from euploid and selfed monosomic tall fescue embryos. Theor. Appl. Genet. 79: 39-44.

Dahleen, L.S.,, Stuthman, D.D., and Rines, H.W. 1991. Agronomic trait variation in oat lines derived from tissue culture. Crop Science 31: 90-94.

Daub, M.E. and Jenns, A.E. 1989. Field and greenhouse analysis of variation for disease resistance in tobacco somaclones. Phytopathology 79: 600-605.

De Paepe, R., Prat, D. and Huguet, T. 1982. Heritable nuclear DNA changes in doubled haploid plants obtained by pollen culture of *Nicotiana sylvestris*. Plant Sci. Lett. 28: 11-28.

Dennis, E.S., Brettell, R.I.S. and Peacock, W.J. 1987. A tissue culture induced Adhi null mutant of maize results from a single base change. Mol. Gen. Genet. 210: 181-183.

Detrez, C., Sangwan, R.S. and Sangwan-Norreel, B.S. 1989. Phenotypic and karyotypic status of *Beta vulgaris* plants regenerated from direct organogenesis in petiole culture. Theor. Appl. Genet. 77: 462-468.

Donn, G., Tischer, E., Smith, J.A. and Goodman, H.M. 1984. Herbicide-resistant alfalfa cells: An example of gene amplification in plants. J. Mol. Appl. Genet. 2: 621-635.

Drake, J.W. 1969. Comparative rates of spontaneous mutation. Nature 221: 1132.

Drew, R.A. and Smith, M.K. 1990. Field evaluation of tissue cultued banans in south-eastern Queensland. Aust. J. Expt. Agri. 30: 569-574.

Dugdale, L.J., Collin, H.A., Issac, S. and Gill, J.J.B. 1993. Leaf Aust. J. Expt. Agri. 30: 569-574.

Eapen, S., Abraham, V., Gerdemann, M. and Schieder, O. 1989. Direct somatic embryogenesis, plant regeneration and evaluation of plants obtained by mesophyll protoplast of *Brassica juncea*. Ann. Bot., 63: 369-372.

Eastman, P.A.K., Webster, F.B., Pitel, J.A. and Roberts, P.R. 1991. Evaluation of somaclonal variation during somatic embryogenesis of Interior Spruce (Picea-Glauca- Engelmannii-Complex) using culture morphology and isozyme analysis. Plant Cell Reports 10: 425.

Edallo, S., Zucchinali, C., Perenzin, M.l and Salamini, F. 1981. Chromosomal variation and frequency of spontaneous mutation associated with *in vitro* culture and plant regeneration in maize. Maydica 26: 39-56.

Encheva, J., Ivanov, P., Tsvetkova, F. and Nikolova, V. 1993. Development of new initial breeding material in sunflower (Helianthus annuus L.) using direct organogenesis and somatic embryogenesis. Euphytica 63: 181-185.

Endo, T., Komiya, T., Mino, M., Nakanishi, K, Fajita, S. and Yamada, Y. 1988. Genetic diversity among sublines originating from a single somatic hybrid cell of *Duboisia hopwoodii* and *Nicotinana tabacum*. Theor. Appl. Genet. 76: 641-646.

Evans, D. 1986. Case histories of genetic variatbility *in vitro*: tomato. In: I.K. Vasil (ed.) Cell culture and Somatic Cell Genetics in Plants. Vol. 3. Plant regeneration and genetic variability. Academic Press, Orlando, pp. 419-434.

Evans, D.A. and Paddock, E.F. 1976. Comparision of somatic crossing over frequency in *Nicotiana tabacum* and three other crop species. Can. J. Genet. Cytol. 18: 57-65.

Evans, D.A. and Sharp, W.R. 1983. Single gene mutation in tomato plants regenerated from tissue culture. Science 221: 949-957.

Evans, D.A., Sharp, W.R. and Medina-Filho, H.P. 1984. Somaclonal and gametoclonal variation. Amer. J. Bot. 71: 759-774.

Fassuliotis, G. 1990. Somaclonal variation for nematode resistance. In: Y.P.S. Bajaj (ed.) Biotechnology in Agriculture and Forestry Vol. II. Somaclonal Variation in Crop Improvement. Springer-Verlag Berlin Heidelberg, pp. 258-268.

Frampton, L.J.J. and Jett, J.B. 1989. Juvenile wood specific gravity of loblolly pine tissue culture plantlets and seedlings. In: D.I. Dunstan (ed.) 4th meeting of the International Conifer Tissue Culture Work Group. Saskatoon, Canada, August 8 to 12, 1988. Canadian J. Forest Res. 19: 1347-1350.

Freytag, A.H., Rao, A., Anand, S.C., Wrather, J.A. and Owens, L.P. 1989. Somaclonal variation in soybean plants regenerated from tissue culture. Plant Cell Reports 8: 199-202.

Fukui, K. 1983. Sequential occurrence of mutations in a growing rice callus. Theor. Appl. Genet. 65: 225-230.

Galiba, G. and Sutka, J. 1989. Frost resistance of somaclonal derived from Triticum aestivum L. winter wheat calli. Plant Breeding 102: 101-104.

George, L. and Rao, P. 1983. Yellow-seeded variants in *in vitro* regenerants of mustard (*Brassica Juncea* Coss var. Rai-5), Plant Sci. Lett. 30: 327-330.

Griesbach, R.J. 1989. Selection of dwarf Hemerocallis through tissue culture. Hortsci. 24: 1027-1028.

Haines, R.J. 1993. *In-vitro* selection in plant breeding. In: B.C. Imrie and J.B. Hacker (eds.). Focused plant improvement: towards responsible and sustainable agriculture. pp. 210-219.

Hammerschlag, F.A. and Ognjanov, V. 1990. Somaclonal variation in peach: Screening for resistance to *Xanthomonas campestris* pv. *pruni* and *Pseudomonas syringe* pv. *syringae*. Acta Horticulturae. 280: 403-408.

Hashim, Z.N., Campbell, W.F. and Carman, J.C. 1990. Morphological analysis of spring wheat (CIMMYT cv. PCTY-10) somaclones. Plant Cell, Tissue and Organ Culture 20: 95-99.

Heath-Pagliuso, S., Pullman, J. and Rappaport, L. 1989. UC-T3 Somaclone: Celery germplasm resistant to *Fusarium oxysporium* f. sp. apil, race 2. Hort. Sci. 24: 711-712.

Heinz, D.J., Krishnamurthi, M., Nickell, L.G. and Maretzki, A. 1977. Cell, tissue and organ culture in sugarcane- improvement. In: J. Reinert and Y.P.S. Bajaj (eds.). Applied and Fundamental Aspects of Plant Cell, tissue, and Organ Culture. Springer, Berlin, pp. 3-17.

Hwang, C.N. and Zimmerman, J.L. 1989. The heat shock response of carrot. Plant Physiology 91: 552-558.

Jain, R.K., Sharma, D.R. and Chowdhury, J.B. 1989. High frequency regeneration and heritable somaclonal variation in Brassica juncea. Euphytica 40: 75-81.

Jain, S., Narnawatee, H.S., Jain, R.K. and Chowdhury, J.B. 1991. Proline status of genetically stable salt tolerant Brassica juncea L. Somaclones and their parent cv. Prakash. Plant Cell Reports. 9: 684-687.

Jalaja Sreenivasan, Sreenivasan, T.V. and Alexander, K.C. 1987. Somaclonal variation for rust resistance in sugarcane. Ind. J. Genet. 47: 109-114.

Johnson, R.T. 1980. Gamma irradiation and *in vitro* induced separation of chimeral genotypes in carnation. Hort. Sci. 15: 605-606.

Kaltsikes, P.J. and Bebeli, P.J. 1993. Somaclonal variation causes changes in the interrelationship between traits in hexaploid triticale. Jap. J. Breed. 43: 45-51.

Kamo, K. and Griesbach, R. 1993. Ploidy changes in Mitchell petunia. Acta Horticulturae No. 336: 307-314.

Katiyar, R.K. and Chopra, V.L. 1990. Somaclonally induced earlyness in *Brassica juncea* germplasm accession with field resistance to important diseases. Plant Breeding. 104: 262-264.

Kidwell, K.K. and Osborn, T.C. 1993. Variation among alfalfa somaclones in copy number of repeated DNA sequences. Genome. 36: 906-912.

Krishnamurthi, M. 1981. Sugarcane improvement through tissue culture and review of progress. In: A.N. Rao (ed.) Tissue culture of Economically Important Plants, Asian Network for Biological Sciences, Singapore pp. 70-77.

Kumar, A. and Cocking, E.C. 1982. Restriction endnuclease analysis of chloroplast DNA in interspecies somatic hybrids of *Petunia*. Theor. Appl. Genet. 60: 197-214.

Landsmann, J. and Uhrig, H. 1985. Somaclonal variation in *Solanum tuberosum* detected at molecular level. Theor. Appl. Genet. 71: 500-505.

Larkin, P. 1987. Somaclonal variation: history, method and meaning. Iowa State J. Res. 61: 393-434.

Larkin, P., Ryan, S., Brettell, R. and Scowcraft, W. 1984. Heritable somaclonal variation in wheat. Theor. Appl. Genet. 67: 443-455.

Larkin, P. and Scowcraft, W. 1981. Somaclonal variation- a novel source of variability from cell culture for plant improvement. Theor. Appl. Genet. 60: 197-214.

Larkin, P.J., Banks, P.M., Bhati, R., Brettell, R.I.S., Davies, P.A., Ryan, S.A., Scowcraft, W.R., Spindler, L.H. and Tanner, G.J. 1989. From somatic variation to variant plants: mechanisma nd applications. Genome 31: 705-711.

Lentini, Z., Earle, E.D.,a nd Plaisted, R.L. 1990. Insect resistant plants with improved horticultural traits from interspecific potato hybrids grow *in vitro*. Theor. Appl. Genet. 80: 95-104.

Levall, M.W. and Bornman, J.F. 1993. Selection *in vitro* for UV-tolerant sugarbeet (*Beta vulgaris*) somaclones. Physiologia plantarum. 88: 37-43.

Lorz, H. and Scowcraft, W.R. 1983. Variability among plants and their progeny regenerated from protoplasts of Su/su heterozygotes of *Nicotiana tabacum*. Theor. Appl. Genet. 66: 67-75.

Maddock, S., Risiott, R., Parmar, S., Jones, M. and Shewry, P. 1985. Somaclonal variation in the gliadin patterns of grains of regenerated wheat plants. J. Exp. Bot. 36: 1976-1984.
Marcotrigiano, M. 1986. Origin of adventitious shoots regenerated from cultured tobacco leaf tissue. Am. J. Bot. 73: 1541-1547.
Marcotrigiano, M., Morgan, P.A., Swartz, H.J. and Ruth, J. 1987. Histogenic instability in tissue culture proliferated strawberry plants. J. Am. Soc. Hortic. Sci. 112: 583-587.
Marcotrigiano, M. and Stewart, R.N. 1984. All variegated plants are not chimeras. Science 223: 505.
Mathews, V.H., Rao, P.S. and Bhatia, C.R. 1985. Somaclonal variation in cotyledonary plants of mungbean. Z. Pflanzenzucht. 96: 169-173.
McCoy, T.J., Phillips, R.L. and Rines, H.W. 1982. Cytogenetic analysis of plant regenerated from oat (*Avena sativa*) tissue culture: frequency of partial chromosome loss. Can. J. Genet. Cytol. 24: 37-50.
McHughen, A. 1987. Salt tolerance through increased vigor in a flax line (STS-II) selected for salt tolerance *in vitro*. Theor. Appl. Genet. 74: 727-732.
McPbheeters, K. and Skirvin, R.M. 1989. Somaclonal variation among ex vitro Thornless Evergreen trailing blackberries. Euphytica 42: 155-162.
Meins, F. 1983. Heritable variation in plant cell culture. Ann. Rev. Plant Physiol. 34: 327-346.
Mitra, R. and Bhatia, C.R. 1986. Repeated DNA sequences and polyploidy in cereal crops. In: S.K. Datta (ed.), DNA systematics Vol. II CRC Press, Boca Raton, Florida pp. 21-43.
Mohmand, A.S. and Nabors, M.W. 1990. Somaclonal variant plants of wheat derived from mature embryo explants of three genotypes. Plant Cell Reports. 8: 556-558.
Morrish, F.M., Hanna, W.W. and Vasil, I.K. 1990. The expression and perpetuation of inherent somatic variation in regenerants from embryogenic cultures of *Pennisetum glaucum* (L.) R. Br. (Pearl millet). Theor. Appl. Genet. 80: 409-416.
Muller, E., Brown, P.T.H., Hartke, S. and Lorz, H. 1990. DNA variation in tissue cultured derived rice plant. Theor. Appl. Genet. 80: 675-679.
Nafziger, E., Widholm, J., Steinrucken, H. and Killmer, J. 1984. Selectiona nd characterization of a carrot cell line tolerant to glyphosate. Plant Physiol. 76: 571-574.
Nelson, E.H., Defrancesco, J.T., Martin, L.W. and Skirvin, R.L. 1989. Evergreen blackbery screening for genetic thornlessness. Acta Horticulturae. 262: 113-118.
Peerbolte, R., Ruigrok, P., Wullems, G. and Schilperoort, R. 1987. T-DNA rearrangements due to tissue culture: somaclonal variation in crown gall tissues. Plant Mol. Biol. 9: 51-57.
Peschke, V. and Phillips, R.L. 1992. Genetic implication of somaclonal variation in plants. Advances in Genetics. 30: 41-75.
Peschke, V., Phillips, R. and Gengenbach, B. 1987. Discovery of transposable elements activity among progeny of tissue culture-derived plants. Science. 238: 804-807.
Pickering, R.A. 1989. Plant regeneration and variants from calli derived from immature embryos of diploid barley (*Hordeum vulgare* L.) and *H. vulgare* X *H. bulbosum* L. Crosses. Theor. Appl. Genet. 78: 105-112.
Preil, W. and Engelhardt, M. 1982. *In vitro* separation of chimeras by suspension cultures of *Euphorbia pulcherima* Wild. Gartenbauwissenschaft 47: 241-244.
Renfroe, M.H. and Berlyn, G.P. 1985. Variation in nuclear DNA content in *Pinus taeda* L. tissue culture of diploid origin. J. Plant Physiol. 121: 131-139.
Reuveni, O., Golubowicsz, S. and Israeli, Y. 1993. Factors influencing the occurrence of somaclonal variation in influencing the occurrence of somaclonal variation in micropropagated banana. Acta Horticulturae No. 336: 357-364.
Rietveld, R.C., Bressan, R.A. and Hasegawa, P.M. 1993. Somaclonal variation in tuber disc derived population of potato. II. Differential effect of genotype. Theor. Appl. Genet. 87: 305-313.
Rietveld, R.C., Hasegawa, P.M. and Bressan, R.A. 1991. Somaclonal variation in tuber disc-derived populations of potato. Theor. Appl. Genet. 82: 430-440.
Rode, A., Hartmann, C., Benslimane, A., Picard, E. and Quetier, P. 1987. Gametoclonal variation detected in the nuclear ribosomal DNA from double haploid lines of a spring wheat (*Triticum aestivum* L. cv. Cesar.). Theor. Appl. Genet. 74: 31-37.
Roest, S., Van Berkel, M., Bokelmann, G. and Broertjes, C. 1981. The use of an *in vitro* adventitious bud technique for mutation breedings of *Begonia hiemalis*. Euphytica 30: 381-388.

Rosati, P., Mezzetti, B., Aucherani, M., Foscolo, S., Predieri, S. and Fasolo, F. 1990. *In vitro* selection of apple root stock somaclones with *Phytopthora cactorum* culture filtrate. Acta Horticulturae. 280: 409-416.

Roth, E.J., Frazier, B.L., Apuya, N.R. and Lark, K.B. 1989. Genetic variation in inbred plants: Variation in tissue cultures of soybean (*Glycin max* (L.) Merrill). Genetics 121: 359-368.

Sala, F., Baratta, G., Todesco, G., Consonni, G., Govazzi, G., Tonelli, G. and Vecchio, F. 1990. Somaclonal variation and chemically induced mutagenesis for the production of salt and drought tolerance in tomato. Acta Horticulturae No. 280: 353-360.

Schwartz, D. and Dennis, E. 1986. Transposase activity of the Ac controlling elements in maize is regulated by its degre of methylation. Mol. Gen. Genet. 205: 476-482.

Scowcroft, W.R. 1984. Genetic variability in tissue culture: Impact on germplasm conservation and utilization. IBPGR report AGPG: IBPGR/84/152 Oct. 1984: 1-41.

Scowcroft, W.R. and Larkin, P.J. 1981. Chloroplast DNA assorts randomly in interspecific somatic hybrids of *Nicotiana debneyi*. Theor. Appl. Genet. 60: 179-184.

Scowcroft, W.R., Larkin, P.J. and Brettell, R.I.s. 1983. Genetic variation from tissue culture. In: Use of Tissue Culture and Protoplasts in Plant Pathology. Academic Press. pp. 139-162.

Seeni, S. and Gnanam, G. 1981. *In vitro* regeneration of chlorophyll chimeras in tomato (*Lycopersican esculentum*). Can. J. Bot. 59: 1941-1943.

Shepard, J.F. 1981. Protoplasts as sources of disease resistance in plants. Ann. Rev. Phytopathol. 19: 145-166.

Sibi, M., Biglary, M. and Demarly, Y. 1984. Increase in the rate of recombinants in tomato (*Lycopersicon esculentum* L.) after *in vitro* regeneration. Theor. Appl. Genet. 68: 317-321.

Sita, G.L. 1991. tissue cultured sandalwood. Curr. Sci. 61:12.

Skirvin, R.M. 1978. Natural and induced variation in tissue culture. Euphytica 27: 241-266.

Skirvin, R.M., Nortom, M. and McPheeters, K.D. 1993. Somaclonal variation: has it proved useful for plant improvement. Acta Hort. no. 336: 333-340.

Smith, M.K. and Drew, R.A. 1990. Growth and yield characteristics of dwarf off-types recovered from tissue-cultured bananas. Aust. J. Expt. Agric. 30: 575-578.

Smith, R.H. and Bhaskaran, S. 1988. Sorghum cell culture: Somaclonal variation/screening. Iowa State J. Res. 62: 571-585.

Sorenson, J.C. 1984. the structure and expression of nuclear genes in higher plants. Adv. Genet. 22: 109-144.

Sree Ramulu, K. 1986. Case history of genetic variability: Potato. In: P.V. Ammirato, D.A. Evans, W.R. Sharp, Y. Yamada (eds.), Cell Culture and Somatic Cell Genetics of Plants. Vol. 3 Macmillan, New York, pp. 449-473.

Sree Ramulu, K., Dijkhuis, P. and Roest, S. 1983. Phenotypic variation and ploidy level of plants regenerated from protoplasts of tetraploid potato (*Solanum tuberosum* L. cv. Bintje.). Theor. Appl. Genet. 65: 329-338.

Stelly, P.M., Altman, D.W., Kohel, R.J., Rangan, T.S. and Commiskey, E. 1989. Cytogenetic abnormalities of cotton somaclones from callus cultures. Genome. 32: 762-770.

Stephens, P.A., Nickell, C.D. and Widholm, J.M. 1991. Agronomic evaluation of tissue culture derived soybean plants. Theor. Appl. Genet. 82: 633-635.

Sun, Z., Zhao, C., Zheng, X., Qi, and Fu, Y. 1983. Somaclonal genetics of rice, *Oryza sativa* L. Theor. Appl. Genet. 67: 67-73.

Swanson, E.B., Herngesell, M.J., Arnolds, M., Sippell, D.W. and Wong, R.S.C. 1989. Microspore muttagenesis and selection: Canola plants with field tolerance to the emidazolinones. Theor. Appl. Genet. 78: 525-530.

Torok de, D. 1968. The cytologic and growth characteristics of tumor and normal clones of *Picea glauca*. Cancer Res. 28: 608-614.

Toth, K.F. and Lacy, M.L. 1991. Increasing resistance in celery to *Fusarium-oxysporum* F Sp Apii Race 2 with somaclonal variation. Plant Disease. 75: 1034-1037.

Vajranabhaiah, S.N. and Reddy, P.C. 1993. *In-vitro* developed desirable somaclone of upland rice (*Oriza sativa* L.) cultivar Halubbalu. Curr. Sci. 65: 174-176.

Van Harten, A.M., Bouter, H. and Broertjes, C. 1981. *In vitro* adventitious bud techniques for vegetative propagation and mutation breeding of potato (*Solanum tuberosum* L.). Euphytica 30: 1-8.

Wang, W.C., Myers, J.R. and Collins, G.B. 1991. Selection of atrazin-resistant tobacco variants from photomixotrophic cultures. Plant Science. 73: 199-209.

Waskom, R.M., Miller, D.R., Hanning, G.E., Duncan, R.R., Voigt, R.L. and Nabors, M.W. 1990. Field evaluation of tissue culture derived sorghum for increase tolerance to acid soils and drought stress. Can. J. Pl. Sci. 70: 997-1004.

Wenzel, G. and Uhrig, H. 1981. Breeding for nematode and virus resistance in potato via anther culture. Theor. Appl. Genet. 59: 333-340.

Whelan, E.D.P. 1990. Meiotic abnormalities in primary regenerants from callus culture of immature embryos of Norstar winter wheat. Genome 33: 260-266.

Williams, M.E., Hepburn, A.G. and Widholm, J.M. 1991. Somaclonal variation in maize inbred line is not associated with changes in the number of location of Ac-homologous sequences. Theor. Appl. Genet. 81: 272-276.

Winton, L. 1971. Tissue culture propagationof European aspen. For. Sci. 17: 348-350.

Witherspoon, W.D. Jr., Wernsman, E.A., Gooding, G.V. Jr. and Rufty, R.C. 1991. Characterization of a gametoclonal variant controlling virus resistance in tobacco. Theor. Appl. Genet. 81: 1-5.

Ye, X.R. and Yu, Y.J. 1989. Study on variation in wheat generated plants II. Cytological and morphological variations of R_1 generation. Acta Genetica Sinica 16: 105-110.

Yung, C.H. and Wernsman, E.A. 1990. The value of gametoclonal variation in breeding for qualitative traits in fluecured tobacco (*Nicotiana tabacum* L.). Theor. Appl. Genet. 83: 381-384.

In Vitro Response as Aid to Measure Genetic Diversity

Sumitra Sen

The study of genetic diversity is an important line of research in programmes on biological diversity. Several parameters have been applied by different authors from time to time to identify cryptic genetic differences which often remain phenotypically indistinguishable (Stebbins, 1971). In addition to the finer details of chromosome characteristics (Sharma, 1974; Sen, 1975; Bennett, 1987), the amount of DNA and repeat DNA sequences do provide good parameters (Bennett and Smith, 1976). Over and above these criteria, extensive works have been carried out in this laboratory on differential behaviour *in vitro* in several taxa of monocotyledons. In addition to recalcitrance and positive response, several factors which control *in vitro* growth and regeneration are species specific. The present review is principally based on the assessment of these results in order to ascertain the importance of *in vitro* response as an indicator of genetic diversity. In view of the contributions of this laboratory the discussion has been centered principally around some liliaceous taxa.

Genus *Allium*

Allium cepa and *A. sativum*, commonly known as onion and garlic are cultivated wildly for their horticultural and economical value.

In respect of callus initiation and organogenesis, the different explants of *A. cepa* and *A. sativum* namely, leaf, bulb scale and root-tips showed differences in their response to callus inducing medium (Chakraborty and Sen, 1988). The bulb scale explants containing the basal disc in *A. cepa* and axillary bud in *A. sativum* were found suitable for formation of callus in all the cultivars studied. A sativum was more responsive than *A. cepa* in relation to *in vitro* growth and both species have different requirements for growth under *in vitro* condition. In *A. sativum*, the addition of IAA and kinetin along with 2,4-D showed increased callusing of the explants.

The different cultivars of *A. cepa* have been noted to show differences in their response to hormone combinations which are consistent in different replication trials. The cultivar I showed response in presence of NAA and kinetin, cultivar II was more responsive in presence of NAA and BAP, whereas cultivar III did not show favourable response in any of the hormone combinations tried. Thus genotypic control is marked in this species at intraspecific level. In general, *A. cepa* did not show profuse callusing in any of the three cultivars studied. This is in contrast to *A. sativum* which shows profuse callus formation in both the cultivars studied. The interspecific differences in this respect is rather distinct.

The cultivar II of *A. cepa* could be maintained till 240 days whereas cultivars I and III could not be maintained beyond 150 days. In *A. sativum* on the other hand, cultivars I and II could be maintained upto 360 and 390 days respectively, indicating no marked difference in response. Thus, as far as maintenance of callus lines is concerned, the differential response of cultivars of two species is quite evident at interspecific level. At the intraspecific level, varietal differences are more prominent in *A. cepa* than in *A. sativum*.

In respect to regeneration, genotypic difference could be recorded as well. In *A. cepa* cultivar I showed regeneration from callus on transfer to MS medium with IAA and kinetin supplement, whereas cultivar II needed transfer of callus to MS with NAA and BAP supplements. The cultivar III on the other hand, needed 15% coconut milk for development of shootbuds. The distinct difference of genotypes in their requirements for regeneration are indeed good parameters of identifying the genetic difference in this species.

In *A. sativum*, regeneration of organ could be noted in both the cultivars on transfer to MS medium with IAA and kinetin supplements. The minimum time of appearance of shoot buds however, differed in the two cultivars. Excepting this minor difference, in respect of regeneration, genotypic influence in *A. sativum* could not visibly be demonstrated.

In *A. cepa* the meristems were noted to develop from the surface tissue of the callus whereas in *A. sativum*, both shoot and root meristem developed from deep seated tissue of the callus which are further reflections of genetic differences.

In general, the genetic control of diversity in respect of callusing, organogenesis and regeneration are marked in both the species, though in *A. sativum*, it is less demonstrable than in *A. cepa* (Chakravorty and Sen, 1988).

Genus *Asparagus*

This genus is well known for its medicinal, food and horticultural value. For induction of callus, different species, namely *A. asperagoides*, *A. cooperi*, *A. officinalis*, *A. plumosus*, *A. racemosus*, *A. verticillatus* were tried and variable response was recorded (Ghosh and Sen, 1991). The combination and concentration of the cytokinin and auxin needed for callusing by different explants differed in different species. For induction of callus, shoot tip explant of *A. cooperi*, *A. verticillatus*, *A. officinalis*, *A. asperagoides* needed NAA and kinetin, whereas in *A. plumosus*, Kn and 2iP were needed. For node explants, requirements were 2,4-D, NAA and Kn for *A. cooperi*, 2,4-D and Kinetin for *A. verticillatus*, 2,4-D and BAP for *A. asperagoides* and NAA and Kinetin for A. plumosus. In A. racemosus callus could be induced from both spear sections and hypocotyl segments. But initiation was noted with 2,4-D + Kn/NAA (Kar and Sen, 1985) only.

Similar behaviour in relation to the type of explant, hormonal combination and individual species was noted to be highly significant in this genus. This may be due to the fact that in addition to different genomic constitution of the species, the explants too often represent different degrees of endoreplicated state of the genome.

The organogenic response of calli of the species also varied depending on species, source of callus line and supplements of basal medium. It has been noted that, the requirement of the amount of L-artinine, as well as its combination with other growth regulators for triggering shoot induction, varied from species to species.

For induction of shoot, cytokinin, L-arginine and low level of NAA (0.02 mg) were essential for all the five species. But the type of cytokinin needed was found to be species specific and depended on callus source. In *A. cooperi*, BAP was effective for induction of shoot and callus from all types of explants (Ghost and Sen, 1992a). In *A. verticillatus*, calli from shoot, phyllode and root responded in BAP but BPA was needed for node and internode explant. Similarly in *A. officinalis*, calli from node, internode and root, needed BPA, whereas BAP was suitable for shoot tip and phyllode. In *A. plumosus*, calli from spear responded to BAP and others required BPA supplement. In *A. asperagoides*, calli from shoot and phyllode needed BAP whereas BPA was effective in others.

For rooting, auxin conjugates were more effective than free acid in species of this genus. But the nature of conjugates needed was not similar in all the species studied. For *A. cooperi*, KIBA (Ghosh and Sen, 1992a), for *A. verticillatus*, IAA-L-alanine (Ghosh and Sen, 1992b), for A. plumosus and *A. asperagoides*, IAA-L-aspartic acid and for *A. officinalis*, IAA-L-phenylalanine were effective (Ghosh, 1992).

As far as shoot multiplication is concerned, high rate of multiplication could be secured from shoot and node explants. However, cytokinin requirements were rather specific. For *A. cooperi*, *A. officinalis*, *A. verticillatus* and *A. asperagoides* it was BAP, whereas for A. plumosus it was DPU (Ghosh and Sen, 1994). Induction of rooting too, was species specific, namely KIBA for *A. cooperi*, IAA-L-Alanine for *A. verticillatus*, IAA-L-aspartic acid for A. plumosus and *A. asperagoides* and IAA-L-phenylalanine for *A. officinalis* (Ghosh, 1992).

In fact, the wide difference in specific requirements for response *in vitro* as manifested in this genus, in callusing and organogenesis including shoot and root development, is a clear reflection of genetic diversity. However, the requirement of MS basic medium for growth in general more or less remains the same at interspecific level. This may also be due to the homogeneity of this taxon. But the requirements of supplements differ markedly at the interspecific level due to possibly cryptic genetic differences. The genus Asparagus however, is quite rich in species diversity. Leaving aside the common basic medium needed for all species in general, the differential response of different genotypes to different factors *in vitro* is quite distinct.

Genus *Kniphofia*

The members of this genus/cultivated in temperate regions as ornamentals. Morphogenesis has been studied *in vitro* in *K. nelsoni* and *K. uvaria*. A comparative analysis of *in vitro* response, shows that α-naphthalene acetic acid exerts an inhibitory effect on shoot formation in both the species, whereas low concentration of BAP accelerates the formation of shoots. With increase in BAP, there is inhibition of shoot response, which is specially manifested in *K. nelsoni*. In Kniphofia, too, species differences could not be brought out markedly in their response *in vitro* (Nayak and Sen, 1990), which is also reflected in the gross homogeneity in karyotypes both in number and structure of chromosomes.

Genus *Lilium*

The genus *Lilium* is confined mostly to the temperate and subtropical regions of the Northern hemisphere. It is popularly known as lilies and its species are of high ornamental value.

A comparative analysis carried out on five species of *Lilium* namely, *L. tigrinum*, *L. longiflorum*, *L. nepalenses*, *L. philippenses* and *L. wallichianum* in relation to *in vitro* response, indicates that in general, bulb scale explants were more suitable for culture work. However, response requirements varied in respect of combinations of growth regulators (Priyadarshi, 1992). In majority of species, for induction of callus, a low concentration of BA with high dose of 2,4-D was more effective. Whereas in *L. tigrinum*, only a low dose of BA was effective (Priyadarshi and Sen, 1989). In *L. nepalenses*, BA was not needed with 2,4-D for callus formation. The callus once formed, could be maintained in the callusing medium under complete darkness. In *L. longiflorum*, replacement of 2,4-D with NAA in the medium, was found to be more effective in maintaining the callus as well as regeneration potential even after long period of culture (Priyadarshi and Sen, 1992). Regeneration of callus could be obtained by complete removal of hormones from the medium as well as incubating the culture under varying light and dark photoperiod.

Direct differentiation of bulblets from the bulb scales were obtained with BA supplement in all the species. However, BA and kinetin were needed in case of *L. tigrinum* and *L. wallichianum* respectively.

Rapid mass propagation of bulblets through intermediary callus was effective in *L. tigrinum*, *L. longiflorum* and *L. philippenses*. However, in *L. nepalenses* and *L. Wallichianum*, the direct differentiation of bulblets from bulb scale explants could be demonstrated.

As regards the influence of chemical constituents of basal medium on bulb growth in *L. tigrinum*, *L. longiflorum* and *L. philippenses*, the growth index of bulb scales was satisfactory in 1/5th MS medium with only 0.5% sucrose followed by 1% activated charcoal. In *L. nepalenses*, the growth index was optimum in 1/2 MS medium followed by 1/4th MS medium with 0.5% charcoal. Three percent sucrose was adequate for leaf development in *L. longiflorum*, whereas it was 6% in *L. tigrinum* and 9% for *L. nepalenses*. In *L. wallichianum*, leaf development was satisfactory in 1/2 to 1/4th MS medium. Charcoal supplement was effective for leaf development in *L. philippenses*. Development of root was optimum in 9% sucrose supplement in all the species except in *L. philippenses* where 6% sucrose was required. However, *L. tigrinum* needed 1% charcoal in MS medium for satisfactory growth.

In general, it has been observed that for growth of bulblets, reduction of strength of MS salts, and for development of root and leaf, higher sucrose concentration in the medium were required. The gross similarity in growth pattern of different species of Lilium of similar age under more or less similar cultural condition, may be a reflection of their broadly similar genetic makeup, in gross features of chromosomes as well. However, in this general setup, specific difference in chemical requirements for different species may be attributed to the finer differences in their genetic architecture indicating distinct genetic diversity.

In this genus, low dose of gamma radiation was also observed to promote the regenerating capacity of the bulb scales in all species which exhibited better growth potential as compared to their control. On the other hand, high dose of gamma radiation led to complete loss of regeneration capacity of the bulb scales (Priyadarshi, 1992). This similarity in *in vitro* response against gamma radiation in all species of Lilium is evidently due to the genetic control of this behaviour exhibited at the generic level.

Genus *Ornithogalum*

The plants are cultivated in temperate regions of Himalayas as ornamentals. In *in vitro* studies of three species namely *O. virens*, *O. thyrsoides* and *O. umbellatum*, it was noted that organogenesis could be obtained from explant tissues through callus formation. Of the three different types of explants such as leaf, bulb scale and root, used for induction of callus, root explant did not show any response, leaf explant was effective only in *O. thyrsoides* whereas bulb scale explant was effective in all the three species. In *O. virens*, callusing needed MS with 2,4-D and BAP supplement, whereas *O. thyrsoides* and *O. umbellatum* required MS with NAA supplement (Nayak and Sen, 1992, 1993). Despite general similarity in response indicating affinity, species differences are reflected in nutrient requirements for callusing and regeneration. This consistent differential response is an index of genotypic control.

Genus *Scilla*

The species *Scilla indica* grows wild in coastal regions of Bombay, monsoon affected regions of Western ghats as well as Central parts of India. The plant is important for its horticultural value, specially because of ornamentation on leaves. The species shows different phenotypes mostly differing from each other in relation to the spotting on the leaf. The bulbs contain sterols and are of commercial importance (Jha and Sen, 1980).

For organogenesis, Murashige and Skoog's medium supplemented with α-naphthalene acetic acid and kinetin, and for initiation of root and growth, basal medium supplemented with α-naphthalene acetic acid yielded optimum results. Leaf tips do not show any response in hormone free medium. Leaf tip of *S. indica* is highly totipotent and has been demonstrated to serve as suitable materials for the study of regeneration of shoot bud (Chakraborti and Sen, 1979).

Different populations of *S. indica* also show their differential response to the artificial medium indicating clear genotypic control (Chakraborti, 1980). The intraspecific differences which are very consistent in their differential response are a clear index of genetic diversity in this species. Such difference is profoundly reflected in the expression of totipotency. Suspension culture has also been utilized for initiation of callus from leaf cells in modified Murashige and Skoog's medium followed by immediate transfer to liquid suspension medium containing α-naphthalene acetic acid. However, no genotypic difference could so far be recorded. Except in suspension culture, genetic diversity at intraspecific level could be demonstrated in this species.

Genus *Urginea*

Diploid, tetraploid, triploid and hexaploid populations of *Urginea indica* were collected from Western, Southern and Northern States of India. Scale leaves were used as explants. Murashige and Skoog's medium was used with certain modifications of vitamins (Jha *et al.*, 1984; Jha and Sen, 1986). Heavy callus formation was observed in the medium containing 2,4-D and coconut milk (Chakraborti and Sen, 1980). Growth of callus was found to increase with age of the tissue upto 180 days after which there was a sudden fall in the growth. Population wise difference in diploid *U. indica* could not be recorded. However, genotypic diversity is very distinct in this species in relation to numerical differences of chromosomes (Jha and Sen, 1983a & b).

CONCLUSION

In general, it is inferred that genetic diversity in relation to *in vitro* response in these genera is reflected both at inter- and intraspecific level to a varying extent. The genotypes may differ in explant preference, requirement of media supplement and regeneration potential *in vitro*. The genetic control of these parameters is indicated in consistent results in repeated trials. However, not all parameters are equally applicable to all the taxa studied. The extent of manifestation of parameters is also under genetic control. In addition to chromosomal characteristics (Sen, 1973), *in vitro* response may be utilized as good indicator of genetic diversity in these genera. The results so far obtained show that, this criterion can be applied as additional evidence to ascertain homogeneity as well as diversity in debated cases. As understanding of the genetic control of response *in vitro*, exhibited either at inter- or intraspecific level, may prove to be of use in framing strategies, for utilizing *in vitro* method for crop improvement.

REFERENCES

Bennett, M.D. 1987. Variation in genomic forms in plant and its ecological implications. New Phytol. 106 (Suppl.): 177-200.

Bennett, M.D. and Smith, J.B. 1976. Nuclear DNA amounts in angiosperms. Philos, Trans. Soc. London, Ser. B. 274: 227-274.

Chakraborti, S.P. and Sen, S. 1979. *In vitro* culture of leaf cells of *Scilla indica* as means of rapid propagation. Cell & Chromosome Newsletter 2: 16-19.

Chakraborti, S.P. and Sen, S. 1980. Polyploidization of chromosomes of *Urginea indica* in *in vitro* callus. Ind. J. Exp. Biol. 18: 1324-1325.

Chakraborti, S.P. 1980. Tissue culture of Indian species of *Urginea* and *Scilla*, their cytological behaviours following *in vitro* growth and chromosome analysis of different populations of *Scilla indica*. Ph.D. Thesis.

Chakravorty, B. and Sen, S. 1988. Comparison of karyological behaviour during *in vitro* growth in *Allium cepa* L. and *Allium sativum* L. Ind. J. Exp. Biol., 26: 828-831.

Ghosh, B. and Sen, S. 1991. *In vitro* Asparagus research- A short report. Asparagus Research Newsletter 8: 1-3.

Ghosh, B. and Sen, S. 1992a. Stable regeneration in *Asparagus cooperi* Baker as controlled by different factors. Plant Science 82: 119-124.

Ghosh, B. 1992. In vivo and *in vitro* study of species of *Asparagus* with special reference to chromosome characteristics and other cell biological parameters. Ph.D. Thesis.

Ghosh, B. and Sen, S. 1994. Effect of explant light intensity and growth regulators on stable regeneration of Asparagus plumosus Baker. Nucleus 37: 24-29.

Jha, S. and Sen, S. 1980. A search for Scilladienolides in *Scilla indica* Roxb. Curr. Sci. 49: 273-274.

Jha, S. and Sen, S. 1983a. Chromosome study of diploid Indian squill. Cytologia 48: 79-86.

Jha, S. and Sen, S. 1983b. Chromosome study of polyploid Indian squill. Cytologia 48: 407-418.

Jha, S. and Sen, S. 1986. Mass propagation of Indian squill *Urginea indica* Kunth *in vitro*. Acta Horticulturae 188: 105-109.

Jha, S., Mitra, G.C. and Sen, S. 1984. In vitro generation from bulb explants of Indian squill *Urginea indica* Kunth. Plant Cell Tissue Organ Culture 3: 91-100.

Kar, D.K. and Sen, S. 1985. Propagation of *A. racemosus* through tissue culture. Plant Cell Tissue Organ Culture 5: 89-95.

Nayak, S. and Sen, S. 1990. Differential requirements for organogenesis in two species of *Kniphofia*. Ind. J. Exp. Biol. 28: 174-178.

Nayak, S. and Sen, S. 1992. Propagation of *Ornithogalum umbellatum* L. through tissue culture. Nucleus 35: 74-77.

Nayak, S. and Sen, S. 1993. Growth, chromosome numbers and DNA content in callus of *Ornithogalum thyrsoides* as influenced by different auxins. Cytobios. 76: 209-216.

Priyadarshi, S. and Sen, S. 1989. In vitro propagation and DNA content in *Lilium tigrinum* Ker-Gawl. Perspectives in Cytology and Genetics. Eds. G.K. Manna and U. Sinha. 6: 137:141.

Priyadarshi, S. and Sen, S. 1992. A revised scheme for mass propagation of Easter lity. Plant Cell Tissue and Organ Culture 30: 193-197.

Priyadarshi, S. 1992. Tissue culture, mutagenesis and chromosome studies of some ornamental species. Ph.D. Thesis.

Sen, S. 1973. Structural hybridity intra- and interspecific level in Liliales. Folia Biologica 21: 183-197.

Sen, S. 1975. Cytotaxonomy of Liliales. Fed. Rep. 86: 255-305.

Sen, S. and Chakroborty, B. 1988. Structural heterozygosity and response *in vitro* in two species of *Allium*. Proceedings Eucarpia 4th Allium Symposium, 6-9 September, Warwick, U.K.

Sharma, A.K. 1974. Plant Cytogenetics. The Cell Nucleus (ed. H. Busch), Vol. II, 263-291. Academic Press, London.

Stebbins, G.L. 1971. Chromosomal evolution in higher plants. Edward Arnold, London.

Phytochrome, Flowering, Genes and Biotechnology[1]

S.C. Maheshwari

INTRODUCTION

Friends: It is a matter of very great pleasure and an honour to be invited to deliver this plenary and inaugural lecture for the first scientific session of this conference being held here in the city of Jaipur. I am delighted for more than one reason. Firstly, I am reminded of some of my boyhood days as this is also the city from where my parents came. Secondly and more importantly, it is a privilege to have this opportunity of paying my own personal tribute to Professor Naresh Chandra, whom we respect not only for his scholarship, original contributions to science and standing as a noted teacher, but also for many personal qualities the most prominent of which is his great modesty.

Turning to the subject of my talk, right at the outset, I should confess and clarify that I am no expert on phytochrome in the international sense. By and large, work in our country is in the initial phase. Further, as many people in the audience know, in my laboratory work has also been done in other areas such as Tissue Culture, Haploids, Biotechnology and Gene Expression. Thus, I have been somewhat of an Intellectual Gypsy, and necessarily therefore my expertise in any one field is diluted. However, I chose the topic since this is an area which is academically very challenging and in any case many of you heart me only about a year ago on the subject of Plant Biotechnology. Early in life, I also happened to meet Harry Borthwick and Sterling Hendricks. Now no more, yet they did pioneering work on the subject and left an indelible impression on me.

In the time I have, I shall try to give you an overview of developments internationally in the area of phytochrome research with a few slides thrown in or our own work here in India. And, I would like to present all researches in a historical perspective so that everyone can understand and appreciate the fascinating subject. I start with the pioneering contribution of James Bonner (with whom both Mrs. Maheshwari and I worked at California Institute of Technology in USA some three decades ago), and who has been one of my great gurus of Plant Molecular Biology. Even though in his laboratory, our concern then was RNA polymerase -- which we helped discover in plants with other friends-- in his own youth, Bonner along with Karl Hamner (1) of Chicago started the chain of discoveries leading to the isolation of phytochrome. Of course, a little earlier, the phenomenon of photoperiodism had been discovered. With *Xanthium pennnsylvanicum*, a short day (SD) plant, Bonner and Hamner found that if an otherwise inductive dark period was

[1] Plenary lecture

interrupted with even a very weak and short flash of light (of only a couple of minutes duration), the inductive effect of the critical night period was completely abolished.

HISTORY AND BASICS OF PHYTOCHROME

The work of Bonner and Hamner was carried forward by Harry Borthwick and Sterling Hendricks (Parker *et al.*, 1945; Parker *et al.*, 1946; Borthwick *et al.*, 1952 for a historical review see Sage, 1992) of the U.S. Department of Agriculture in Beltsville in USA. They set up one of the largest spectrographs ever assembled to determine the action spectrum of inhibition of flowering by light. The night-break effect observed earlier on flowering of *Xanthium* clearly implied that some special photoreceptor may be involved in sensing light interruption. The reason that the USDA workers decided to mount a major research effort to determine the action spectrum of this effect on flowering was to enable them to obtain an idea of the absorption spectrum and thus a clue to the identity of the concerned pigment. Normally, such as spectrum for a certain physiological action must closely match the absorption spectrum of a putative photoreceptor. Both workers were outstanding men. Borthwick who initiated these researches was botanist-- in fact he also did some noteworthy work in plant embryology. Very wisely, he sought the collaboration of Hendricks, a physical chemist (the choice could not be better as he obtained his Ph.D. under the famous Nobel Prize winning chemist, Linus Pauling, who was also then at Caltech). With the technical assistance also of Jack Downs (interestingly, he had served in the US Navy during World War II and had acquired expertise in optics and repair of periscopes of submarines), they set up a spectrograph that threw a spectrum nearly 2 metres wide Sage, 1992). In addition to *Xanthium,* they employed soybean (also a SD plant). Potted plants were brought in the spectrograph room and placed at various wavelength stations. The spectrograph lamp was then switched on briefly in the middle of the night and flowering scored later.

It was soon observed that at Ca. 660 nm, the energy requirement to inhibit flowering was the lowest. That meant that the putative pigment must absorb maximally at 660 nm in the red region of the spectrum. Another remarkable discovery was that the action spectrum was almost identical not only for the inhibition of flowering of other SD plants but for a whole lot of diverse processes such as promotion of seed germination, inhibition of stem growth etc., which were also known to be affected by light. Therefore, the pigment had the place of a master switch or a regulator. Moreover, while the night-break flashes inhibited flowering in SD plants, they promoted flowering in long-day (LD) plants indicating that light absorption by the same pigment could lead to effects that could be both positive or negative, depending on the plant and the particular physiological phenomenon.

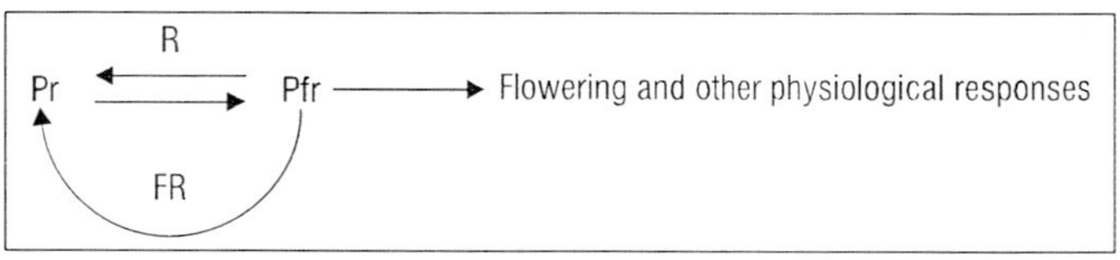

An interesting discovery that followed soon was that action or red-light was negated not only by darkness but also by far-red region of the spectrum with a peak at 730 nm. Far-red light acted instantaneously. Indeed, one could alternately switch on red and far-red lights for a night-break experiment -- in such experiments what mattered was the last treatment indicating that there was a single pigment which has however two forms as given below (Hendricks and Borthwick, 1959).

This pigment was later named as phytochrome. Since this pigment was conceived originally only on theoretical grounds, as is common in science, many were highly sceptical, and it was thus dubbed by many as the Pigment of Imagination. But the pigment is now a reality.

It is necessary briefly to restate that FR and darkness work in the same direction. Since normal daylight has a high R/FR ratio, it was proposed that ultimately daylength brought about its effect by changing ratio of Pr to Pfr. The role of FR in nature, under an open sky, is still debatable-- despite its high level at certain times of the diurnal cycle, for example at sunset. But even without any special effect of FR light, the more energized and biologically active Pfr form reverts back slowly to the Pr form in darkness and that is most likely the timing mechanism in photoperiodic phenomena. A plant cell, at a particular location, obviously should have just the right amount of Pfr to bring about a certain effect.

With the action spectrum established, one could now actually isolate the pigment with the characteristic of a matching absorption spectrum. For this, a biophysicist and a biochemist joined the team, namely Warren Butler and Harold Siegelman. The pigment was finally identified spectrally in maize extract in the summer of 1959 and partially purified (Butler *et al.*, 1959). But if you are ready to believe, with Bruce Bonner (then a Research Associate) in the Galston laboratory at Yale and where I too began my postdoctoral studies the same year, we constituted the second team in the world to have attempted purification of the pigment. This work was taken up on pea, but since Professor Galston had other plans for me, I could not continue. Nonetheless, Bruce made several important contributions (Bonner, 1960, 1961; Taylor and Bonner, 1967) including the first isolation of a phytochrome from an alga, *Maesotenium*. Incidentally, it may amuse the audience to know that my adventure even for the first night with phytochrome caused a serious, though a rather transient, domestic problem which comes to my mind. A worried Mrs. Maheshwari, who had no advance information on my clandestine project to isolate phytochrome after dinner, reached the Yale Security Police to locate as to where I was, until she found me-- in the early hours of the morning-- still in the coldroom with Bruce (because phytochrome is so labile all work has to be done at 4°C). The absorption spectra of purified phytochrome in both Pr and Pfr forms are now given in almost all text-books. The chromophore was identified by Siegelman *et al.* (1996) as an open-chain tetrapyrrole. But as we shall see later, essentially phytochrome is a dimeric protein of a fairly large size, with more than a thousand amino acid residues in each monomer.

RELATIONSHIP OF PHYTOCHROME TO HORMONES

Before going on further with phytochrome, I would like to devote a few minutes on the relationship of light absorption by phytochrome to hormones. Since phytochrome is a rather large protein which cannot move from one cell to another, the consummation of flowering, as also other physiological phenomena, must involve special signal transduction chain(s) involving certain low molecular weight second messengers. In other words, ultimately, some hormone-like substance may be produced which must translocate or diffuse through the plant to shoot apices or other parts. Long ago, on the basis of grafting experiments carried out between SD and LD plants -- one photoinduced and the other not and yet resulting in flowering in both-- it was hypothesized that there is a universal flowering hormone and which was named as florigen (Chailakhyan, 1936). Some of the first grafting experiments were carried out by the distinguished German botanists, Georg

Melchers and Anton Lang. Many plant scientists have looked for florigen since then and in the rosette type of LD plants, a gibberellin may be the candidate. However, no one has any real idea of the identity of florigen, and for the SD or DN (day-neutral) plants there are no clues whatsoever. However, clever and yet simple experiments, like excision of a photoinduced leaf after varying intervals of time, have allowed us to measure the rate of translocation of the putative florigen. And this takes only a few hours.

I would like to take a few minutes to tell you about our work on the flowering of Lemnaceae which is one of the topics that I took up for research in my early days as a faculty member in Delhi University after returning from Caltech. The Botany Department in those days had no biochemical facilities of any real standard. And catapulted from USA, back to India, work on Lemnaceae (the family comprises some of the minutest plants which can be grown aseptically in an ordinary air-conditioned room) or on tissue culture seemed to be the only feasible projects. With the help of several students (I would like to mention in particular Dr. Jitendra Paul Khurana who has been my longest collaborator in the area of photobiology), we discovered the flower-inducing effect of cytokinins and iron-chelates in 1960s and 1970s. But more recent work, that began in 1980s (Tanot *et al.*, 1987) and is still going on, has shown that even more potent are salicylic acid (SA) and its derivative acetylsalicylic acid (the latter compound is the principal ingredient of the common headache relieving drugs such as aspirin). We do not have much idea of how SA works. But SA has been found in recent years to have other important roles in the physiology of a plant, such as in termogenesis and plant defense, and is now meriting many articles (Raski, 1992; Raskin *et al.*, 1987; Malamy *et al.*, 1990). Another compound which is of great interest is cyclic AMP or cAMP, which too induces flowering in certain Lemnaceae, e.g. Lemna paucicostata, strain 6746 (Khurana *et al.*, 1987, 1988).

In the last few years, we have demonstrated the presence of cAMP in the Lemnaceae by use of HPLC, radioimmunoassay and other techniques (Gangwani *et al.*, 1991). True, at present cAMP cannot be regarded an established signalling molecule in higher plants, but we have to keep an open mind since, as I shall inicate later, another cyclic nucleotide, cGMP, of the same family is highly effective in evoking certain other responses. To summarize the results of our work on members of the family *Lemnaceae* and on various other plants in laboratories around the world, at this moment, a universal flowering hormone has not yet been found. But this is a very important area on which work needs to be started anew with more modern techniques and with greater vigour since it is reasonable to assume that the signalling mechanism may have been conserved and may be the same in flowering plants generally (see ref. O'Neill, 1992 for a review).

CHEMISTRY OF PHYTOCHROME

Now, let me come back to phytochrome. As mentioned above, quite early in phytochrome research, the chromophore was found to be a tetrapyrole (unlike in chlorophyll, this is an open chain-type) which was deduced on the basis of absorption spectra of phycocyanins that are common pigments in blue-green algae. Later studies, for example, of Wolfhart Rudiger in Germany and Pill-Soon Song in USA have confirmed this (for references see Page, 1992 Kendrick and Kronenberg, 1994). Some kind of cis-trans type of isomerization occurs when phytochrome converts from one form to another.

However, far more difficult has been the task of determining the structure of the apoprotein and the precise nature of the conformational change induced as a result of absorption of photons by the chromophore. One reason has been that even though phytochrome was isolated some 30 years ago, native phytochrome proved difficult to purify. Since it readily splits into two fragments nearly half the original, for most of its history, we have lived unknowingly with an fallacious molecular weight (of ca. 60 kD) of the polypeptide. But now we know through our own work in wheat (Mukesh Malik, Vijay Sharma, Ph.D. Theses, Delhi University, 1990 and 1994, see also Malik *et al.*, 1992 as also many other studies that the true molecular weight generally ranges between j120 and 127 kD, depending upon the species.

The new recombinant technology has come as an extremely powerful tool for the advancement of research on phytochrome. Quail's group in the USA was the first to clone the cDNA (Hershey *et al.*, 1984). Later, genomic library clones were also obtained and now we have complete sequences available of not only of amino acids of the phytochrome molecule but of nucleotides of the gene itself which has typical introns and exons as in eukaryotes. The phytochrome polypeptide from dark-grown *Avena* seedlings consists of 1,128 amino acids. The chromophore is attached to a cysteine residue in the N-terminal half of the molecule. Unfortunately most investigators, who have tried to determine directly the amino acid sequence, have found the N-terminal end blocked. But, in wheat, we have determined the N-terminal end blocked. But, in wheat, we have determined the N-terminal amino acid sequence where it is apparently free and one characteristic feature of this end is the cluster of serine residues (Vijay Sharma, Ph.D. Thesis, cited above). This confirms that what we have isolated is indeed phytochrome. As is well-known, in proteins, serine residues are of great importance since, like tyrosine, they can be phosphorylated changing not only the conformational status but biological activity as well.

Many studies have been conducted by diverse spectroscopic techniques like ORD, CD, NMR and so on. Even electron microscopy has been done of phytochrome molecules (Sage, 1992). The native phytochrome molecule appears to be a dimer (total m.w. ca. 250 kD) with a shape and organization rather like that of an antibody molecule i.e. C-terminal ends close together but the N-terminal ends somewhat free from each other). Various data indicate that during Pr Pfr transitions, the N-terminal ends of the molecule may open or close up a little.

HOW DOES PHYTOCHROME FUNCTION?- STUDIES ON MECHANISM OF ACTION

Localization

Leaving aside pure chemistry, let us now look into the manner in which phytochrome may function in a cell. Unfortunately, chemical studies on the structure of phytochrome have so far given little insight about its localization or mode of its action. But some elegant studies have been done in Germany on the alga *Mougeotia* where the chloroplast reorients itself to adjust to prevailing light conditions and which was found to be under control of phytochrome fairly early. Using microbeams of R and FR lights it was shown by Haupt and co-workers that phytochrome is most likely localized in and near the plasma membrane both in the Pr and Pfr forms (Hampt and Weisenseel, 1976). In these cells, the microbeam needs to pass only through the edges to bring about chloroplast rotation. But red and far-red lights have their maximal action only if light vibrates in a certain plant as found by use of a polarizer. And to reverse a R effect by FR, the ploarizer needs to be turned at 90°! Obviously, such results cannot be had unless the photoreceptor is located in or attached to membranes in a definite orientation.

Some recent work in our laboratory on wheat also shows that phytochrome, at least in its Pfr form, is largely associated with the plasma membrane though in the Pr form it may remain soluble (Malik, Ph.D. Thesis, cited above). But it is important to state that localization of phytochrome is highly controversial since many other studies on higher plants employing other techniques (such as immunochemical localization) are at variance with this idea. What is clear from the amino acid sequence data is that there are no typical hydrophobic helices as in rhodopsin, ATPase and other integral membrane proteins. For this reason, we need to further extend our work. It is possible also that phytochrome may only transiently associate with the inner face of the plasma membrane in the Pfr form.

Membrane Action

While direct association with plasma membrane has yet to be proved unequivocally, there is a wealth of physiological evidence to indicate that phytochrome action may indeed be closely related to membranes. Fondeville, Borthwick and hendricks (1966) found in the mid-sixties that leaf movement in *Mimosa pudica* is under the control of phytochrome. These movements are so rapid that you have to think of a change in membrane permeability which is best explained if phytochrome, too, is associated with membranes.

In the seventies and early eighties, work in Galston's laboratory at Yale, particularly by Ruth Satter (Satter and Galston, 1973), has provided similar evidence for pulvinar cells of leaves of two legumes, namely *Samanea saman* and *Albizzia julibrissin*. Their experiments employing electron-microprobe X-ray analysis are the most extensive till date and it seems considerable changes in ion-fluxes occur diurnally between two sides of a pulvinus so that turgor is lost or regained. The most prominent changes have been shown for K^+ ions. Again, if phytochrome is associated with membranes, these results are relatively easy to explain.

Several other investigations have shown that irradiation of plant tissues or organs like roots or coleoptiles by R or FR lights causes rapid changes in the direction and magnitude of electric potentials across cell membranes and which may be in the order of a few millivolts. Such experiments were first reported by Shrank in USA in 1950s (Shrank, 1950) for coleoptiles in relation to phototropism as also geotropism, but largely forgotten or ignored. In subsequent years interest in them was revived by the work of Tanada (1968), Jaffe (1968) and then by Newman and Briggs (1972). These experiments show that generation of such potentials is reversibly controlled by R and FR lights. Although more complicated explanations for the potential changes can be devised, it is simplest to assume that phytochrome is membrane localized and Pr Pfr transitions may alter the activity of ion channels of a certain class or possibly even of certain type of ion pumps, say Ca^{++} ATPase.

Changes in Ca^{++} Flux

We could now turn to some work in our own laboratories at Delhi. For some years we have been engaged in studies changes in Ca^{++} flux across membranes of cells of wheat (coleoptiles and enclosed leaf primordia). The work was started after Stanley Roux and coworkers in USA (Roux, 1986) reported that in oat coleoptiles R light drives Ca^{++} out of cells and FR in. The direction of change however was opposite of that in algae where Haupt and his pupils first reported an effect on Ca^{++} flux. In a bid to confirm the general phytochrome effect and in particular verify the direction in which Ca^{++} flows, we have employed a variety of techniques and using intact coleoptile or isolated protoplasts. As found by us (for technical details see Minnie Mehta, Ph.D. Thesis, 1990, Suveena Sharma,

M. Phil. Thesis, 1990, Delhi University), Ca^{++} ions can readily enter or leave a cell depending whether R or FR light exposure has been given. But we found that with R pulses they move into a cell and with FR, they move out equally rapidly. Calculations show that Ca^{++} levels can transiently increase by about 1000-fold by a single R light pulse of only 1 or 2 min duration (Mehta *et al.*, 1993). Similar results were reported a few years ago by Professor Sopory's group at JNU in New Delhi in maize (Das and Sopory, 1985) using somewhat different techniques. Indeed, while our work was in press, in England, Trevawas and colleagues (Shacklock *et al.*, 1992) also reported similar results in wheat employing Ca^{++} binding dyes that fluoresce inside a cell. Despite the use of different approaches, all recent work leads to the same conclusion that a R light pulse causes an influx of Ca^{++} ions and a FR pulse an efflux.

How Ca^{++} May Work?

In the last decade, in animal cells, Ca^{++} has gained considerable importance as a second messenger switching off or on various metabolic pathways (Morgan, 1989). Ca^{++} is now known to bind with calmodulin which is a small protein and either alone or bound with calmodulin it may activate a large number of protein kinases by complexing to these enzymes. There is another class of protein kinases, where Ca^{++} as also a lipid-like molecule DAG (diacylglycerol) are required simultaneously for full activation. As many in the audience may already be aware and might readily appreciate, phosphorylation by kinases has turned out to be among the two or three most important discoveries in modern biochemistry. Activated protein kinases may in turn work on many target enzymes-phosphorylating and changing their conformational state- and consequently turning them on or off depending on the enzyme. But in some cases it is dephosphorylation which turns on a particular gene or an enzyme and change in Ca^{++} levels can also lead to changes in the activity of phosphatases.

In our laboratories, some work has been done recently on changes in profiles of phosphorylated proteins in various sub-cellular fractions of wheat coleoptile cells (Vijay Sharma, cited above). Such experiments can be done by the use of either P^{32} labelled ATP or inorganic P^{32} depending on whether the phosphorylation reaction is being studied *in vitro* or *in vivo*, i.e. in the intact plant tissue. A number of protein bands are either phosphorylated or dephosphorylated in response to R or FR light treatments as also presence or absence of Ca^{++}. However, we have no clues concerning the identity of the proteins which will demand enormous work in future and use of sophisticated techniques like microsequencing.

Before we switch to the next topic, let me also say that in animals, Ca^{++} level can rise in a cell not only because of entry from the extracellular bathing fluid, but also by release from internal stores or pools such as mitochondria or more importantly specialized regions of the endoplasmic reticulum, often called calciosomes. There are not many investigations in plants concerning release of Ca^{++} from internal pools, but we have some unpublished evidence that mitochondria may indeed take up or release Ca^{++} if irradiated *in vitro* with R and FR light (Minnie Mehta, Ph.D. Thesis, cited above).

Phytochrome and the PI Pathway

A major established pathway for the release of Ca^{++} from internal stores is the PI (phosphatidylinositol) pathway. Unfortunately, there will be no time for me to go into a description of the PI cycle or work that led to its discovery. Let me just state that it is already well established in animals (36) and recent work in many laboratories shows that plants too contain various inositol phospholipids like PI, PIP, PIP_2 and IP_3. A key enzyme in this pathway is phospholipase C or PLC and according to current concepts most stimuli like hormones work on PLC (through the intermediary of G-proteins) so that IP3 and DAG are released after hydrolysis of PIP_2. IP3 then combines with a certain class of receptors on membranes of mitochondria or the ER. These receptors are in fact integral parts of the Ca^{++} channel-cum-receptor complexes. And the binding causes the release of Ca^{++} to the cytosol.

In plants, the first evidence for activation of PLC by light came from Ruth Satter-- Galston's former associate-- just before her untimely death (Morse *et al.*, 1987 see also Morse *et al.*, 1990). German workers have also provided evidence for this idea through their work on mosses (for reference see Morse *et al.*, 1990). In recent years, my former associates Professor Sudhir Sopory and Professor Sipra Guha-Mukherjee have also been working on this problem at JNU in New Delhi. Overall, they are supportive of the idea of phytochrome mediated changes in PI metabolism (Guron *et al.*, 1992; Acharya *et al.*, 1991) but their results raise some doubt as to whether PLC is the primary target of phytochrome action. Instead of finding an increase in IP3 level, they find an increase in PIP2 or PIP levels and think that the primary target may be another enzyme in the cycle. More work will be necessary to determine the precise mechanism of phytochrome action.

Regulation of Gene Expression

At this point we may now turn to DNA and genes. Although rapid responses such as influx of Ca^{++} ions are of great interest and can certainly explain modulation of metabolism, there is extensive evidence that phytochrome does control gene activity as well.

The idea of gene regulation by phytochrome was proposed long ago by Hans Mohr (1996). However, it is only recently after the advent of recombinant DNA methods, that definitive evidence and insight have come. Interestingly, the first studies were made by Elaine Tobin (1981) at UCLA on the weed *Lemna* already referred to by me (I think this is a good example to remind ourselves that scientific discoveries are often made more easily on weeds and that plants like *Lemna, Datura* and *Arabidopsis* all have a place in scientific research even in a country such as ours). Similar studies as by Tobin were almost simultaneously made in Germany by Klaus Apel (1979). Expression of certain genes like *rbcS* (encoding the small subunit of Rubisco) is tightly coupled to phytochrome transformations. More sophisticated work has since been done by groups of Nam-Hai Chua (Gilmartin *et al.*, 1990; Sarokin and Chuan, 1992; Krategiri *et al.*, 1989) and Anthony Cashmore (Dutta and Cashmore, 1989) at the Rockefeller and Pennsylvania State Universities. By delection analysis of promoters of concerned genes (cloned in *E. coli* plasmids and then used for making transgenic plants), novel cis-acting sequences called Light Responsive Elements or LREs have been identified in plant promoters. These enhancer-like sequences are recognized by *trans*-acting proteinaceous factors which somehow stimulate transcription of genes by RNA polymerase. Currently, the focus of attention in many laboratories are these factors.

Indeed, an interesting recent discovery is that many of these *trans*-acting factors phosphorylate or dephosphorylate under the influence of specific protein kinases or phosphatases. In animals, extensive evidence exists for such a role in respect of both cAMP and Ca^{++} dependent protein kinases to regulate the phosphorylation of trans-acting factors (Alberts *et al.*, 1994). Very recently similar evidence has come for plants including some from our lab on *Vigna* (Arati Sakhuja, M.Phil. Thesis, 1994, Delhi University). Phytochrome can thus control the phosphorylation of these factors via Ca^{++} or calmodulin.

And this is a finding of special importance because flux of ions, like Ca^{++}, can explain not only electrical phenomena such as generation of voltage gradients, or movement of leaves, or activation of entire sets of cytosolic enzymes, but also of release/binding of trans-acting factors to RNA polymerase and thus control of gene activity itself. In contrast to the situation in animal cells, cAMP has not yet been shown to be a signalling agent for genes in higher plants. However, it is worthy of note that some recent work by Chua and coworkers does indicate that cGMP, another cyclic nucleotide, may indeed play such a role (Bowler *et al.*, 1994).

THE LATEST DEVELOPMENTS

In the last part of this talk, I touch on a few of the latest developments. Of course, at this stage, it looks as if, instead of resolving the many mysteries that have long existed, we are further adding to the confusion. Thus, even though one phytochrome was already taxing our minds, the fact is that use of recombinant DNA technology has thrown up entirely novel findings. And the most important among them is that now we have multiple phytochromes to contend with! There is extensive evidence now that in addition to the classical phytochrome (now being called phytochrome A or PHYA isolated from dark-grown seedlings in various laboratories), there is another class of phytochrome, PHYB, also called green phytochrome, which is more abundant in light.

Multiplicity of Phytochromes

An example of the multiplicity of phytochromes can be had in some work done in our laboratory on wheat (Vijay Sharma, Ph.D. Thesis, cited above) in collaboration also with Professor Rameshwar Sharma of Hyderabad. Using certain antibodies specific to either phytochrome A or phytochrome B and obtained originally from Japan from Masaki Furuya, we have provided evidence for two distinct phytochromes in wheat seedlings. In darkness, we have a phytochrome of about 124 Kd; but light-grown seedlings show another phytochrome of a somewhat lower molecular weight typical of the green phytochromes. The original dark phytochrome begins to decrease rapidly on exposure to light.

The first evidence of such multiple phytochromes was obtained by Merten Jabben and Gerald Deitzer (1978) in USA some years ago in oat employing dual-wavelength difference spectroscopy of norflurazon-grown seedlings (such treatment bleaches chlorophyll and thus removes interference in spectrophotometry). Later, Masaki Fuiruya (1989) in Japan found that the peptide digests of phytochrome also turned out to be slightly different in etiolated and green plants. But, now, recombinant DNA technology has allowed isolation of two types of genes for the two phytochromes (Hanelt *et al.*, 1992; Quail, 1994). It seems that phytochrome B i.e. the green phytochrome is more primitive since in mosses such as *Ceratodon purpureus*, this is the only type present (I suppose that Nature, in its wisdom, decided to evolve later the dark type phytochrome).

What is surprising, however, is that the number of different phytochromes is much larger. The most intensive analysis so far has now been done on *Arabidopsis thaliana* and there are as many as 5 genes for phytochrome in this plant (Quail, 1994). The situation in the living plant may actually be even more complex. Work on pea shows and even for the same phytochrome gene, e.g. phyA, there may be multiple transcripts (Sato, 1988). These arise probably because of the existence of multiple TATA boxes. No one yet clearly understands as to why multiple transcripts of different lengths are produced.

Is Phytochrome a Protein Kinase?

Perplexing as the new findings are, as scientists we continue to believe in the parsimony of Nature. I am sure that clarity and some unifying theme will soon emerge. At least in one instance, the moss *Ceratodon*, the use of the recombinant DNA technique has paid a rich dividend. Comparison of the putative amino acid sequence data shows that phytochrome of this moss is most probably a novel, two-in-one, photoreceptor-cum-protein kinase which is directly regulated by absorbed photons (Algarra *et al.*, 1993; Schneider-Poetsch, 1992). The C-terminal end has a structure typical of kinases whereas the N-terminal end has the attached chromophore for sensing light. In animals, kinases are generally activated in response to ligands or hormones. However, plants appear to have evolved a novel mechanism-- instead of ions or ligands they use R and FR light photons (or merely darkness) to control kinase activity in at least certain members of this class of enzymes.

In higher plants, we are not sure as to what happens. Claims have indeed been made that phytochrome has kinase activity (Wong *et al.*, 1986). But others think that the kinase is a separate entity and merely co-purifies with phytochrome. Maybe, in higher plants, the light-absorbing and kinase functions have separated into two distinct genes and proteins. The protein products may, however, associate with each other in the living cell.

Function of Multiple Phytochromes-- Transgenic Plants

At this point no one precisely knows the role of different phytochromes. One idea is that PHYA may function in germinating seeds or dark-grown seedlings and then-- as development proceeds other phytochromes may play their roles: for example PHYB in leaf expansion or PHYC and D in flower initiation and so on. But this seems to be a rather simplistic notion. It is more likely that different phytochromes may be linked to different fluence responses- for example to various phenomena controlled by VLF (Very Low Fluence) or LF (Low Fluence) or HER (High Energy Reaction) responses.

One approach towards understanding the function of different phytochromes is the construction of transgenics which overexpress one or the other type of phytochrome (Cherry and Vierstra, 1994). Such experiments have been done though till date in almost all studies only phytochrome A has been overexpressed since this is the only phytochrome gene that is widely available so far.

The Approach of Mutant Analysis

Another approach employs mutants as pioneered, for example, by Maarten Koornneef (Koornneef and Kendrick, 1994) in Holland and expanded more recently by Joanne Chory and Xing-Wang Deng in USA and Garry Whitelam in UK. *Arabidopsis thaliana* has proved to be the material of choice and various mutants have been generated some lacking PHYA or PHYB or all phytochromes (the latter is the result when chromophore synthesis fails). Double mutants have also been made lacking simultaneously more than one phytochrome. An interesting outcome of experiments with some of these mutants is that even a single physiological response, and interestingly flowering itself, may be controlled by more than one phytochrome (Halliday *et al.*, 1994). In LD *Arabidopsis*, flowering occurs rather early and prematurely if simultaneously two phytochromes are lacking.

Biochemical analyses of various mutants shows that in some cases the phenotypic defect is not due to a mutation of *phy* gene, but of some other gene(s) involved in the synthesis of components of the signal transduction chain for phytochrome action. It appears that in certain responses phytochrome action. It appears that in certain responses phytochrome may directly affect transcription since some of the recently discovered mutant loci like *det* (de-etiolated) or *cop* (constitutive photomorphogenesis) appear to code for proteins known to possess Zn-finger domains or nuclear localization signals. Both *det* and *cop* mutants are a new class of mutants in which a seedling displays the phenotype of a light-grown seedling even in darkness with precocious expression of many photosynthesis related genes (of both nuclear and plastid origin) as also inhibition of hypocotyl elongation (Reed *et al.*, 1992).

PHO MUTANTS

At this point, I would also like to mention some recent work of my colleague Dr. J.P. Khurana who is pioneering in India the use of *Arabidopsis* for analysis of photomorphogenesis via the mutant approach (Khurana *et al.*, 1993). Khurana obtained some of the first mutants showing lack of normal phototropic response in Kenneth Poff's laboratory in U.S.A. After returning to India, he has been working on certain newer mutants, e.g. the *pho* (plumular hook open, even in dark) mutants earlier known as JK mutants. These mutants, like the *det* and *cop* mutants, seem to affect photoreceptor function. But pho seems to be different from the two classes discussed earlier. Three of the *pho* mutants characterized flower in complete darkness! Physiological and biochemical analyses show that the defect is not in the synthesis of phyA or even phyB. However, further work remains to be done to determine whether another phytochrome (C, D or E) has been mutated or, else, some key link of the signal transduction chain is missing.

CONCLUSION

It is time now to conclude. I am afraid the working of phytochrome is still shrouded in mysteries. Maybe, we can illustrate the confusion in present day research though the example of several blind men trying to recognize an elephant. Thus, phytochrome is some animal or a beast whose contours we are beginning to recognize only now, though not too well. On one side, absorption of photon occurs through the chromophore but then on the other we have a bewildering array of various manifestations of the photoconversion process. Conversion of Pr to Pfr may thus result in increase of its kinase activity. Or, it may lead to conformational change in a neighbouring protein such as a Ca^{++} ion channel or a quanylate kinase, or an ion-pump. Or, it primarily affects a G-protein. No one knows the truth at this moment. In fact, with the discovery of multiple phytochromes, the scene has become a bit more complex: we have now a herd of elephants or a number of different animals to deal with. Only more work should clarify the picture.

I wish to close with a reference to plant biotechnology. Studies by several groups, leading to transgenic plants with phytochrome gene of another plant (and consequent over-expression), have already been mentioned. But much more may be expected in future. Control of flowering or of other developmental processes is of great importance to agriculture. Once we know of the precise mechanism of action of various phytochromes and have genes available with us, new avenues would have opened up in plant biotechnology. Clearly, the next decade will be an exciting one and surely tissue culture and biotechnological approaches for gene delivery will have a crucial role.

ACKNOWLEDGMENTS

My thanks are due to my former pupils Dr. J.P. Khurana and Dr. Vijay Sharma (now in the Department of Plant Molecular Biology at Delhi University) for many helpful discussions. Apart from Vijay, my thanks are also to Dr. Minnie Mehta, Dr. Mukesh Malik, Ms Suveena Sharma and Ms Arati Sakhuja who worked for their Ph.D. and M. Phil. Theses. Through all of them I have been able to keep in touch with this fascinating subject. Lastly, a word of sincere appreciation is due to Mrs. Shashi Mehta and Mrs. Neelam Bedi for help in various ways and Mr. Satyendra Patwal and Mrs. Renu Dhawan for processing the raw material through computers in this shape.

REFERENCES

Acharya, M.K., Munjal, I.D. and Guha-Mukherjee, S. 1991. Light-induced rapid changes in inositol phospholipids and phosphatidylcholine in *Brassica* seedlings. Phytochemistry 30: 2895-2897.

Alberts, B., Bray, D., Lewis, J., Raff, M., Roberts, K. and Watson, J.D. 1994. Molecular Biology of the Cell. Garland Publishing, New York.

Algarra, P., Linder, S. and Thmmler, F. 1993. Biochemical evidence that phytochrome of the moss Ceratodon purpureus is a light regulated protein kinase. FEBS Letts 315: 69-73.

Apel, K. 1979. Phytochrome-induced appearance of mRNA activity for the apoprotein of the light-harvesting chlorophyll a/b protein of barley (*Hordeum vulgare*). Eur. J. Biochem. 97: 183-188.

Berridge, M.J. 1993. Inositol trisphosphate and calcium signalling. Nature 361: 315-325.

Bonner, B.A. 1960. Partial purification of the photomorphogenic pigment from pea seedlings. Plant Physiol. 35 (Suppl. xxxii).

Bonner, B.A. 1961. Properties of phytochrome from peas. Plant Physiol. 36 (Suppl. xiii).

Borthwick, H.A., Hendricks, S.B. and Parker, M.W. 1952. The reaction controlling floral initiation. Proc. Natl Acad. Sci. USA 38: 929-934.

Bowler, C., Neuhaus, G., Yamagota, H. and Chua, N.-H. 1994. Cyclic GMP and calcium mediate phytochrome phototransduction. Cell 77: 73-81.

Butler, W.L., Norris, K.H., Siegelman, H.W. and Hendricks, S.B. 1959. Detection, assay, and preliminary purification of the pigment controlling photoresponsive development of plants. Proc. Natl. Acad. Sci. USA 45: 1703-1708.

Chailakhyan, M.Kh. 1936. On the hormonal theory of plant development. C.R. (Dokl.) Acad. Sci. USSR 12 (Part III): 443-447.

Cherry, J.R. and Vierstra, R.D. 1994. The use of transgenic plants to examine phytochrome structure/function. In: Photomorphogenesis in Plants. R.E. Kendrick and G.H.M.. Kronenberg (Eds.), Kluwer Academic Publishers, Dordrecht, pp. 271-297.

Das, R. and Sopory, S.K. 1985. Evidence of regulation of calcium uptake by phytochrome in maize protoplasts. Biochem. Biophys. Res. Commun. 128: 1455-1460.

Datta, N. and Cashmore, A.R. 1989. Binding of a pea nuclear protein to promoters of certain photoregulated genes is modulated by phosphorylation. Plant Cell 1: 1069-1077.

Fondeville, J.C., Borthwick, H.A. and Hendricks, S.B. 1966. Leaflet movement of *Mimosa pudica* L. indicative of phytochrome action. Planta 69: 357-364.

Furuya, M. 1989. Molecular properties and biogenesis of phytochrome I and II. Adv. Biophys. 25: 133-167.

Gangwani, L., Tamot, B.K., Khurana, J.P. and Maheshwari, S.C. 1991. Identification of 3'5'-cyclic AMP in axenic cultures of *Lemna paucicostata* by high-performance liquid chromatography. Biochem. Biophys. Res. Commun. 178: 1113-1119.

Gilmartin, P.M., Sarokin, L., Memelink, J., and Chua, N.-H. 1990. Molecular light switches for plant genes. Plant Cell 2: 369-378.

Guron, K., Chandok, M.R. and Sopory, S.K. 1992. Phytochrome mediated rapid changes in the level of phosphoinositoles in etiolated leaves of *Zea mays*. Photochem. Photobiol. 56: 691-695.

Halliday, K.J., Koorneef, M. and Whitelam, G.C. 1994. Phytochrome B and at least one other phytochrome mediate the accelerated flowering response of *Arabidopsis thaliana* L. to low red/far-red ratio. Plant Physiol. 104: 1311-1315.

Hamner, K.C. and Bonner, J. 1938. Photoperiodism in relation to hormones as factors in floral initiation and development. Bot. Gaz. 100: 388-431.

Hanelt, S., Braun, B., Marx, S. and Schneider-Poetsch, H.A.W. 1992. Phytochrome evolution: A phylogenetic tree with the first complete sequence of phytochrome from a cryptogamic plant (*Selaginella marensii* Spring). Photochem. Photobiol. 56: 751-758.

Haupt, W. and Weisenseel, M.W. 1976. Physiological evidence and some thoughts on localized responses, intracellular localization and action of phytochrome. In: Light and Plant Development. H. Smith (Ed.), pp. 63-74. Butterworths, London.

Hendricks, S.B. and Borthwick, H.A. 1959. Photocontrol of plant development byt he simultaneous excitations of two interconvertible pigments. Proc. Natl Acad. Sci. USA 45: 344-349.

Hershey, H.P., Colbert, J.T., Lissemore, J.L., Barker, R.F. and Quail, P.H. 1984. Molecular cloning of cDNA for Avena phytochrome. Proc. Natl Acad. Sci. USA 81: 2332-2336.

Jabben, M. and Deitzer, G. 1978. Spectrophotometric measurements in light-grown *Avena sativa* L. Planta 143: 309-313.

Jaffe. M.J. 1968. Phytochrome-mediated bioelectric potential changes in oat seedlings. Plant Physiol. 50: 687-693.

Katagiri, F., Lam, E. and Chua, N.-H. 1989. Two tobacco DNA-binding proteins with homology to nuclear factor CREB. Nature340: 727-730.

Kendrick, R.E. and Kronenberg, G.H.M. (Eds.). 1994. Photomorphogenesis in Plants. Kluwer Academic Publishers, Dordrecht.

Khurana, J.P., Tamot, B.K. and Maheshwari, S.C. 1988. Role of adenosine, 3', f'-cyclic monophosphate in flowering of a short-day duckweed *Lemna paucicostata* 6746. Plant Cell Physiol. 29: 1023-1028.

Khurana, J.P., Tamot, B.K. and Maheshwari, S.C. 1987. Role of catecholamines in promotion of flowering in a short-day duckweed, *Lemna paucicostata* 6746. Plant Physiol. 85: 10-12.

Khurana, J.P., Kochhar, A., Nayyar, M., Sharma, V.K., Tyagi, A.K. Maheshwari, S.C. and Poff, K.L. 1993. *Arabidopsis* mutants in the study of photomorphogenesis. In: Proceedings of Symposium on Photosynthesis and Plant Molecular Biology organized at Jawaharlal Nehru University, by Department of Atomic Energy, Government of India, Bombay.

Koornneef, M. and Kendrick, R.E. 1994. Photomorphogenic higher plants. In: Photomorphogenesis in Plants. R.E. Kendrick and G.H.M. Kronenberg (Eds.). Kluwer Academic Publishers, Dordrecht. pp. 601-624.

Malamy, J., Carr, J.P., Klessig, D.J. and Raskin, I. 1990. Salicylic acid: a likely endogenous signal in the resistance response of tobacco to viral infection. Science 250: 1002-1004.

Malik, M.K., Sharma, V.K., Khurana, J.P. and Maheshwari, S.C. 1992. Purification and immunological detection of phytochrome from wheat coleoptiles. J. Plant Biochem. Biotechnol. 1: 27-31.

Mehta, M., Malik, M.K., Khurana, J.P. and Maheshwari, S.C. 1993. Phytochrome modulation of calcium fluxes in wheat (*Triticum aestivum* L.) protoplasts. Plant Growth Reg. 12: 293-302.

Mohr, H. 1966. Differential gene activation as a mode of action of phytochrome 730. Photochem. Photobiol. 5: 469-483.

Morgan, N.G. 1989. Cell Signalling. Open University Press, Milton Keynes.

Morre, D.J., Boss, W.F. and Loewus, F.A. 1990. Inositol metabolism in plants. Wiley-Liss. New York.

Morse, M.J., Crain, R.C. and Satter, R.L. 1987. Light-stimulated inositol phospholipid turnover in *Samanea saman* leaf pulvini. Proc. Natl Acad. Sci. USA 84: 7075-7078.

O'Neill, S.D. 1992. The photoperiodic control of flowering: Progress towards understanding the mechanism of induction. Photochem. Photobiol. 56: 789-802.

Parker, M.W., Hendricks, S.B., Borthwick, H.A. and Scully, N.J. 1945. Action spectrum for the photoperiodic control of floral initiation in Biloxi soybean. Science 102: 152-155.

Parker, M.W., Hendricks, S.B., Borthwick, H.A. and Scully N.J. j1946. Action spectrum for the photoperiodic control of floral initiation of short-day plants. Bot. Gaz. 108: 126.

Quail P.H. 1994. Phytochrome genes and their expression. In: Photmorphogenesis in Plants. R.E. Kendrick and G.H.M. Kronenberg (Eds.), Kluwer Academic Publishers, Dordrecht, pp. 71-104.

Raskin, I. 1992. Role of salicyclic acid in plants. Annu. Rev. Plant Physiol. Plant Mol. Biol. 43: 439-463.

Raskin, I., Ehmann, A., Melander, W.R. and Meeuse, B.J.D. 1987. Salicyclic acid: a natural inducer of heat production of *Arum* lilies. Science 237: 1601-1602.

Reed, J., Nagpal, P. and Chory, J. 1992. Searching for phytochrome mutants. Photochem. Photobiol. 56: 833-838.

Roux, S.J. 1986. Role of calcium ions in phytochrome responses: An update. Physiol. Plant. 66: 344-348.

Sage, L. 1992. Pigment of the Imagination: A History of Phytochrome Research. Academic Press, New York.

Sarokin, L. and Chua, N.-H. 1989. Binding sites for two novel phosphoproteins, 3AF5 and 3AF3, are required for rbcS-3A expression. Plant Cell 4: 473-483.

Sato, N. 1988. Nucleotide sequence and expression of the phytochrome gene in *Pisum sativum*. Differential regulation by light of multiple transcripts. Plant Mol. Biol. 11: 697-710.

Satter, R.L. and Galston, A.W. 1973. Leaf movements: Rosetta stone of plant behaviour? BioScience 23: 407-416.

Schneider-Poetsch, H.A.W. 1992. Signal transduction by phytochrome: Phytochromes have a module related to the transmitter modules of bacterial sensor proteins. Photochem. Photobiol. 56: 839-846.

Shacklock, P.S., Read, N.D. and Trewavas, A.J. 1992. Cytosolic free calcium mediates red light-induced photomorphogenesis. Nature 358: 753-755.

Shrank, A.R. 1950. Electrical polarity and auxins. In: Plant Growth Substances. F. Skoog (Ed.). University of Wisconsin Press, Madison. pp. 123-140.

Siegelman, H.W., Turner, B.C. and Hendricks, S.B. 1966. The chromophore of phytochrome. Plant Physiol. 41: 1289-1292.

Tamot, B.K., Khurana, J.P. and Maheshwari, S.C. 1987. Obligate requirement of salicyclic acid for short-day induction of flowering in a new duckweed, *Wolffiella hyalina*. Plant Cell Physiol. 28:349-353.

Tanada, T. 1968. A rapid photoreversible response of barley root tips in the presence jof 3-indoleacetic acid. Proc. Natl Acad. Sci. USA 59: 376-380.

Taylor, A.O. and Bonner, B.A. 1967. Isolation of phytochrome from the alga *Mesotaenium* and liverwort *Sphaerocarpus*. Plant Physiol. 42: 762-766.

Tobin, E.M. 1981. Phytochrome mediated regulation of messenger RNAs for the small subunit of ribulose 1,5-bisphosphate carboxylase and the light harvesting chlorophyll a/b-protein in *Lemna gibba*. Plant Mol. Biol. 1: 35-51.

Wong, Y.S., Cheng, H.C., Walsh, D.A.a nd Lagarias, J.C. 1986. Phosphorylation of *Avena* phytochrome *in vitro* as a probe of light induced conformational changes. J. Biol. Chem. 61: 12089-12097.

Tissue Culture of Aquatic Plants and Its Applications

H.Y. Mohan Ram and Anuradha Agrawal

Plant tissue culture that began as an intellectual curiosity has presently attained a respectable status because of its significant contributions to basic knowledge and biotechnology. Its use has led to several practical gains to agriculture and industry. The established benefits of plant tissue culture have been in micropropagation, production of haploids, disease detection and eradication, elimination of breeding barriers, biosynthesis of secondary metabolites, generation of variability and in conservation of germplasm (Mohan Ram and Agrawal, 1990). Availing the advantage of cellular totipotency and somatic embryogenesis and the ability to insert genes have been the key factors in the promising area of plant biotechnology in the past few years.

Inspite of the commendable progress made with land plants, utilization of results of tissue culture of aquatic plants has not been impressive. Published accounts on the tissue culture of water plants are limited (see Mohan Ram, 1978, 1991; Mohan Ram & Kakkar, 1983; Kane, 1989a,b). The present paper reviews the work done in aquatic plant tissue culture and outlines the prospects of its utility. It takes into account the numerous biological aspects unique to aquatic plants. In covering the broader points of view, we pay our tribute to Professor Naresh Chandra, whose vision and botanical interests have been wide and whose contributions have encompassed several facets of botany.

AQUATIC PLANTS: DEFINITION, DIVERSITY AND PRESENT STATUS

Being the most successful and dominant group of plants, with a broad spectrum of distribution, diversity and adaptability angiosperms have perennially fascinated botanists. Approximately one per cent of all the flowering plants occupy fresh water habitats. By definition aquatic plants are interpreted as all seed-bearing plants whose photosynthetically active parts are permanently or, at least, for several months each year are submerged in water or float on the surface of water (Cook, 1990). the plants may grow in permanent or seasonally semi-permanent water. The systematic relationships of aquatic flowering plants are varied. Yet they share many similar characteristics necessitated by a common habitat (Sculthorpe, 1967). These include presence of air cavities, development of special structures for buoyancy, reduction in mechanical and vascular tissue, adaptation in the mode of pollination and fruit/seed dispersal. However, depending on whether the plants are amphibious, attached, free-floating, emergent or submersed, they differ in the degree of adaptation to their surroundings.

Besides regulating key ecological processes, aquatic plants serve a variety of purposes such as sources of food, fibre, paper pulp, green manure and for mineral recycling and control of pollution (Boyd, 1968, 1970, 1972; Gujral *et al.*, 1986). However, in disturbed aquatic ecosystems some of the alien species such as *Eichhornia*, *Salvinia* and *Pistia* have turned into pernicious and unmanageable weeds which have dried up water bodies used by rural people for their daily needs, choked irrigation channels, blocked hydroelectric installations, hampered navigation and have seriously affected fishing. With explosion in human population, pressure on land is mounting and the compulsive demand for the study and utilization of marine and freshwater ecosystems is not difficult to realize.

About 470 species of flowering plants constitute the aquatic flora of the Indian subcontinent, accounting for nearly half of the world's known aquatic angiosperms (Biswas and Calder, 1937; Subramanyam, 1962; Deb, 1976; Gopal, 1990). The vast potential of water plants is only marginally utilized and no scientific approach has been developed for deriving greater economic and ecological benefits (Mohan Ram, 1991). Modern technology is seldom applied to the cultivation of aquatic food plants which are mostly grown in native cultures by traditional methods (Anonymous, 1976).

THE USE OF *IN VITRO* TECHNIQUES IN AQUATIC PLANTS

Growing whole plants rather than only their parts under precisely controlled and aseptic conditions provides extensive opportunities to understand problems of organization, integration and interaction (Mohan Ram, 1978). Aquatic plants are specially favourable for this purpose because of the ease with which they can be propagated, grown and multiplied in liquid culture (Mohan Ram, 1991). Studies have demonstrated that certain aquatic plants are amenable to *in vitro* propagation by proliferation from pre-existing buds (Doreswamy and Mohan Ram, 1969; Kane, 1989a; Mohan Ram, 1991).

The convenience with which nutrients can be supplied makes the axenic culture of whole hydrophytes ideally suited for understanding chemical ecology, to do critical analytical work dealing with metabolism and to visualize fine structure. Another advantage is the general absence of a callus phase in cultures of submerged aquatic plants, which often becomes irksome in the case of tissue from land plants (Mohan Ram, 1978).

Establishment of Cultures and Nutrition

A serious obstacle that could have dissuaded many investigators from using hydrophytes as experimental material is the inherent difficulty in establishing aseptic cultures from plants growing in natural surroundings. Hillman (1961) has reviewed the various procedures to overcome this problem with regard to duck weeds. For emergent macrophytes, Kane *et al.* 1988a, 1990a, 1991a), Kane and Albert (1989b), Kane and Sheehan (1988) used defoliated aerial shoots of *Myriophyllum*, *Cryptocoryne* and *Proserpinaca* to raise aseptic cultures. The stem segments were surface sterilized with 1.05% solution of sodium hypochlorite containing 0.01% (v/v) Tween 20 for 5-20 min. and rinsed thrice with sterile deionized water. Explants of *Nymphaea* epiphyllous plantlets and rhizome tips were first immersed in 50°-70° (v/v) ethanol for 1 min. followed by the above treatment (Jenks *et al.*, 1990, Lakshmanan, 1994).

Whereas these techniques may be suitable for duckweeds and emergent hydrophytes, our experience with other aquatic plants has been different. Surface-sterilized fruits which are mature but not dehisced and fresh or dry seeds provide good starting material. The presence of a hard fruit wall or seed coat protects the embryo from the adverse effects of sterilizing agents. Successful cultures have thus been raised in *Utricularia* (Mohan Ram and Doreswamy, 1966; Mohan Ram and Dutta, 1966), *Vallisneria* (Uma and Mohan Ram, 1972), *Ceratophyllum* (Mohan Ram and Kapoor, 1974) *Limnophila* (Rao and Mohan Ram), *Neptunia oleracea* (Kakkar and Mohan Ram, 1986), lotus (Kane *et al.*, 1988b; 1990b) and *Trapa* (Agrawal, 1993). Recently members of the Podostemaceae have also been raised *in vitro* (Vidyashankari and Mohan Ram, 1987; Sehgal *et al.*, 1993) starting with seeds obtained from undehisced fruits.

Once initial axenic cultures are established, aquatic plants can be grown on a defined mineral medium containing sucrose. Experiments designed to find out whether or not an energy source is required for the growth of these chlorophyllous plants in light have demonstrated that without sucrose growth is rather poor (Hillman, 1959; Doreswamy and Mohan Ram, 1969; Sehgal, 1976). In the presence of 2 per cent sucrose, however, the amount of light required seems to be rather low, suggesting that light probably exercises a morphogenetic role (Hillman, 1957). Vitamins do not appear to be necessary, although in their presence growth is markedly better (Doreswamy and Mohan Ram, 1969; Uma and Mohan Ram, 1972; Sehgal, 1976). The pH of natural waters is between 4.0 and 10.2 and uptake of iron by plants is difficult. Iron supplied as ferric chloride is not effective in supporting normal growth *in vitro*. The requirement of chelating agents such as EDTA (ethylene-diamine-tetraacetic acid) and EDDHA (ethylene-diamine-di-o-hydroxyphenylacetic acid) for a large number of water plants cultivated *in vitro* emphasizes the role of natural chelates in fresh waters (Fogg, 1959; Hillman, 1959; Doreswamy and Mohan Ram, 1969; Gupta and Maheshwari, 1970; Venkataraman *et al.*, 1970; Uma and Mohan Ram, 1972; Sehgal, 1976).

Insectivores are one of the most curious groups of plants, which are endowed with special devices to trap insects. There were claims that the nitrogen derived from the insect prey is an obligate requirement for some of these plants (Lloyd, 1942; Pringsheim and Pringsheim, 1962, 1967; Harder, 1963). *Utricularia* is a large carnivorous genus with several species occurring in water. Pringsheim and Pringsheim (1967) reported that peptone or beef extract is essential for the cultured plants of *U. minor* and *U. ochroleuca*. These substances promoted good vegetative growth in *U. exoleta* and were absolutely essential for flowering (Harder, 1963). In axenic cultures of *U. gibba* (Mohan Ram and Dutta, 1966) beef extract and tryptone supported maximum vegetative growth and delayed the abscission of bladders. In *U. inflexa* var. *stellaris*, growth occurs even on a medium containing nitrate, thus demonstrating that there is no absolute requirement for organic nitrogen or protein (Doreswamy and Mohan Ram, 1969). However, the presence of organic nitrogen sources such as casein hydrolysate and yeast extract stimulated higher rates of growth. It is natural to question the specific function of the bladders. Bladders probably resulted from a morphological mutation, which during evolution must have become secondarily useful in an environment deficient in nitrogen. This view is strengthened by the fact that under the same ecological conditions, numerous other flowering plants exist without any special contrivances for deriving extra sources of nitrogen.

Regeneration and Micropropagation

The propensity for regeneration and vegetative propagation in the water plants is extraordinarily high, probably as an adaptation to turbulence and other mechanical disturbances in water which break up the plant body. This is commonly observed in rivers where portions and small islands of weeds like the water hyacinth float along the current. Fragmentation of the plant body followed by regeneration from a part bearing a bud is a common phenomenon. This has been amply substantiated by studies carried out *in vitro* with *Utricularia* (Doreswamy and Mohan Ram, 1969), Ceratophyllum (Sehgal, 1976), *Vallisneria spiralis* (Uma and Mohan Ram, 1972), *Limnophila aquatica* (Rao and Mohan Ram, 1981), *Neptunia oleracea* (Kakkar and Mohan Ram, 1986), *Cryptocoryne lucens* (Kane *et al.*, 1990) and *Myriophyllum aquaticum* (Kane *et al.*, 1991a). In *Utricularia*, excised internodes that lacked buds failed to regenerate. Lobes of isolated leaves, however, readily gave rise to new plants by initiating adventitious buds (Doreswamy and Mohan Ram, 1969). Regeneration through adventitious shoot buds was also observed in the excised root tips of *Limnophila indica* (Rao and Mohan Ram, 1981) cultured in Nitsch's liquid medium with 2% sucrose. Regeneration of adventitious shots and root regeneration from aerial leaf explants was successfully accomplished in the heterophyllous aquatic *Myriophyllum heterophyllum* (Kane and Albert, 1989a). These results indicate that like land plants, hydrophytes are capable of organogenesis from non-meristematic tissues. Kane and Albert (1989) have demonstrated a high regeneration potential (2600 shoots from a single leaf explant in one month's time).

Commercial micropropagation of aquatic plants has been limited to two end uses: aquaria and horticulture. These include *Cryptocoryne lucens, Echinodorus*, lotus (*Nelumbo lutea*), water lily (*Nymphaea* Daubeniana, *Nymphaea* hybrid James Brydon), *Myriophyllum* and *Potenderia* (Kane *et al.*, 1988, 1990a,b, 1991a,b; Jenks *et al.*, 1990; Lakshmanan, 1994).

A rapid micropropagation procedure has been recently developed for water chestnut (*Trapa* sp.), a minor food crop with recalcitrant seeds (Agrawal, 1993). A high rate of post-transplantation establishment is the outcome of this work.

Patterns of Development

Heterophyll (formation of several kinds of leaves on the same individual) is a characteristic feature of many rooted aquatic angiosperms. The submerged laminae are thin and highly dissected; they lack cuticle and stomata, and have chloroplasts in epidermal cells. In contrast, the aerial leaves have broad, thick, laminae with entire margins, well-developed cuticle and mesophyll, and numerous stomata. In these leaves the epidermal cells lack chloroplasts. In some plants heterophylly is a manifestation of heteroblastic development with juvenile leaves arising in an orderly sequence before adult ones are formed, regardless of environmental influence (Sculthorpe, 1967). However, in most hydrophytes factors such as depth of water, relative humidity, light intensity, photoperiod, temperature and variations in salt concentration seem to be involved in determining leaf shape. Dissected water leaves can be initiated by submerging aerial stems (Bostrack and Millington, 1962) but inducing aerial leaves under water is more tricky. Jones (1955, 1956), however, was able to obtain aerial leaves under water in *Callitriche* by increasing the osmotic concentration of the water.

Using the heterosporous fresh water fern *Marsilea*, Allsopp (1955) was able to control heteroblastic development by manipulating sucrose concentration. He considered that heteroblastic development *in vitro* is regulated by the effect of nutritional factors such as sucrose and nitrogen on the shoot apex, a critical minimal nutritional status being necessary for the production of adult leaves (Allsopp, 1965, 1967). Heterophylly has been investigated to a considerable extent in our laboratory using the classical amphibious material *Limnophila*, in which the factors that influence the shoot apex in determining leaf form have been identified. For instance, when the nodal explants of submerged shoots of *L.*

indica were cultured in Nitsch's liquid medium containing abscisic acid at 10^{-7}M and 10^{-6}M, they bore regenerated shoots with typical aerial leaves, even under submerged conditions (Mohan Ram and Rao, 1982). ABA induced floating/aerial leaf form in *Potamogeton nodosus* (Anderson, 1978), *Myriophyllum* (Kane and Albert, 1989b), *Proserpinace palustris* (Kane and Albert, 1982, 1987a) and *Hippuris vulgaris* (Kane and Albert, 1987b). Thus heteroblasty is a fundamental morphological problem which has been observed but not explained. The *in vitro* technique coupled with molecular probes should enable us to understand the nature of the signals which reach the apical meristem and regulate the pattern of leaf development.

In experiments dealing with the effect of cytokinins on the development of leaves in *Ceratophyllum*, an unusual feature came to our attention. The degree of dichotomy and the length of the segments of the leaf were found to be directly related to the concentration of benzyladenine. The fact that the delimitation of the species within this genus is based on the degree of leaf dichotomy (Fassett, 1953) suggests that the different species must be producing varied amounts of endogenous cytokinins. This aspect, however, needs verification. Other interesting responses to cytokinins were the retardation of growth rate, acute internodal condensation, resulting in exposed apical meristems and release of dormant lateral buds (Sehgal, 1976).

Flowering

The organ culture technique has been successfully employed to investigate floral evocation, to study the persistence of the induced state, and also to determine the factors that support flower morphogenesis. As far as the aquatic flowering plant are concerned, the main thrust has been in probing the mechanism of floral induction especially using cultures of duckweeds. These tiny monocotyledons offer distinct advantages. They multiply rapidly and produce a large, genetically homogenous, population. They are simple (lack vasculature), with no part of the plant being more than a few cells away from the medium. They grow easily and require only a few minutes of red light a day to maintain normal growth (Hillman, 1976).

Kandeler (1955) was the report flowering in the culture of *Lemna gibba*, (a long-day plant). Hillman (1957) induced flowering in *L. perpusilla* under short days by incorporating EDTA into the medium. It has now been established that chelates like EDTA and EDDHA are essential for flowering in other members of the Lemnaceae with the exception of *Wolffia papulifera* (Maheshwari and Seth, 1965). Czygan (1962) brought *L. minor* plants to flower with oestrogens under short days.

Cytokinins induce flowering in the duckweeds under non-inductive conditions (Hillman, 1957 and Maheshwari and Venkataraman, 1966). Surprisingly, ABA induces flowering in *L. perpusilla* (Higham and Smith, 1969). Flowering has been induced in the duckweeds under non-inductive conditions using salicylic acid, acetyl salicylic acid, tannic acid, benzoic acid, 8-hydroxyquinoline (a Cu chelating agent) and c-AMP (3', 5'-cyclic monophosphate) (Khurana and Maheshwari, 1978, 1980, 1983a,b, 1984, 1986; Khurana *et al.*, 1988).

The factors regulating flowering in *Utricularia inflexa* var. *stellaris* has been reviewed by Mohan Ram (1989). This plant is photosensitive and flowering can be induced by exposing 4-week-old cultures to short-day regimes (Mohan Ram and Doreswamy, 1966). Although initiation of inflorescence primordia occurred with 10 cycles of 8 hours light and 16 hours dark, fruit and seed set took place only after 20 or more cycles were given. Organic nitrogen supplied as yeast extract inhibited flowering, while beef extract, tryptone, peptone, casamino acids and casein hydrolysate depressed the percentage of cultures flowering (Doreswamy and Mohan Ram, 1971). Flowering could not be induced by cytokinins under non-inductive conditions. Under short days, kinetin and benzyladenine enhanced flowering and zeatin depressed it. Gibberellic acid (GA_3) also lowered the percentage of flowering cultures. A commercial sample of ethephon (2-chloroethyl phosphonic acid) induced flowering in all cultures kept under long days at 10^{-8}M. However, a highly purified sample caused induction but not flower development (Mohan Ram *et al.*, 1972).

Eichhornia crassipes flowers under both long-day and short-day conditions. However, detailed studies on the effects of different day lengths on flowering have not been carried out (Pieterse, 1985). Importantly plants do not flower below 21 °C but respond positively to gibberellic acid treatment (Pieterse *et al.*, 1976).

Pollination

The pollination biology of aquatic angiosperms is a fascinating field of enquiry. The study of pollination is basic to the understanding of studies of sexual reproduction. There are molecular mechanisms which can intervene to bring about male sterility (Goldberg *et al.*, 1993). Curiously nearly all aquatics have retained the pollination mechanisms of their terrestrial ancestors (Cook, 1990). There are, however, notable exceptions. At least 28 genera from nine families have evolved pollination mechanisms using water as an agent. However, the submerged pollination (Hypohydrophily) with wettable pollen and stigmas has been shown in only five genera. Additionally six genera may perhaps also be pollinated underwater (Cock, 1990).

Sehgal (1976) has studied pollination in two species of monoecious *Ceratophyllum in vitro*. The male flowers bear a large number of spirally arranged stamens. Each stamen has a float at its apex and produces an enormous amount of pollen. Owing to a continuous system of lacunae, gases accumulate in the floats. At maturity the basal cells of the stamens give way, and the stamens rise to the surface of water due to buoyancy. Some pollen grains germinate even in intact stamens. Abscission of the stamens stimulates further pollen germination (about 10 per cent in nature and almost 100 per cent *in vitro*). After floating on the surface for a day, the anters dehisce liberating all the pollen, both ungerminated and germinated. These gradually sink, and one of the tubes effects pollination when it meets a female flower with its protruding stigma.

Mohan Ram and Doreswamy (1966) have shown that in the *in vitro* raised plants *Utricularia inflexa* var. *stellaris*, flowers undergo self-pollination. The pollen grains germinate in situ and the germinated grains are deposited *en masse* by the inward bending of the stamens and the dehiscence of the anthers close to the funnel shaped stigma. A good seed set has been recorded *in vitro*, emphasizing that the plants are self-compatible.

Evan though pollination in many hydrophytes is accomplished above water, there is often a post-fertilization bending of the peduncle so that the development of the fruits and seeds takes place under water.

Somaclonal Variation

A markedly different type of plant form was noted in the cultures of *Utricularia inflexa* var. *Stellaris* kept under long days. In contrast to the normal habit, fine thread-like stolons with profuse branches with shorter and sparingly dissected leaves were noted. This abnormal condition was termed bushy clone (Mohan Ram *et al.*, 1972). The bushy habit could be maintained by subculture. Two other characteristics of these bushy plants were the failure to flower under any photoperiodic treatment (the normal plants flower under short days) and the tendency to sink. In nature, plants showing this habit have not been recorded. It is likely that it is a spontaneous bud mutation in culture.

Bushy seedlings have also been observed in the cultures of *Ceratophyllum demersum* and *C. echinatum* along with normal plants (Sehgal, 1976). These abnormal plants were produced both from excised embryos and embryos enclosed in fruits, collected from plants growing *in vivo* and *in vitro*. The incidence of production of these abnormal seedlings was higher in immature fruits, or those produced early in the season. Seedlings were characterized by the absence of chlorophyll, slow growth rate, distorted fleshy cotyledons, condensed internodes, short leaves and small white shoot apices. Excised portions of these seedlings failed to regenerate. High (36°C) and low 6°C) temperature treatments, and growth regulators such as kinetin, benzyladenine, GA_3, and NAA could not induce normal growth. Incorporation of growth retardants, TIBA (triiodobenzoic acid),. 2,4,5-T (2,4,5-trichloroacetic acid) and the morphactin chlorflurenol also failed to induce the morphological features unique to the bushy habit in normal plants. Surprisingly, some of the bushy seedlings reverted to normal growth by the production of a normal axillary but or by transformation of the shoot apex. Once transformed, the normal branches even produced flowers and fruit.

Such bushy plants have never been observed in nature, even though the massive cotyledons could have supported their growth for a short period. Understandably, severe competition in nature would have prevented their survival. The significance of the bushy habit needs to be critically examined.

CONSERVATION THROUGH PLANT TISSUE CULTURE: THE CASE HISTORY OF THE FAMILY PODOSTEMACEAE

The strange family Podostemaceae comprises of 49 genera and 240 species of aquatic plants. Of these 18 are endemic to India (Nagendran, 1975). The podostemads have an astonishingly reduced thalloid plant body (lacking true root-shoot axis), resembling an alga, lichen or a bryophyte. They are found growing in gushing mountain streams, rivers or cataracts and live in submerged conditions attached to rocks and boulders by means of a gum secreted by their rhizoids. Flowering is initiated when the water level recedes (Vidyashankari and Mohan Ram, 1987; Mohan Ram and Sehgal, 1992).

Early attempts to germinate the seeds of podostemads and raise plants were unsuccessful. Pannier (1960) had observed that the lack of research on this interesting group of plants was mainly because they could not be maintained alive out of their native habitat for any period of time.

When the seeds of some members of the Podostemaceae such as *Griffithella hookeriana, Indotristicha ramosissima, Polypleurum stylosum, Hydrobryopsis sessilis* and *Dalzellia Zeylanica* were transferred to a nutrient medium aseptically, they failed to germinate. However, a novel technique of planting seeds on a cube of thermocole (Polystyrene foam) and floating the cubes on the sterile medium, led to high rates of germination (Vidyashankari and Mohan Ram, 1987). For sustained growth and development, the constituents of Murashige and Skoogs medium (1962) (excepting sucrose) were diluted from 1/5 to 1/20th of the original concentration (Sehgal *et al.*, 1993). thus for the first time podostemads have been grown under controlled conditions. Simple, rapid, reproducible protocols have been established to complete the seed-to-flowering cycle in *Griffithella hookeriana* (Vidyashankari and Mohan Ram, 1987), *Indotristicha ramosissima* (Vidyashankari, 1988; Sehgal and Mohan

Ram, unpublished), *Polypleurum stylosum* (Sehgal *et al.*, 1993), *Hydrobryopsis sessilis* and *Dalzellia zeylanica* (unpublished). Very recently, *Willisia selaginoides*, a highly endemic and endangered species has also been successfully established *in vitro*.

The Podostemaceae are embryologically unique on account of the presence of pseudo-embryo sacs, presence of paired pollen (Podostemoideae), absence of double fertilization, lack of endosperm etc. As the plants inhabit water falls and rapidly flowing water, their early developmental morphology has not been understood. Using whole plant cultures it has become possible to investigate germination, origin of the thallus, its homology and development, flowering, pollination, seed set, anatomy and cytology. Our knowledge about these incompletely understood group of plants has been extended. More importantly, *in vitro* culture of the podostemads has opened possibilities of its ex situ conservation, since their natural habitats are threatened by human activities such as construction of dams, installation of hydroelectric stations etc. Research is now being directed at developing protocols for micropropagation of the podostemads and their transfer to appropriate natural sites. It is also envisaged to design short-term and long-term strategies for the *in vitro* storage of podostemads by either slow-growth or cryopreservation techniques to conserve their diversity.

PROSPECTS

Having reviewed the work already done, it is ventured to direct attention to certain aspects in which future research may be carried out.

Depollution and Recovery of Nutrients from Waste Waters

Aquatic plants are excellent indicators of environmental contaminants such as sulphur dioxide, ozone, pesticides and chemicals, as they present rapid toxic symptoms. They are also useful in accumulating heavy metals in large amounts. Arsenic is taken up by *Ceratophyllum* up to 26 ppm with no adverse effects. Among the duckweeds *Spirodela* is reported to have yielded zinc at a concentration factor of about 25,000 (Hutchinson, 1975). Other duckweeds, notably *Lemna* concentrate aluminium 660,000-fold, iron 307,000-fold, titanium 102,000-fold, copper 79,000-fold and cobalt 26,000-fold (Anonymous, 1976). There is thus a specific advantage in using aquatic plants to depollute streams and lakes into which effluents from industrial establishments are discharged in the urban areas. Aquatic weeds are able to survive in water with high BOD and toxic chemicals (Dinges, 1976). Water hyacinth accumulates a large number of heavy metals and has been explored for treating waste waters from dairies, piggeries, textile mills, oil mills and metal work industries (Abbasi, 1987).

Jamil (1990) has presented data on the accumulation of heavy metals such as Cd, Hg and Zn by water hyacinth from solutions containing cadmium chloride, mercuric chloride and zinc sulphate at 25, 50, 75 and 100 ppm, after a period of 1,3,4 and 7 days. In most cases maximum concentration was noted after 24 hr treatment. In another study water hyacinth plants were grown in solution with salts of metals such as Hg, Cd, Pb and Cr, at concentrations ranging from 25 to 100 ppm. The maximum uptake of these toxic metals was found at 25 ppm.

Various groups of plants have been studied for their ability to cycle heavy metals from a pond with post-sewage water containing Fe, Mn, Zn, Cr, Cu, Ni and Pb (Ozimek, 1985). Of the plant groups, pleustonic plants (comprising *Lemna minor* and *L. gibba*) showed a higher efficiency for removal of metals as they thrived in polluted waters, absorbed the metals efficiently, were not limited by light and did not set the sediment in motion. Plants such as *Lemna, Myriophyllum, Scirupus* and *Pistia* have been used in the treatment of industrial waste waters and nutrient-rich agricultural drainage effluents.

Plants that are tolerant to heavy metals synthesize heavy metal binding polypeptides referred to as cadystins and phytochelatins. The enzyme that catalyzes their biosynthesis, phytochelatin synthase, is constitutively expressed in plants. These polypeptides apparently sequester and detoxify excess metal ions and play a role in homeostasis. The nature and role of these polypeptides, which possess a general structure (-Glu-Cys)$_n$-Gly, (where n=2-7), have been recently reviewed by Steffens (1990).

The utilization of aquatic weeds for this purpose will require well-designed programmes to pick up such native species which can selectively take up heavy metals and also carcinogens. For accurate, quick and dependable results, the *in vitro* technique should prove invaluable.

Intensive *in vitro* selection of clones of aquatic weeds should also be undertaken to serve as tools for bioassay of other pollutants such as chromates, cyanides, thiocyanates, phenols, chlorides and detergents.

Determination of the Effectiveness of Herbicides and Insecticides

Eradication of aquatic weeds and controlling mosquitoes that live on them are becoming a major problem for many developing countries. This is understandable because environmental conditions for weed growth are most congenial in the tropics. All possible methods are being tested with the objective of maximizing effectiveness and minimizing the hazards. The merit of the *in vitro* system lies in the fact that the efficacy of a very large number of herbicidal compounds could be tested on great many weed species, cultured under laboratory conditions, even in seasons when they do not occur in nature.

Possible Medical Uses

In general hydrophytes are poor in secondary metabolites. Very few medicinal principles have been isolated and identified from them (Sculthorpe, 1967). Yet traditional uses of aquatic plants have been reported from India (*Acorus calamus, Hygrophila spinosa, Bacopa monnieri*, etc.). It is also likely that sufficient scientific attention has not been paid to them.

Basic Biological Studies

Knowledge of the biology of aquatic weeds has been built largely on observations. However, valuable this approach has been, the causative aspects can be understood only by experimentation. Many angiosperms come to flower only when their shoot apices rise above water or the water level falls substantially. In what ways water modifies the photoperiodic or temperature conditions to prevent flowering is worthy of enquiry.

Chromatographic analysis of extracts of closely resembling species of Lemnaceae, grown under identical conditions in aseptic culture, has shown striking differences in the phenolic compounds and flavonoids (McClure and Alston, 1964; Alston, 1966). Chemotaxonomy may be important in the delimitation of other species of hydrophytes also.

Sehgal (1976) has reported that she was able to isolate and identify from aspetic cultures of *Ceratophyllum, C. echinatum*, an American species hitherto unreported from India.

Sircar and co-workers have extracted growth hormones from the roots of *Eichhornia* (see Bhanja *et al.,* 1967). If this finding can be developed into a commercial process, horticultural practices in the country can be greatly augmented. Mutation, selection of high yielding clones, their propagation and hormonal extraction will be undoubtedly facilitated by the *in vitro* techniques.

Conservation of Biodiversity

In vitro conservation of germplasm is being increasingly used for such species in which conventional seed storage in gene banks is problematic. These include plants which do not produce seeds, or are vegetatively propagated, or genetically heterogeneous or produce recalcitrant seeds. Aquatic plants such as *Trapa* are reported to produce seeds which become non-viable if their moisture content is reduced (Chin and Roberts, 1980). Conservation of such plants would be possible if *in vitro* methods are employed. Moreover, as stated earlier, tissue culture system may prove to be the only means of conserving plants such as *Willisia selaginoides* and *Indotristicha tirunelvelliana* of the Podostemaceae, which have a highly restricted distribution and are found in fragile ecosystems vulnerable to anthropogenic activities.

Aquarium Plants

A vast potential exists for the commercial production of aquarium plants such as *Cryptocoryne, Myriophyllum, Cabomba* and others, by micropropagation. Plants raised *in vitro* are preferred over plants collected from natural waters as they are free from pathogens that attack domesticated fish. However, presently only France and Holland are engaged in this practice (Pierik, 1987). A tropical country like India which abounds in a rich diversity of potential aquarium plants should take advantage of this endowment. As already stated, aquatic plants can be multiplied at a low cost, are easy to handle and have no problem of post-transhipment hardening. Recent techniques, such as the use of cultuSAK™, a commercially available heat sealable *in vitro* culture system, would enhance the production and marketing of aquarium plants (Kane *et al.,* 1990c).

In vitro raised water ornamentals can be used for mutation breeding to generate variants as conventional breeding is difficult in them.

CONCLUSION

In the above account on attempt has been made to outline the immensely fascinating aspects of water plants and the use of tissue culture techniques in visualizing and understanding some aspects of their biology and biotechnology. *In vitro* culture provides on ideal system to carry out investigations on hormonal and environmental mechanisms controlling plant growth and development of hydrophytes, which may otherwise remain neglected due to inaccessibility in their natural surroundings. There is a vast potential for development of commercial tissue culture of aquatics.

REFERENCES

Abbasi, S.A. 1987. Cited in Smith, W.H. and K.R. Reddy (Eds.) Proceedings of Aquatic Macropytes for Water Treatment and Resource Recovery, Magnolia Publishing Orlando, 175 pp.

Agrawal, A. 1993. Some aspects on the biology of *Trapa*, PhD Thesis, Univ. of Delhi, Delhi.
Allsopp, A. 1951. *Marsilea* spp. Materials for experimental study of morphogenesis, Nature 168: 301.
Allsopp, A. 1955. Experimental and analytical studies of pteridophytes. 27. Investigations on *Marsilea*: 5. Cultural conditions and morphogenesis with special reference to the origin of land and water forms, Ann Bot 19: 247-264.
Allsopp, A. 1965. Land and water forms: Physiological aspects, HandbPfl Physiol, 15: 1236-1355.
Allsopp, A. 1967. Heteroblastic development in vascular plants, *Adv. Morphogenesis*, 6: 127-171.
Alston, R.e. 1966. Chemotaxonomy or biochemical systematics? In Swain, T. (Ed.) comparative Phytochemistry Academic Press, London.
Anderson, L.W.J. 1978. abscisic acid induces formation of floating leaves in the heterophyllous *aquatic* angiosperm *Potamogeton nodosus*, Science N. Y. 201: 1135-1138.
Anonymous. 1976. Making Aquatic Weeds Useful: Some Perspectives for Developing Countries National Academy of Science Washington D.C. 171 pp.
Bhanja, A., Mukherjee, R.K., Roy, N.K., Banerjee, S., Sircar, P.K. and Sircar, S.M. 1967. Isolation and identification of growth substances in the root of water hyacinth (Eicchornia crassipes Mort. Solms.), Proc. Int. Sym. Plant Growth Substances Botany Dept., Calcutta Univ. pp. 47-56.
Biswas, K. and Calder, C.C. 1937. Handbook of Common Water and Marsh Plants of India and Burma, 1936: Health Bulletin No. 24, Manager of Publications Delhi, 140 pp.
Bostrack, J.M. and Millington, W.F. 1962. On the determination of leaf form in an aquatic heterophyllous species of *Ranunculus*, Bull Torrey bot Club 89: 1-20.
Boyd, C.D. 1968. Fresh water plants: A potential source of protein, Econ Bot. 22: 359-368.
Boyd, C.D. 1970. Vascular aquatic jplants for mineral nutrient removal from polluted waters, Econ. Bot. 24: 1-95.
Boyd, C.D. 1972. A bibliography of interest in the utilization of vascular aquatic plants, Econ Bot. 26: 74-84.
Chin, H.F. and Roberts, E.H. (Eds.) 1980. Recalcitrant Crop Seeds, Tropical Press Sdn Bnd Malaysia.
Cook, C.D.K. 1990. Aquatic Plant Book SPB Academic Publishing The Hague The Netherlands.
Czygan, F.C. 1962. Blutenbildung bei Lemna minor nach zusatz von Oestrogenen, Naturwissenschaften 49: 285-286.
Deb, D.B. 1976. A study on the aquatic vascular plants of India, Bull Bot Soc. Bengal 29: 155-170.
Dinges, R. 1976. Water hyacinth cultures for waste water treatment, Texas Department of Health Texas Austin 143 pp.
Doreswamy, R. and Mohan Ram, H.Y. 1969. Studies on growth and flowering in axenic cultures of insectivorous plants. I. Seed germination and establishment of cultures in *Utricularia inflexa* Forsk., *Phytomorphology*, 19: 363-371.
Doreswamy, R. and Mohan Ram, H.Y. 1971. Studies on the growth and flowering in axenic cultures of insectivorous plants. II. Induction of flowering and development of flowers in *Utricularia inflexa*, Z Pflphysiol, 65: 315-325.
Fassett, N.C. 1953. North American Ceratophyllum (synopsis en Lengua espanda), Communications del Instituto de Investigationes cientificas, 2: 25-45.
Fogg, G.E. 1959. Dissolved organic matter in oceans and lakes, New Biol. 29: 31-48.
Goldberg, R.B., Beals, T.P. and Sanders, P.M. 1993. Anther development: jBasic principles and practical applications, Plant Cell, 5: 1217-1229.
Gopal, B. (Ed.) 1990. Ecology and management of aquatic vegetation in the Indian subcontinent, Kluwer Academic Publishers. The Netherlands 246 pp.
Gujral, G.S., Jain, S.K. and Vasudevan, P. 1986. Aquatic biomass: Production, potential and scope for exploitation, Res. & Ind. 31: 41-47.
Gupta, S. and Maheshwari, S.C. 1970. Growth and flowering of *Lemna paucicostata*. I. General aspects and role of chelating agents in flowering. Pl Cell Physiol. 11: 83-95.
Harder, R. 1963. Blutenbildung durch tierische zusatznahrung und andere Faktoren bei *Utricularia exoleta* R. Braun, Planta 59 459-471.
Higham, B.M. and Smith, H. 1969. The induction of flowering by abscisic acid in *Lemna perpusilla* 6746: Life Sci. 8: 1061-1065.
Hillman, W.S. 1957. Photoperiodic control of flowering in *Lemna perpusilla*, Nature 181: 1275.

Hillman, W.S. 1959. Experimental control of flowering in *Lemna*. i. General methods. Photoperiodism in L. perpusilla 6746, Am J. Bot. 46: 466-473.

Hillman, W.S. 1961. The Lemnaceae or duckweeds. A review of the descriptive and experimental literature, Bot Rev. 27: 221-281.

Hillman, W.S. 1976. Calibrating duckweeds; light, clocks, metabolism, flowering, Science 193: 453-458.

Hutchinson, G.E. 1975. A Treatise on Limnology. Vol. III. Limnological Botany, Wiley Interscience, New York.

Jamil, K. 1990. Environmental Biology of Water Hyacinth, Avichal Science Foundation, Vallabh Vidyanagar, 104 pp.

Jenks, M., Kane, M., Marousky, F., McConnell, D. and Sheehan, T. 1990. *In vitro* establishment and epiphyllous plantlet regeneration of Nymphaea Daubeniana, Hort Science 25: 1664.

Jones, H. 1955. Further studies on heterophylly in *Callitriche intermedia*: Leaf development and experimental induction of ovate leaves, Ann Bot 19: 369-388.

Jones, H. 1956. Morphological aspects of leaf expansion especially in relation to change sin leaf form, In Milthorpe F Ed. The Growth of Leaves, Butterworth Scientific Publications, London.

Kakkar, M. and Mohan Ram, H.Y. 1986. Regeneration of whole plants from tissue cultures of the tropical aquatic legume *Neptunia oleracea*, J Pl Physiol. 126: 83-91.

Kandeler, R. 1955. Uber die Blutenbildung bei *Lemna gibba* L. I. Kultur bedingungen und Tageslangenabhangigkeit, Zeit Bot 43: 36-71.

Kane, M. 1989a. Micropropagation of freshwater plants, Restoration and Management Notes, 7: 98-99.

Kane, M.E. 1989b. Aquatic plant tissue culture: A useful tool for the propagator and hybridizer? Water Gdn J. 5(1): 22-24.

Kane, M.E. and Albert, L.S. 1982. Environmental and growth regulator affects on heterophylly and growth of *Proserpinaca intermedia* Mark. (Haloragaceae), Aquat Bot, 13: 78-85.

Kane, M.E. and Albert, L.S. 1983. Effects of ABA on heterophylly and stomatal development on submerged shoots of *Myriophyllum heterophyllum* Michx., Am J. Bot. 70: 23.

Kane, M.E. and Albert, L.S. 1987a. Abscisic acid induces aerial leaf morphology and vasculature in submerged *Hippuris vulgaris* L., Aquat Bot. 28: 81-88.

Kane, M.E. and Albert, L.S. 1987b. Integrative regulation of leaf morphogenesis by gibberellic and abscisic acids in the aquatic angiosperm Proserpinaca palustris L., Aquat Bot. 28: 89-96.

Kane, M.E. and Albert, L.S. 1989a. Comparative shoot and root regeneration from juvenile and adult aerial leaf explant of variable- leaf milfoil, J. Aquat Plant Manage, 27: 1-10.

Kane, M.E. and Albert, L.S. 1989b. Abscisic acid induction of aerial leaf development in *Myriophullum* and *Proserpinaca* species cultured *in vitro*, J. Aquat Plant Manage 27: 102-111.

Kane, M.E., Gilman, E.F. and Jenk, M.A. 1991a. Regenerative capacity of *Myriophyllum aquaticum* tissues cultured *in vitro*, J. Aquat Plant Manage, 29: 102-109.

Kane, M.E., Gilman, E.F., Jenks, M.A. and Sheehan, T.J. 1990a. Micropropagation of the aquatic plant *Cryptocoryne* lucens, HortScience 25: 687-689.

Kane, M.E., Jenks, M.A. and Sheehan, T.J. 1990b. Propagation studies in the Nymphaeacea: American lotus, Water Gdn J. 6: 31-33.

Kane, M.E., McConnell, D.B., Sheehan, T.J. and Dehgan, B. 1988a. A laboratory exercise to demonstrate adventitious shoot formation using stem internodes of parrot-feather, HortScience 23: 408.

Kane, M.E., Philman, N.L., Bartuska, C.A. and McConnell, D.B. 1993. Growth regulator effects on *in vitro* shoot regeneration of *Crassula helmsii*, J. Aquat. Plant Manage, 31: 59-64.

Kane, M.E., Philman, N.L. and Lee, T.M. 1990c. Evaluation of CultuSAK™ *in vitro* plant cell culture system, Pro Fla State Hort Soc. 103: 182-186.

Kane, M.E., Philman, N.L., Lee, T.M. and Jenks, M.A. 1991b. Micropropagation and transplant growth performance of wetland plants: *Potenderia cordata*, Hortscience 26: 756.

Kane, M.E. and Sheehan, T.J. 1988. Aquatic plant tissue culture; Fluffing up the parrot feather, Water Garden J. 4: 19-23.

Kane, M.E., Sheehan, T.J. and Ferwerda, F.H. 1988b *in vitro* growth of American lotus embryos, HortScience 23: 611-613.

Khurana, J.P. and Maheshwari, S.C. 1978. Induction of flowering in *Lemna paucicostata* by salicylic acid, Pl Sci Lett. 12: 127-131.

Khurana, J.P. and Maheshwari, S.C. 1980. Some effects of salicylic acid on growth and flowering in Spirodela polyrrhiza SP_{20}, Pl Cell Physiol. 21: 923-927.

Khurana, J.P. and Maheshwari, S.C. 1983a. Floral induction in *Wolffia microscopica* by salicylic acid and related compounds under non-inductive long days, Pl Cell Physiol. 24: 24: 902-912.

Khurana, J.P. and Maheshwari, S.C. 1983b. Effect of 8-hydroxyquinoline on flowering and endogenous levels of iron and copper in Lemna paucicostata strain LP_6, Pl Cell Physiol 24: 1251-1254.

Khurana, J.P. and Maheshwari, S.C. 1984. Floral induction in short-day. *Lemna paucicostata* 6746 by 8-hydroxyquinoline, under long day, Pl Cell Physiol. 25: 77-83.

Khurana, J.P. and Maheshwari, S.c. 1986. A comparision of the effects of chelates, salicylic acid and benzoic acid on growth and flowering of *Spirodela polyrrhiza*, Pl Cell Physiol. 27: 919-924.

Khurana, J.P., Tomat, B.K. and Maheshwari, S.C. 1988. Role of adenosine 3'5'-cyclic monophosphate in flowering of a short day duckweed *Lemna paucicostata* 6746, Pl Cell Physiol. 29: 1023-1028.

Kumazawa, M. 1967. An experimental study on the seedling of *Utricularia pilosa* Makino, *Phytomorphology*, 17: 494-498.

Lakshamanan, P. 1994. *In vitro* establishment and multiplication of *Nymphaea* hybrid James Brydon, Pl Cell Tissue and Organ Cult. 36: 145-148.

Llyod, W.F. 1942. The Carnivorous Plants, Chronica Botanica Co, Waltham.

Maheshwari, S.C. and Seth, P.N. 1965. Photoperiodic control of flowering in *Wolffia papulifera*, Pl Cell Physiol. 7: 163-165.

Maheshwari, S.C. and Venkataraman, R. 1966. Induction of flowering in a duckweed- *Wolffia microscopica*- by a new Kinin, Zeatin, Planta 70: 304-306.

McClure, J.W. and Alston, R.E. 1964. Patterns of selected chemical component of *Spirodela oligorhiza* formed under various conditions of axenic culture, Nature 201: 311-313.

Mohan Ram, H.Y. 1978. *In vitro* culture of aquatic flowering plants; Achievements and prospects, 10th Philip R. White Mem Lec., Ahmedabad, 17 pp.

Mohan Ram, H.Y. 1989. *Utricularia inflexa* var. *stellaris*, In Halevy, A.H. (Ed.) CRC Handbook of Flowering Vol. VI CRC Press Boca Raton, Florida, pp. 655-660.

Mohan Ram, H.Y. 1991. Biology of aquatic flowering plants, Proc Indian Natn. Sci., Acad. B57: 95-108.

Mohan Ram, H.Y. and Agrawal, A. 1990. Plant tissue culture: Achievements and prospects, Biology Education 7: 182-197.

Mohan Ram, H.Y. and Dutta, S. 1966. *In vitro* culture of utricularia Curr, Sci. 35: 48-50.

Mohan Ram, H.Y. and Kakkar, M. 1983. Role of tissue culture in the study of aquatic plants, Bull Bot. Surv. India. 25: 26-34.

Mohan Ram, H.Y. and Kapoor, A. 1974. *In vitro* growth and development of hornwort In 3rd Int. Conf. Plant Tissue and Cell Culture Leicester, U.K.

Mohan Ram, H.Y. and Rao, S. 1982. *In vitro* induction of aerial leaves and of precocious flowering in submerged shoots of *Limnophila indica* by abscisic acid, Planta, 155: 521-523.

Mohan Ram, H.Y. and Sehgal, A. 1992. Podostemaceae- The strange family of aquatic angiosperms, Palaeobotanist, 41: 192-197.

Mohan Ram, H.Y., Harada, H. and Nitsch, J.P. Studies on the growth and flowering in axenic culture of insectivorous plants. III. Effects of photoperiod, ethrel, morphactin and a few other growth substances and metabolic inhibitors on *ultricularia inflexa*, Z Pfl physiol. 68: 235-253.

Mohan Ram and Doreswamy, R. 1966. Growth and Flowering of *Utricularia inflexa* Forsk. Var. stellaris Taylor in axenic culture, Naturwssenchaften, 53: 387.

Murashige, J. and Skoog, F. 1962. A revised medium for rapid growth and bioassays with tobacco tissue cultures, Physiologia pl 15: 473-479.

Nagendran, C.R. 1975. Studies on posostemaceae, PhD thesis univ mysore, mysore 173 pp.

Ozimek, T. 1985. Heavy metal content in macrphytes from pond wuplied with post-sewage water, Symposia Biologica Hungarica 29" 41-70.

Pannier, F. 1960. Physiological response of podostemaceae in their natural habitat, Int. Rev. Ges Hydrobiol 45: 347-354.

Pierik, R.L.M. 1987. *In vitro* culture of higher plants, Martinus Nijhoff Publishers, Dordrecht.

Pieterse, A.H. 1985. *Eichhornia crassipes*, In Halevy, A.H. Ed. CRC Handbook of Flowering Vol. 2; CRC Press, Inc. Boca Raton, Florida, 462-465 pp.

Pieterse, A.H., Aris, J.A.M. and Butter, M.E. 1976. Inhibition of float formation in water hyacinth by gibberellic acid, Nature, 260: 388-389.

Pringhsheim, E.G. and Pringhseim, O. 1967. Kleiner Bertrag Zur Physiologie von *Utricularia*, Z Pfl Physiol 57: 1-10.

Pringsheim, E.G. and Pringsheim, O. 1962. Axenic cultue of Utricularia, Am. J. Bot., 49: 898-901.

Rao, S. and Mohan Ram, H.Y. 1981. Regeneration of whole plants from the cultured root tips of *Limnophila indica*, Can J. Bot. 59: 969-973.

Sculthorpe, c.D. 1967. The biology of aquatic vascular plants, Edward Arnold, London, 610 pp.

Sehgal, A. 1976. Some Aspects of Developmental Biology of *Ceratophyllum*, PhD Thesis Delhi, 99 pp.

Sehgal, A., Mohan Ram, H.Y. and Bhatt, J.R. 1993. *In vitro* germination, growth, morphogenesis and flowering of an aquatic angiosperm *Polypleurum stylosum* (Podostemaceae), Aquat Bot. 45: 269-283.

Steffens, J.C. 1990. The heavy metal-binding peptides of plants, Ann Rev Plant Physiol Plant Mol. Biol. 41: 553-575.

Subramanyam, K. 1962. Aquatic angiosperms, Botanical Monograph No. 3, Council of Scientific and Industrial Research, New Delhi, 190 pp.

Uma, M.C. and Mohan Ram, H.Y. 1972. *In vitro* culture of *Vallisneria spiralis*, Phytomorphology, 22: 121-124.

Venkataraman, R., Seth, P.N. and Maheshwari, S.C. 1970. Studies on the growth and flowering of a short-day plant *Wolffia microscopica*. I. General aspects and induction of flowering by cytokinins. Z Pflphysiol 17: 721-731.

Vidyashankari, B. 1988. Seed germination and seedling morphology in *Indotristicha ramossimma* (Podostemaceae) grown *in vitro*, Curr Sci. 57: 369-373.

Vidyashankari, B. and Mohan Ram, H.Y. 1987. *In vitro* germination and origin of thallus in *Griffithella hookeriana* (Podostemaceae), Aquat Bot 28: 161-169.

Auxin-Calcium Interaction in Plant Growth and Development

S.C. Bhatla, Geetika Kalra and Sangeeta Kapoor

According to the classical theory of hormonal regulation of plant growth and development, modulation of hormonal levels and interactions between different hormones are responsible for the regulation of different aspects of plant development. Recent investigations suggest interesting similarities between plants and animals in terms of reception, transduction and amplification of the hormonal signals (Guern, 1987). Due to their lipophilic character, plant hormones have the potential to interact with the hydrophobic phase of the membranes. Most of the recent studies on the mechanism of action of plant hormones have been concerned with the modification of membrane properties, such as transmembrane electrical potential differences, permeability to ions, calcium binding to membranes or phosphorylation of membrane proteins. Binding of the hormones to the discrete protein components, supposed to have receptor function, has also received much attention. Most of the membrane receptors that bind hormone molecules, are thought to undergo a conformational change when they bind to a ligand at the cell exterior. This change leads to the generation of an intracellular signal that alters the behaviour of the target cell. The intracellular signaling molecules are often referred as second messengers, the first messengers being the hormones themselves.

If calcium is acting as a second messenger in mediating the physiological responses evoked by the hormonal stimuli, then it must meet the following criteria: 1. Cytoplasmic calcium levels must change in response to hormonal stimuli from the environment and the neighbouring cells; and these changes must precede a physiological response; 2. It must be possible to evoke a physiological response by inducing a change in the cytoplasmic calcium level in the absence of hormonal stimulus; 3. Cells must posses the mechanism to sense the changes in cytosolic calcium levels, and translate them into physiological responses; 4. Blocking the operation of a second messenger sensing system must block the physiological response to the hormonal stimulus. Keeping these criteria in view, several investigations have been undertaken in the recent past to demonstrate the role of calcium as a second messenger in the action of plant hormones. In the present work, an attempt has been made to critically analyse the work done so far with regard to auxin-calcium interaction in plant growth and development.

CALCIUM AND AUXIN TRANSPORT

As early as in 1956, Andreae and Van Ysselstein provided evidence for the involvement of Ca^{2+} in the uptake of auxin by roots. In 1970, Stahly and Benson indicated an interrelationship between the movement of calcium and the polar basipetal movement of IAA. Transport of auxins in plant cells shows all the characteristics of active transport system. Basipetal transport of IAA in sunflower stem sections is suppressed by washing in EDTA solution, and it is restored by subsequent application of calcium solutions (DelaFuente and Leopold, 1973). This shows that EDTA inhibition is due to elution of Ca^{2+} ions and infers that Ca^{2+} is actually removed by the treatment. In calcium-depleted state, it appears that the intensity of transport is lowered but the velocity remains unchanged. This shows that calcium effect on auxin transport is principally through maintaining the structural integrity of the membranes. Plasmalemma enriched membrane preparations from the etiolated soybean hypocotyls posses an active ATP-dependent calcium pump which leads to calcium accumulation (Kubowica *et al.*, 1982). Their observation indicates that calcium efflux pump is altered in its activity by IAA pretreatment of the tissue.

DeGuzman and DelaFuente (1984) presented evidence for the polar flux of Ca^{2+} in short hypocotyl segments of sunflower (*Helianthus annuus* L.), that is enhanced by IAA but inhibited by the auxin transport inhibitor, TIBA. It was suggested that calcium requirement for auxin transport might be the result of stimulus-secretion coupling. The concept states that Ca^{2+} acts as a second messenger and its entry at the basal end activates basipetal secretion of IAA. With the initial entry of Ca^{2+} into the cytoplasm, the internal concentration of Ca^{2+} may rise above a threshold level leading to Ca^{2+} enhanced secretion of IAA. The increased secretion of IAA causes more Ca^{2+} to enter the cell. The cycle continues until either of the substance reaches a low threshold level unable to stimulate the secretion or uptake of the other. Cohen and Lilly (1984) studied the changes in $^{45}Ca^{2-}$ concentration following auxin treatment of protoplasts isolated from etiolated soybean hypocotyls. Their results are consistent with other reports in saying that one rapid effect of auxin is a change in calcium flux which may result in a lower internal calcium concentration. Much of the calcium in plant tissues is extracellular. There seem to be two pools of extracellular calcium in intact plant cells- a small pool of easily displaceable calcium directly involved in auxin transport and a large pool of insoluble and strongly adsorbed calcium not involved in auxin transport. The high capacity of cell wall to bind calcium may prevent the available pool of Ca^{2+} in the hypocotyls from building up. Incubation of hypocotyl segments in calcium solution probably causes an increase in the available calcium in the extracellular space. Basipetal auxin transport is dependent upon the supply of Ca^{2+} which comes from the small pool of available calcium in the free space. Thus, according to chemiosmotic hypothesis of IAA transport, the polar movement starts with the passive entry of the undissociated IAA mostly at the apical end of the cell. The molecule ionizes in the neutral cytoplasm and binds to the putative carrier on the inside surface of the basal plasmalemma. As the internal concentration of IAA reaches a certain level, membrane depolarization ensues followed by thermodynamically downhill Ca^{2+} influx. Rise in the calcium concentration in the cytoplasm may lead to the binding of this element to the calmodulin which may then activate the IAA carrier complex, causing the secretion of IAA.

CALCIUM MODULATION OF AUXIN-MEMBRANE INTERACTION

High resolution electron microscopy and fluorescence probe experiments have demonstrated that the plasma membrane exhibits a direct conformational response to auxin *in vitro* and *in vivo* (Morre and Bracker, 1976). Concentration of $CaCl_2$, which inhibits elongation of hypocotyl segments, causes the membrane profiles to thicken, whereas IAA, at concentration which promotes elongation, causes the membranes to appear thinner. The dimensional changes induced by $CaCl_2$ and IAA are partially reversible, i.e. membranes can be made alternatively thick or thin by successive treatments with $CaCl_2$ and IAA. The response of cell membranes *in situ* is similar to that of the isolated membrane vesicles. Binding of Ca^{2+} to plasmamembrane leads to lateral compaction of the membrane and auxin-induced thinning of the membrane results from calcium displacement and subsequent relaxation of the membrane to its preferred packaging density in the absence of Ca^{2+} (Buckhout *et al.*, 1980).

PROTEIN PHOSPHORYLATION

Isolated membranes of soybean incorporate ^{32}P from Y-(^{32}P) ATP *in vitro* (Morre *et al.*, 1984). The incorporation is rapid and does not require added calcium. SDS-PAGE revealed several protein bands. Auxin (2,4-D) stimulation of ^{32}P incorporation into the material from membrane vesicles is reflected by the enhanced incorporation of protein bands with apparent molecular weight of 45,000 and 50,000. The kinetics of protein phosphorylation seems much too rapid to result secondarily from the primary auxin effect on phosphatidyl inositol (PI) breakdown. In 1985, Varnold and Morre reported the *in vitro* phosphorylation of membrane proteins of particulate fractions from etiolated hypocotyls of soybean (*Glycine max* L. Wagne), in response to calcium and auxin treatments, alone and in combination. Phosphorylation was stimulated by the growth-promoting concentrations of 2,4-D. 2,4-D + Ca^{2+} inhibit phosphorylation of these proteins. The findings suggest the possibility of a complex 2,4-D-proteinkinase-calcium interaction in the regulation of phosphorylation of membrane proteins in plants. To confirm these interactions, Veluthambi and Poovaiah (1986) selected auxin-responsive oat coleoptile segment system, and effect of Ca^{2+} and calmodulin antagonists on the *in vitro* protein phosphorylation was studied by SDS-PAGE. Two polypeptides were phosphorylated both under *in vitro* and *in vivo* conditions, but the pattern of phosphorylation of several other polypeptides was different under the two conditions, indicating that *in vivo* phosphorylation pattern of proteins is not truly reflected by *in vitro* phosphorylation studies. Protein phosphorylation in the control and auxin-treated coleoptiles was same.

Recently, evidence has been presented for the role of auxins in protein phosphorylation in pea epicotyl (Reddy *et al.*, 1987) and the possible involvement of calmodulin in the auxin-induced changes. In the presence of auxins, phosphorylation of 23,000, 82,000, 105,000 and 110,000 mol. wt. polypeptides decreases markedly, whereas phosphorylation of 19,000, 24,000 and 28,000 mol. wt. polypeptides is increased. Auxins have earlier been shown to increase calcium efflux from various tissues and plasma membrane preparations. All findings taken together indicate that auxins lower cytosolic calcium levels. The observed changes in phosphorylation in the presence of IAA could be attributed to decrease in cytosolic calcium thereby inactivating Ca^{2+} and Ca^{2+} calmodulin-dependent enzymes (protein kinases and phosphatases). The results obtained by inhibiting calmodulin activity are consistent with these conclusions, since chlorpromazine-induced changes in protein phosphorylation are similar to auxin-induced changes.

CELL ELONGATION

It has been suggested that auxins induce cell elongation in the stem and coleoptiles by removing calcium from the cell wall (Bennet-Clark, 1955). According to this calcium-bridge hypothesis, the factor which limits cellular expansion is the number of calcium ions which cross link with the pectin chains of the cell wall. Removing such cross-linkages by auxin would result in elongation. It has been suggested that auxin brings about this removal of calcium from the pectin polymers by chelation or by methylation of the carboxyl groups to which the calcium is attached. Calcium-Bridge Hypothesis was criticized on the basis of the following points: 1. IAA is only a weakly complexing agent (Cohen *et al.*, 1958), 2. IAA does not cause a measurable loss of calcium from cell wall (Cleland, 1960), and 3. IAA does not interfere in the binding of calcium to the cell wall (Burling and Jackson, 1965).

According to the acid growth theory (Rayle and Cleland, 1979; Cleland, 1980), auxins stimulate cells to excrete H^+; the resulting decrease in wall pH causes cell wall loosening and subsequent cell elongation. In 1976, Cohen and Nadler examined cationic requirement of IAA-induced acidification. Ca^{2+}-H^+ exchange has been found to be the mechanism for acidification in *Avena* coleoptiles. The acidification is dependent on divalent cations with an order of effectiveness of $Ca^{2+}>S^{2+}>Mn^{2+}>Mg^{2+}$. The Ca^{2+} response is IAA-dependent. It is not only clear that the IAA-induced acidification responds to divalent cations, but that the divalent cations do not stimulate a process occurring in the absence of IAA. Preincubation of tissue in mild acid exchanges away endogenous cations and effects of endogenous cations can be observed. $CaCl_2$ stimulated acidification by acid washed tissue in the presence of IAA.

Role of calmodulin in auxin-induced elongation has been investigated by using its potent antagonists (Raghothama *et al.*, 1985). Addition of calmodulin antagonists, like chlorpromazine (CPZ), trifluoperazine (TFP) and napthalene sulphonamide derivatives, to the incubation medium decreases auxin-induced elongation in oat and corn coleoptiles. The release of free amino acids from the tissue has been used as a parameter for measuring cellular integrity. The release of amino acids at higher concentration of CPZ may be due to irreversible membrane damage. The results suggest the involvement of calmodulin in auxin-mediated coleoptile elongation. The involvement of calmodulin in cellular processes may be due to changes in the actual content of calmodulin and/or by modulation of its activity due to altered calcium levels in the cytoplasm.

Hasenstein and Evans (1986) investigated the interaction of Ca^{2+} and auxin on root elongation in the seedlings of *Zea mays* L. The relationship between Ca^{2+} availability and root growth does not appear to reflect a direct dependence of growth on extracellular Ca^{2+} concentration. Roots of seedlings raised in the absence of applied calcium exhibit about the same growth rate as the seedlings raised with supplemented calcium. This can be interpreted as follows: The level of free calcium in the cytoplasm is low even though the level of extracellular calcium is high. This Ca^{2+} gradient is maintained by active efflux of Ca^{2+} and by sequestering of Ca^{2+}. A sudden influx of Ca^{2+} into the cytoplasm due to hormone-induced release of Ca^{2+} from sequestering sites enhances the activity of Ca^{2+} pumps, which reestablishes low levels of free cytoplasmic Ca^{2+}. If auxins inhibit root growth by causing an increase in cytoplasmic Ca^{2+} either by release from intracellular stores or influx across the plasmalemma, it must inhibit growth most effectively in cells with high Ca^{2+} levels and less effectively in Ca^{2+} depleted cells.

Cunninghame and Hall (1986) studied the effect of calcium antagonists and inhibitors of secretory processes on Auxin-induced elongation and fine structure of *Pisum sativum* stem segments. It is seen that the treatment of dark-grown pea stem segments with EDTA, EGTA, $CaCl_2$, CTC, A-23187 (calcium ionophore) and verapamil reduced elongation in the presence of IAA. Inhibition increased with increasing concentration of these substances. IAA-induced cell wall extension is considered as a biphasic response, initial phase being wall loosening due to H^+ secretion, and final phase involves regulation of supply of new cell wall material. The role of Ca^{2+} in the 2nd phase is also evaluated by examining the ultrastructure of the epidermal cells of the segments following treatment with various compounds. The structure and function of dictyosomes is known to be dependent on specific ion balance within the cell. Dilation and accumulation of vesicles in the vicinity of dictyosomes was seen following treatment with $LaCl_3$ (Lanthanum chloride) and EGTA. This suggested that vesicle fusion with plasma membrane had been impaired. Dilated vesicles may not be capable of fusing with the plasma membrane, perhaps because this is dependent on the prevailing Ca^{2+} concentration. The results suggest that elongation is dependent in part on the functioning of the dictyosomes and that they are controlled by the prevailing ion balance. Findings of Reddy *et al.* (1988) are also in agreement with those of Cunninghame and Hall (1986), that auxin-induced elongation is inhibited in the presence of calcium chelators and channel blockers, indicating that calcium has an important role in auxin-induced cell elongation. They have further shown that calcium plays no role in the induction of early IAA-regulated genes. However, the possibility exists that calcium may regulate the expression of late genes in pea or other non-auxin regulated genes. Furthermore, effect of calcium on the translation of IAA regulated mRNAs is also possible, since calcium and phospholipids play role at the level of translation initiation. Involvement of Ca^{2+} messenger system has been further investigated by Brummell and Maclachlan (1989), in the promotory action of auxins on cell wall synthesis and growth in the stem segments from etiolated seedlings of *Pisum sativum* L. cv. Alaska. Intracellular calcium antagonist, TMB-8, inhibits auxin-stimulated growth and cell wall synthesis which could be reversed by exogenous Ca^{2+}. TMB-8 inhibition is observed only in the presence of auxin.

GRAVITROPISM

Since calcium is essential for the polar transport of auxins, asymmetry in the distribution of calcium at the root tip may play a role in the establishment of the auxin asymmetry in the elongation zone. The movement of calcium does not appear to be simply by diffusion or by movement along the moist surface of the root since: 1. metabolic inhibitors strongly retard the movement; 2. movement across the tip is much greater than the movement across elongation zone; 3. movement across the tip is asymmetric in horizontally placed root, and 4. the pattern and magnitude of calcium movement across the root remains unchanged when the two sides of the root (donor and receiver) are isolated by a layer of vacuum grease along the surface on each side of the root (Lee *et al.*, 1983a). Application of calcium chelating agents (EDTA or EGTA) to the tips of maize roots causes a loss of gravitropic sensitivity (Lee *et al.*, 1983b). When the chelator is replaced with calcium chloride, gravitropic sensitivity is restored. Asymmetric application of $CaCl_2$ near the tips of decapped roots causes curvature towards the calcium source. Thus, calcium may play a key role in linking gravity detection to gravitropic curvature in roots.

Light and electron microscopic studies on the distribution of calcium in gravitropically-responding oat coleoptiles reveals an accumulation of Ca^{2+} in the upper halves of the horizontal, gravistimulated coleoptiles within 10 min of stimulus onset (Slocum and Roux, 1983). After 30 min of gravistimulation, a marked lateral asymmetry in the distribution of Ca^{2+} is developed at the region where maximal bending will occur. Following 3 hrs of gravistimulation, both gravitropic curvature and Ca^{2+} asymmetry are well developed, Ca^{2+} staining being particularly intense along the upper, concave region of the bend. Subcellular distribution of calcium in control and gravistimulated organs by electron-microscopic localization of antimonate precipitates, indicates that Ca^{2+} asymmetry results primarily from changes in the distribution of Ca^{2+} within the apoplastic compartment, as large accumulation of this cation occurs at the cuticle. It is not clear what type of motive force might be involved in the directional displacement of apoplastic Ca^{2+}, one possibility being the exchange of Ca^{2+} for H^+ along a proton gradient within the tissue. As auxin concentration gradients are observed to develop across the gravistimulated tissues (Shaw *et al.*, 1973), H^+ secretion, which is coupled to polar auxin transport, would likely follow a similar gradient.

Studies on epidermal cells of maize coleoptiles indicate that auxin triggers oscillations in cytosolic free calcium and cytosolic pH (Brummer *et al.*, 1985; Felle *et al.*, 1986; Felle, 1988). Using a scanning laser confocal microscope, a new visible wavelength Ca^{2+} probe-fluo-3 and the fluorescent pH indicator BCECF, Gehring *et al.* (1990) have recorded rapid light-induced increases in Ca^{2+}, and a lowering of pH, in cells on the shaded side of maize coleoptiles. In horizontally oriented coleoptiles, Ca^{2+} increases and pH decreases in the more rapidly elongating cells on the lower side. The changes detected are consistent with a role for auxin in geo- and phototropism. The changes in Ca^{2+} and pH during exposure to unilateral illumination and gravity indicate either altered sensitivity to auxin in specific regions of the coleoptile or a redistribution of endogenous auxin. The distribution of auxin-regulated RNAs has been found to change when vertically oriented soybean hypocotyls are placed horizontally (McClure and Guilfoyle, 1989). Within 20 min RNAs decrease on the upper side and increase on the lower side of the hypocotyls. The presence of the RNAs is highly correlated with cell elongation. The results of Gehring *et al.* (1990) suggest that increases in Ca^{2+} would precede these changes in RNAs and that Ca^{2+} may, therefore, be a secondary messenger involved in the accumulation of such auxin-regulated RNAs. Such a mechanism may also involve the rapid breakdown of polyphosphoinositides for the mobilisation of intracellular Ca^{2+} and calmodulin-activated or Ca^{2+} activated protein kinases.

MOSS PROTONEMA DIFFERENTIATION

In the moss protonema, a permanent polar IAA transport is necessary in order to maintain the caulonema state of differentiation. Any disturbance in the basipetal transport leads to dedifferentiation to chloronema (Knoop, 1984). There is a significant increase in the pH of the surrounding medium during IAA-induced caulonema differentiation in the suspension cultures of *Funaria hygrometrica* (Johri and D'Souza, 1990). In the liquid medium buffered at pH 5, the differentiation response due to IAA is delayed, but caulonema differentiates even in the control cultures after a lag of 6 ± 1 days. The delay in caulonema differentiation is restricted to the apical cells of the side branches only. There is no delay in the differentiation of primary caulonema in the buffered medium. Similar observations have been made in the protonema raised on media buffered at pH 6 or 7. The observed pattern of caulonema differentiation suggests that the transport of IAA occurs as passive diffusion of undissociated molecules. The stability of a caulonema filament depends on the continued presence of signal(s) emanating from the caulonema apical cell (Knoop, 1984). The inhibition of IAA transport and of caulonema differentiation by PCIB raise the possibility that one of the signals from the apical cells could be auxin. The results, therefore, imply that the signal (i.e. auxin) must be transported basipetally continuously for the differentiation of subapical cells into caulonema. A relationship between IAA transport and the stability of caulonema state has also been proposed earlier (Rose and Bopp, 1983). By reasoning along these lines, the problem of caulonema differentiation can be treated or reduced to a problem of creation of new apical cells *de novo*, either from chloronema apical cells or from side branch initials. Since the primary caulonema formation is not accompanied by an inhibition of chloronema, the formation of primary caulonema apical cells must be under the regulation of extremely low levels of endogenous auxin.

Involvement of calcium in protonema differentiation in *Funaria hygrometrica* has been investigated in the present work with reference to the maintenance of apical cell and caulonema differentiation (Kapoor and Bhatla, unpubl.). These parameters have been investigated by using Ca^{2+} chelator (EGTA) to create Ca^{2+} deficiency in the extracellular medium, by using extra and intracellular Ca^{2+} channel blockers and calmodulin inhibitor. It is clear from these investigations that EGTA and $LnCl_3$ (a competitive inhibitor of Ca^{2+} uptake) result in significant degeneration of the tip cells and also completely suppress caulonema differentiation. The effects of these agents in the wild strain and the auxin-sensitive mutants (87.13 and NAR-2) can not be reversed by auxin treatment. All the strains under investigation (except wild type) require a certain threshold level of Ca^{2+} in the medium, in order to exhibit caulonema differentiation. In wild type, however, some caulonema differentiation can be observed even in the absence of extracellular Ca^{2+}. The response becomes maximum and then gets stabilised at 1.26 mM or higher concentration of Ca^{2+}, in the mutants NAR-2 and 87.13 and 2.1 mM or higher in case of wild type. IAA stimulates the response further. The observations also indicate the essentially of Ca^{2+} at a certain minimum level to evoke caulonema formation by auxin. It is quite likely that the protonema of wild type has a much higher natural status of endogenous free Ca^{2+}, thus the observation that IAA can induce caulonema formation in minus calcium conditions in wild type. Results obtained by the use of verapamil further confirm the primary role of Ca^{2+} in the auxin-induced protonema differentiation. These observations indicate that extracellular Ca^{2+} must be transported through the Ca^{2+} channels to bring the endogenous Ca^{2+} to a physiologically meaningful level in order to trigger protonema differentiation. Verapamil is able to suppress caulonema induction by IAA in the mutants NAR-2 and 87.13 and it also nullifies auxin action in wild type, he degree of sensitivity of the various strains being in the order of NAR- 2>87.13> wild type.

Once within the cell, Ca^{2+} gets mobilised to evoke the morphogenetic response through a cascade of events. The involvement of intracellular Ca^{2+} and the possible mediation of the response through calmodulin have been confirmed by the use of TMB-8 (intracellular Ca^{2+} antagonist) and TFP (calmodulin inhibitor), respectively. TMB-8 suppresses spontaneous as well as auxin-induced differentiation, mutant NAR-2 being the most sensitive. In the mutant 87.13, TFP-suppressed protonema differentiation can be partially overcome by IAA. In the mutant NAR-2, auxin is unable to overcome the inhibitory action of TFP on caulonema differentiation and in wild strain of *Funaria*, the effect of TFP is negligible. These observations indicate that calmodulin interacts with auxin only in the further stimulation of an already induced process. It does not seem to be involved as a primary factor in the induction of caulonema.

Reversion of symptoms due to calcium deficiency (by $LnCl_3$ or EGTA) is possible to varying extents, depending on the strain and the treatment. In the wild type, upto 2 hours of incubation in $LnCl_3$ or EGTA can be reversed. In the mutants 87.13 and NAR-2, upto 4 hours of incubation in these treatments can be reversed. In the mutant 86.1, upto 4 hours of incubation in $LnCl_3$ or EGTA can be reversed. All the three mutants seem to be more resistant to calcium deficiency compared to the wild type. Order of Ca^{2+} deficiency resistance seems to be 86.1>NAR-2 and 87.13>w.t. Once the tip cell degeneration starts in the calcium deficient or minus calcium condition, then no regeneration of normal tip cells is observed even after 24, 48, 72 or 120 hours of reversion treatments. Protonema does exhibit growth due to the branching of subapical cells of the filaments. Even in minus calcium medium, protonema continues to grow as a result of abnormal divisions of the subapical cells. All strains, except mutant 87.13, exhibit an irreversible action due to verapamil, TMB-8 and TFP on protonema differentiation. Complete reversion of TMB-8 treatment is observed in the mutant 87.13.

Calcium ionophore (A-23187) has the potential to substitute for IAA in inducing cell differentiation in moss protonema. It is capable of triggering caulonema differentiation in both auxin-sensitive and auxin-insensitive mutants. This has been observed in the protonema cultures of *Funaria* in the absence of exogenous IAA. Induction of buds by the ionophore treatment, in otherwise non-inductive conditions (absence of cytokinin), proves mediation by calcium in cytokinin-induced morphogenetic response as well (as also observed earlier by Saunders and Hepler, 1982). The fact that calcium ionophore is able to nullify the effect of verapamil (an extracellular calcium channel blocker) to a large extent (depending on the strain), by promoting caulonema differentiation in cultures pretreated with verapamil, further confirms the involvement of extracellular calcium in protonema differentiation. Since the morphogenetic response due to verapamil is reversible by Ca^{2+} ionophore treatment even in the absence of IAA, it indicates that IAA triggers cell differentiation by opening the channels for extracellular calcium, thereby increasing the level of intracellular calcium.

Investigations on the changes in calcium fluxes indicates that: 1. IAA induces quick changes in calcium uptake by the protonema filaments, 2. The observed changes in the uptake potential are in congruence with the morphogenetic response evoked by exogenous IAA. 3. Induction of caulonema in the auxin-sensitive mutant (87.13) as well as further stimulation (by IAA) of spontaneously-induced caulonema in wild type, are accompanied with a quick increase in calcium uptake due to auxin treatment. These uptake changes have been observed to take place within 5 minutes of incubation in IAA-containing bathing solution and the observed changes remain unaltered even after 45 or 60 minutes of treatment. This indicates that the change in uptake potential for calcium ions in the moss protonema is a fast response due to IAA and it precedes much before the actual morphogenetic response can be detected. In other words, auxin-induced caulonema differentiation is accompanied with a quick uptake of more calcium ions from the surrounding medium. Thus, auxin response appears to be mediated through calcium.

CONCLUSIONS

Sufficient evidence has accumulated to convincingly prove the involvement of calcium in the action of auxins on cell elongation (Cunninghame and Hall, 1986, Reddy *et al.*, 1987). In some systems (e.g. maize root tips, sunflower hypocotyls), calcium and IAA interact in an antiport fashion i.e., calcium movement in one direction is tightly linked with IAA movement in the opposite direction (DeGuzmern and DelaFuente, 1984; DelaFuente, 1984, Lee *et al.*, 1984). Perturbing either of the two, affects the other and this can alter the response. Extracellular calcium stimulates IAA accumulation if the protoplasts of *Funaria hygrometrica* are prepared in a Ca^{2+} free medium. Furthermore, Ca^{2+} application lowers the pH of the isolated protoplasts, (Geier *et al.*, 1990). There is an increasing amount of evidence implicating the activation of channels in a number of calcium dependent events in plant cells, including blue light-dependent phototaxis, and a number of growth substance-regulated events involving auxins, cytokinins and gibberellic acid. Calcium channels play a major role in IAA-induced pulvinus growth, but are less essential to gravity-induced growth. Calcium channel blockers greatly inhibit IAA action, more so in the upright pulvini than in gravistimulated ones (Brock *et al.*, 1992). These blockers have little effect on gravistimulated pulvini in the absence of IAA. Varying reports are available with regard to the effect of IAA on calcium uptake. Indoleacetic acid increases Ca^{2+} uptake into vesicles isolated from the elongation zone but has no effect on uptake in the meristematic or maturing zones in soybean hypocotyls (Kubowicz *et al.*, 1982). It promotes calcium release from the membranes of soybean and mungbean hypocotyl and pea epicotyl (Buckhout *et al.*, 1980). IAA has no effect on calcium uptake in wheat protoplasts (Akerman *et al.*, 1983) and hypocotyl segments of *Cucurbita peop* L. (Astle, 1986). Felle (1988) suggested that both cytosolic pH and calcium are the messengers in early auxin action and IAA causes a decrease in cytosolic pH and an increase in cytosolic (Ca^{2+}). Gehring *et al.* (1990a, b) confirmed that IAA rapidly lowers the cytosolic pH and increases the levels of (Ca^{2+}) in the cytoplasm.

ACKNOWLEDGMENTS

Thanks are due to C.S.I.R., New Delhi for financial support.

REFERENCES

Andreae, W.A. and Van Ysselstein, M.W.H. 1956. Studies on 3-Indole Acetic Acid metabolism III. The uptake of 3-Indole acetic acid by pea epoicotyla and its conversion to 3-indole acetylaspartic acid, Plant Physiol 31: 235-240.

Astle, M.C. 1986. Effects of calmodulin antagonists on transmembrane auxin transport in Cucurbita pepo L. hypocotyl segments In: Molecular aspoects of calcium in Plant development Nato ASI Series, Life Sciences 433 pp.

Bennet-Clarke, T.A. 1955. Salt Accumulation and mode of action of auxins. A preliminary hypothesis In: The chemistry of mode of action of plant growth substances Butterworths London, 284 pp.

Brock, T.G., Burg, J., Ghoshen, N.S. and Kaufman, P.B. 1982. The role of calcium in growth induced by indole-3-acetic acid and gravity in the leaf-sheath pulvinum of oat (*Avena sativa*), Bot Gaz (Chicago) 141: 418-421.

Brummell, D.A. and Maclachlan, G.A. 1989. Calcium antagonist TMB-8 inhibits cell wall formation and growth in pea, J. Exp. Bot. 40: 559-565.

Brummer, B., Bertl, I., Potrykus, I., Felle, H. and Parish, R.W. 1985. Evidence that fusicoccin and indole-3-acetic acid induce cytosolic acidification of *Zea mays* cells, FEBS Lett. 189: 109-114.

Buckhout, T.J., Morren, S.J., Young, K.A. and Low, P.S. 1980. Response of isolated plant membranes to auxins: Calcium release, Bot Gaz (Chicago) 141: 418-421.

Burling, E. and Jackson, W.T. 1965. Changes in calcium levels in cell walls during elongation of Oat coleoptiles sections, Plant Physiol. 40: 138-141.

Cleland, R.E. 1960. Effect of auxin upon loss of calcium from cell walls, Plant Physiol. 35: 581-584.

Cleland, R.E. 1980. Auxin and H^+ excretion: the state of our knowledge In: Plant Growth Substances, Academic press, Inc, New York, 71 pp.

Cohen, D., Ginzburg, B.Z. and Heitner-Wirguin, C. 1958. Metal chelating properties of plant growth substances, Nature 181: 686-687.

Cohen, J.D. and Lilly, N. 1984. Change in ^{45}Calcium concentration following auxin treatment of protoplasts isolated from etiolated soybean hypocotyls, Plant Physiol. Suppl. 75: 109.

Cohen, J.D. and Nadler, K.D. 1976. Calcium requirement for indole acetic acid-induced acidification by *Avena* coleoptile, Plant Physiol, 57: 347-350.

cunninghame, M.E. and Hall, J.L. 1986. The effect of calcium antagonists and inhibitors of secretory process on auxin-induced elongation and fine structure of *Pisum sativum* stem segments, Protoplasma 133: 149-159.

De Guzmann, C.C. and Dela Fuente, R.K. 1984. Polar calcium flux in the polar secretion of indoleacetic acid, Plant Physiol. 76: 342-346.

Dela Fuente, R.K. 1984. Role of calcium in the polar secretion of indoleacetic acid, Plant Physiol. 76: 342-346.

Dela Fuente, R.K. and Leopold, A.C. 1973. A role for calcium in auxin transport Plant Physiol. 51: 845-847.

Felle, H. 1988. Auxins causes oscillations of cytosolic free calcium and pH in *Zea mays* coleoptiles, Planta, 174: 495-499.

Felle, H., Brummer, B., Bertl, A. and Parish, R.W. 1986. Indole-3-acetic acid and fusicocein cause cytosolic acidification of corn coleoptile cells, Proc. Natl. Acad. Sci. USA, 83: 8992-8995.

Gehring, C.A., Irving, H.R. and Parish, R.W. 1990a. Effects of auxina nd abscisic acid on cytosolic calcium pH in plant cells, Proc. Natl Acad. Sci USA 87: 9645.

Gehring, C.A., Williams, D.A., Cody, S.H., Parish, R.W. 1990b. Phototropism and geotropism in maize coleoptiles are spatially correlated with increases in cytosolic free calcium, Nature, 345: 528-530.

Geier, U., Werner, O. and Bopp, M. 1990. Indole-3-acetic acid uptake in isolated protoplasts of the moss *Funaria hygrometrica*, Physiol. Plant 80: 584-592.

Hasenstein, K.H. and Evans, M.L. 1986. Calcium dependence of rapid auxin action in maize roots. Plant Physiol. 81: 439-443.

Johri, M.M. and Souza, D. 1990. Auxin regulation of cell differentiation in moss protonema. In: Plant growth substances, Springer-Verlag Heidelberg, 407 pp.

Knoop, B. 1984. Development in bryophytes. In: Experimental biology of bryophytes. Academic Press, London, 143 pp.

Kubowica, B.D., Vanderhoef, L.N. and Hanson, J.B. 1982. ATP-dependent calcium transport in plasmalemma preparations from soybean hypocotyls, Plant Physiol., 69: 187-191.

Lee, J.S., Mulkey, T.J. and Evans, M.L. 1983a. Gravity induced polar transport of calcium across root tips of maize, Plant Physiol. 73: 874-876.

Lee, J.S., Mulkey, T.J. and Evans, M.L. 1983b. Reversible loss of gravitropic sensitivity in maize roots after tip application of calcium chelators, Science, 220: 1375-1376.

Lee, J.S., Mulkey, T.J. and Evans, M.L. 1984. Inhibition of polar calcium movement and gravitropism in roots treated with auxin transport inhibitors, Planta, 160: 536-543.

McClure, B.A. and Guilfoyle, G. 1989. Rapid redistribution of auxin-regulated RNAs during gravitropism, Science, 243: 91-93.

Morre, D.J. and Bracker, C.E. 1976. Ultrastructural alteration of plant plasma membranes induced by auxin and calcium ions, Plant Physiol. 58: 544-547.

Morre, D.J., Morre, J.T. and Varnold, R.L. 1984. Phosphorylation of membrane located proteins of soybean *in vitro* response to auxin, Plant Physiol., 75: 265-268.

Raghothama, K.G., Mizrahi, Y. and Poovaiah, B.W. 1985. Effect of calmoculin antagonists on auxin induced elongation, Plant Physiol. 79: 28-33.

Rayle, D.L. and Cleland, R.E. 1979. Control of plant cell enlargement by hydrogen ions, Current topics in developmental biology, 11: 187-214.

Reddy, A.S.N., McFadden, J.J., Friedmann, M. and Poovaiah, B.W. 1987. Signal transduction in plants: Evidence for the involvement of calcium and turnover of inositol phospholipids, Biochem. Biophys. Res. Commun. 149: 334-339.

Reddy, A.S.N., Koshiba, T., Theologis, A.a nd Poovaiah, B.W. 1988. The effect of calcium antagonists on auxin-induced elongation and on the expression of two auxin-regulated genes in pea epicotyls, Plant Cell Physiol. 29: 1165-1170.

Rose, S. and Bopp, M. 1983. Uptake and polar transport of indoleacetic acid in moss rhizoids, Physiol. Plant. 58: 57-61.

Shaw, S., Gardner, G. and Wilkins, M.B. 1973. The lateral transport of IAA in intact coleoptiles of *Avena sativa*, Planta 115: 97-111.

Slocum, R.D. and Roux, S.J. 1983. Cellular and subcellular localization of calcium in gravistimulated oat coleoptiles and its possible significance in the establishment of tropic curvatures, Planta 157: 481-492.

Stohly, E.A. and Binson, N.R. 1970. Calcium levels of Golden delicious apples sprayed with 2,3,5-Tri-iodobenzoic acid, J. Amer. Soc. Hort. Sci. 95: 726-727.

Varnold, R.L. and Morre, D.J. 1985. Phosphorylation of membrane located proteins of soybean hypocotyls. Inhibition by calcium in the presence of 2,4-dichlorophenoxyacetic acid, Bot. Gaz. 146: 315-319.

Velathombi, K. and Poovaiah, B.W. 1984. Calcium promoted protein phosphorylation in plants, Science 223: 167-169.

Role of Glyoxalase I in Cell Proliferation in Higher Plants

Renu Deswal and Sudhir K. Sopory

Several physiological, biochemical and genetical evidences in microbial and animal systems suggest involvement of the glyoxalase enzyme system in detoxification (Aronsson *et al.*, 1977), cell proliferation (Murata *et al.*, 1988) tumours (Jerzykowski *et al.*, 1978), diabetes (Thornalley *et al.*, 1988), vesicle mobilization (Gillespie, 1979) and microtubule assembly (Gillespie, 1975). In plants, not much attention has been focused on the enzyme system as very few laboratories world wide are engaged in plant glyoxalase research and therefore the available information is very scant.

We first reported the presence of glyoxalase I enzyme in plants and purified it from *Brassica juncea*. Norton *et al.* (1990) have also purified glyoxalase I and II from *Aloe vera* and Paulus *et al.* (1993) purified it from soybean. Glyoxalase II has also been purified from maize (Norton *et al.*, 1989). Earlier, the enzyme activity was reported from Douglas Fir both from callus and needle of the gymnosperm (Smits and Johnson, 1981), however the enzymes were not purified. In this paper we will briefly review the work done in our laboratory to understand the properties and role of glyoxalase I in plants.

WHAT IS GLYOXALASE ENZYME SYSTEM?

Glyoxalase enzyme system comprises of two enzymes glyoxalase I and II (Racker, 1957). The enzyme system converts cytotoxic and cytostatic ketoaldehyde methylglyoxal to nontoxic hydroxy acid thereby relieving the system of the toxic and growth inhibitory effects of this compound.

A non-enzymic adduct of methylglyoxal and glutathione is used as the substrate by glyoxalase I, the first enzyme of the system and is converted to S-D-lactoyl glutathione. Glyoxalase II the second enzyme of the system utilizes the product of the first reaction as its substrate and converts it to D-lactic acid. Glutathione is regenerated to be used again (Fig. 1).

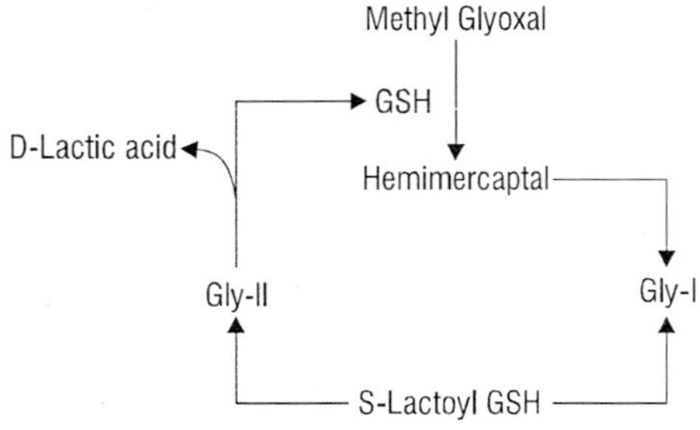

FIG. 1: The glyoxalase system

EXISTENCE OF GLYOXALASE I IN PLANTS

The enzyme glyoxalase I was detected in a number of dicots and monocots using modified assay protocols for the plant system. The activity of the enzyme was checked in a number of plants. The level of enzyme was found to be higher in dicots than in monocots (Table 1), however the exception was *Sorghum* which showed very high activity. Among a number of *Brassica* species screened for the enzyme activity *Brassica juncea* showed reasonably good enzyme levels, in addition, light grown seedlings always showed higher activity in comparison to dark grown seedlings of the same developmental stage. To select for a glyoxalase enriched part of the seedling the activity was checked in cotyledon, root and shoot. Cotyledon showed 3-4 folds more activity than root and shoot. Therefore the enzyme was purified from seven days old light grown seedlings having cotyledons and little hypocotyl.

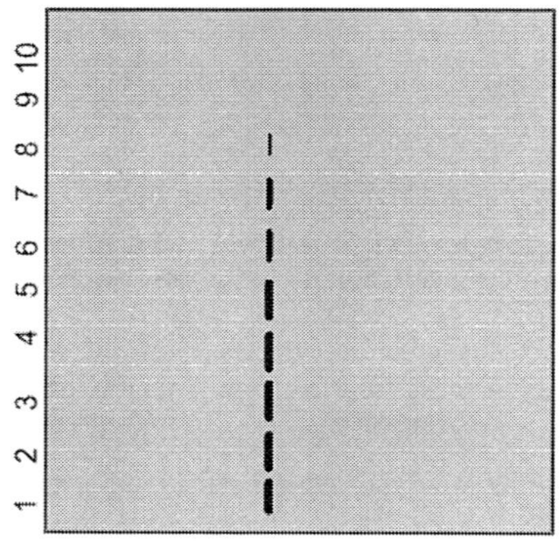

FIG. 2: Test of homogeneity of purified glyxalase I fraction. Different amounts of glyoxalase I protein were loaded in the gel and it was stained with CBB. Well no. 1-10 contain 75, 60, 40, 35, 20, 8, 4, 2, 0.8 and 0.4 μg glyoxalase I in that order.

The purification protocol basically consisted of ammonium sulphate precipitation of 15,000 rpm supernatant followed by affinity and gel filteration chromatography using S-hexyl GSH agarose or sepharose and sephadex G-50 respectively. The gel filtration column was used to remove the inhibitor S- hexyl glutathione and glutathione used in the elution buffer as they interfere in the protein estimation by the Bradford's method.

The homogeneity of purified protein was checked by SDS PAGE and densitometric scanning of the gel. Different amounts of the purified protein were loaded on a 10% SDS gel to check for the contaminants (Fig. 2). The purification of the enzyme from *Brassica* conclusively showed existence of glyoxalase I in plants (Deswal and Sopory, 1991).

TABLE 1: A comparison of glyoxalase I activity in different plants

Name of Plant	Specific Activity of Glyoxalase I (μ/mg protein)
Monocots	
Sorghum	4.9
Wheat	0.6
Maize	0.6
Barley	0.4
Dicots	
Cicer	0.84
Cucumber	1.12
Cajanus cajan	1.00
B. nigra	0.71
B. campestris	0.70
B. carinata	0.80
B. juncea	1.40
B. oleracea	1.10

CORRELATION BETWEEN GLYOXALASE I AND CELL PROLIFERATION AND DIFFERENTIATION

To check if glyoxalase I was involved with cell proliferation, different systems were used. It was observed that the activity of the enzyme glyoxalase I in pea roots was always higher in the dividing zone i.e. in the root-tip (100% more) in comparison to the elongation zones (Ramaswamy *et al.,* 1983). Similarly in highly proliferative.

Brassica calli the levels of the enzyme was 50% more in comparison to differentiating calli (Seth *et al.,* 1987). It was also observed that cell division inhibitors like colchicine (200 mg/ml) and vinblastine inhibited cell growth as well as glyoxalase I. While on the other hand inhibitors of the enzyme glyoxalase I like squaric acid (4 mM) showed a negative effect on cell growth in pea (Ramaswamy *et al.,* 1983b). Suspension culture of *Amaranthus* calli showed inhibition of cell division with concomitant fall in the enzyme levels on treatment with methylglyoxal.

Effect of several plant growth promoting factors like light, hormone and PMA (Phorbol Myristate Acetate, a tumour promoter) was checked on glyoxalase I activity. In *Amaranthus* white light showed 175% more enzyme in light grown calli over the dark controls. Five minutes of red light irradiation of the same duration stimulated the activity by 278% (Chakravarty and Sopory, 1990). In *Brassica* also light grown seedlings always showed

20%-30% more activity than the dark grown seedlings. In soybean cell suspension an auxin dependent culture was obtained which when starved, resulted in the arrest of cell division and decrease in glyoxalase I activity. However, on transferring these cells back to hormone containing medium, an increase in glyoxalase I activity preceded an increase in cell division (Paulus *et al.*, 1993). In *Brassica oleracea* cultures differentiation was observed on basal medium and if hormones were added callusing occurred. Addition of spermidine increased the fresh weight as well as glyoxalase I activity and use of spermidine inhibitor MGBG lead to differentiation and a decrease in glyoxalase I activity (Sethi *et al.*, 1987). PMA, the tumour promoting agent stimulated the enzyme activity in dark as well in light grown seedlings in *Brassica* (Fig. 3).

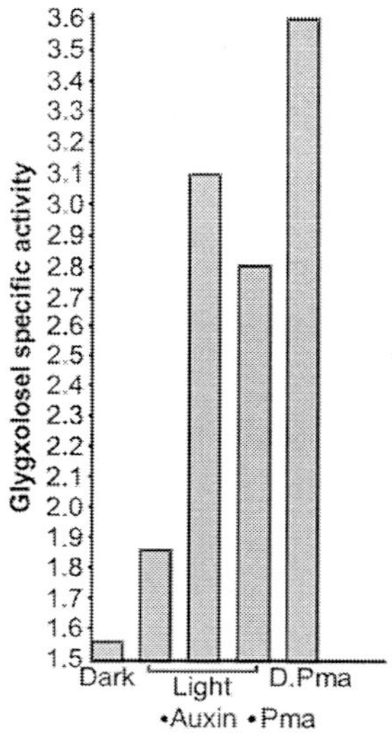

FIG. 3: bar diagram showing the increase in glyoxalase I activity by light, hormone (IAA) and PMA. Glyoxalase I activity was measured in 7 days old dark grown and light grown seedlings. The shoots of light and dark grown seedlings were incubated with 100 ng PMA in 20 ml distilled water for 10 min and activity was checked. For auxin stimulation 10^{-5}M auxin was supplied to intact seedlings growing in light for 7 days and activity was checked on day 7.

We have obtained monospecific polyclonal antibodies for *Brassica* glyoxalase I. *In situ* localization of glyoxalase I protein by tissue printing using antibodies correlated with activity data. Higher level of glyoxalase I protein as observed by tissue printing and westerns was seen in cotyledons. It has been shown that cotyledons have high regeneration capacity if used as an explant in tissue culture (Sharma *et al.*, 1991).

CONCLUSIONS

Our biochemical and immunological studies clearly indicate that glyoxalase I could be used as a marker of cell growth. We observed that generally the poorly regenerating species contained less enzyme. Moreover it has been shown that cotyledons have good regeneration capacity and we observed a high level of the enzyme in this particular part of the seedling. Whether the high level of glyoxalase I makes cotyledons a good explant or glyoxalase has some yet another undiscovered role, needs to be understood. Fast modulation of the enzyme activity by light and PMA indicate post-transational modification of enzyme as observed in case of sexual differentiation of yeast (Inoue *et al.*, 1991a). More work is needed in this direction.

ACKNOWLEDGMENTS

R.D. thanks DST for the financial assistance in form of research associateship from the DST project.

REFERENCES

Aronsson, A.C. and Mannervik, B. 1977. Characterization of glyoxalase I purified from pig erthrocytes by affinity chromatography. Biochem. J. 165: 503-509.

Chakravarty, T.N. and Sopory, S.K. 1990. Light stimulated cell proliferation and glyoxalase I activity in callus culture of *Amaranthus paniculatus*. In Nijikamp, H.J.J., Van Der Plas L.H.W. and Aartrijk, J.V. (Eds.). Progress in Plant Cellular and Mol. Biol. Kluwer Acad. Publ., Derdrecht, 379-384.

Deswal, R.a nd Sopory, S.K. 1991. Purification and partial characterization of glyoxalase I from a higher plant *Brassica juncea*. FEBS Lett. 282: 277-280.

Gillespie, E. 1975. Cell free microtubule assembly: evidence for control by glyoxalase. Fed. Proc. 34: 541.

gillespie, E. 1979. Effects of S-lactoylglutathione and inhibitors of glyoxalase I on histamine release from human leucocytes. Nature 277: 135-137.

Inoue, Y., Choi, B.Y., Murata, K. and Kimura, A. 1990a. Sexual response of *Saccharomyces cerevisiae* in phosphorylation of yeast glyoxalase I by cell extract of mating factor treated cells. J. Biochem. (Tokyo) 108: 4-6.

Jerzykowski, T., Winter, R., Matuszewski, W.a nd Piskorska, D. 1978. A re-evaluation of studies on the distribution of glyoxalases in animal and tumour tissues. Int. J. Biochem. 9: 853-860.

Murata, K., Saikusa, T., Inoue, Y. and Kimura, A. 1988. Molecular cloning of gene enhancing Yeast glyoxalase I activity and its effect on yeast cell size. J. Ferment. Technol. 66: 495-500.

Norton, S.J., Talesa, V., Yuan, W.J. and Principato, G.B. 1990. Glyoxalase I and II from *Aloe vera*. Purification, characterization and comparison with animal glyoxalase, Biochem. Int. 22: 411-418.

Paulus, C., Kollner, B. and Jacobson, H.J. 1993. Physiological and biochemical characterization of glyoxalase I, a general marker for cell proliferation from soybean cell suspension. Planta 189: 561-566.

Racker, E. 1951. Mechanism of action of glyoxalases. J. Biol. Chem. 190: 685-696.

Ramaswamy, O., Pal, S., Mukherjee, S.G. and Sopory, S.K. 1983a. Presence of glyoxalase I in pea. Biochem. Int. 7: 307-318.

Ramaswamy, O., Pal, S., Mukherjee, S.G. and Sopory, S.K. 1984. Correlation of glyoxalase I activity with cell proliferation in *Datura* callus culture. Plant cell Rep. 3: 121-124.

Sethi, U., Basu, A. and Mukherjee, S.G. 1988. Control of cell proliferation and differentiation by regulating polyamine biosynthesis in cultures of *Brassica* and its correlation with glyoxalase I activity. Plant Sci. 56: 167-175.

Sharma, K.K., Bhojwani, S.S. and Thorpe, T.A. 1991. The role of cotyledonary tissue in the differentiation of shoots and roots from cotyledon explants of *Brassica juncea* (L.) czern. Pl. Cell, Tissue and Organ Cult. 24: 55-59.

Smits, M.M. and Johnson, M.A. 1981. Methylglyoxal: Enzyme distibution relative to its presence in Douglas Fir needles and absence in Douglas fir needle callus. Arch. Biochem. Biophys. 208: 431-439.

Thornalley, P.J. 1988. Modification of the glyoxalase system in human red blood cells by glucose in *in vitro*. Biochem. J. 254: 751-755.

Peroxidases in Tissues Cultured *In Vitro*

P. L. Swarnkar and Vibha Tyagi

INTRODUCTION

The initiation of organised development in tissues cultured *in vitro* is associated with the variations in the activities of various enzymes and the levels of cellular metabolites. The genetic information contained in the DNA of the nucleus and in some of the organelles control the structure and function of plants. The genetic information is expressed in the form of enzymes and structural proteins, which vary in the level of their activities spatio-temporally Dougall (1965), using labelled tyrosine in tobacco callus and obtained evidence that protein synthesis occurred during shoot formation. Enzymes may exist in multiple molecular forms also. Markert and Moller (1959), coined the term isozymes for these multiple forms of an enzyme having identical substrate specificity within the same organism. Commission on Biological Nomenclature of IUPAC-IUB (1977) defined isozymes as multiple molecular forms of an enzyme occurring within a single species, as a result of the presence of more than one structural gene. Analysis of various cellular metabolites, activity of various enzymes and their isozymic profiles during organogenesis *in vitro* provide a reasonable and promising approach towards an understanding of the biochemical basis of the developmental pathway. Arnison and Boll (1971) used isozyme patterns as tools to investigate specific traits in cell cultures of bush bean (*Phaseolus vulgaris* cv. contender) and have shown that the study of isozymes was potentially a highly useful biochemical index by which to judge cell and tissue individually. Earlier Mako and Novacky (1969) and Novacky and Hampton (1968) showed that changes in isozymes took place during morphogenesis. Similarly Heinz and Mee (1971) and Noh and Minocha (1990) have reported isozyme variation in regenerated plants from callus cultures.

Involvement of peroxidase in a variety of processes in cell differentiation (Andreev and Shaw, 1965; Shannon, 1968; Srivastava and van Huystee, 1977; Thorpe *et al.*, 1978 and Thorpe, 1980) has attracted the attention of many workers. Peroxidase isozymes are organs/tissue specific (Asins *et al.*, 1982) so different organs from the same plant may show different peroxidase patterns (Hamill and Brewbaker, 1969 and Evans and Alldridge, 1965). Analyses of peroxidase isozymes have been used to study tissue specificity (Bassiri and Carlson, 1979), developmental regulation (Thorpe *et al.*, 1978), repair of damaged cell walls (Birecka and Miller, 1974; Lagrimini, 1991) and in disease resistance (Nessel and Mader, 1977). This enzyme uses hydrogen peroxide to oxidize a wide range of organic and inorganic hydrogen donors such as phenolics, cytochrome C, nitrites, leucodyes, ascorbic acid, indoles, amines, and iodide ions (Saunders *et al.*, 1964). Peroxidase (EC: 1.11, 1.7) is a glycoprotein containing ferri protoporphyrin IX as a prosthetic group (Conroy *et al.*, 1982). They catalyze reactions for lignification (Fry, 1986; Harkin and Obst, 1973; Lagrimini *et al.*, 1987) and is also important in hormone generation and detoxification of H_2O_2 (Conroy *et al.*, 1982), suberization (Espelie *et al.*, 1986), oxidation and polymerization of soluble phenolics (Srivastava and van Huystee, 1977), formation of hydrogen peroxide (Mader *et al.*, 1980), chlorophyll degradation and senescence (Yamauchi and Watada, 1991), IAA metabolism (Coumans-Gilles *et al.*, 1982; Hinnman and Lang, 1965) hormone regulation (Albeles *et al.*, 1989) and cell elongation (McAdam *et al.*, 1992; Goldberg *et al.*, 1986). Thus, because of their ubiquity, multiplicity and ease of detection isoperoxidases have been studied more than any other plant isozyme groups.

METHODOLOGY

To assay and visualise peroxidase enzyme various techniques have been used. Peroxidases are enzymes whose primary function is to oxidize a variety of hydrogen donors at the expense of hydrogen peroxide. Utilizing this character all procedures that are used to assay peroxidase enzyme usually have different substrates and H_2O_2 in appropriate buffer. Siegel and Galston (1967) have used 5 mM guaiacol as the most versatile substrate. Peroxidase activity towards guaiacol was quantitated by following the rate of conversion of this substrate to tetraguaiacol. Certain peroxidases were resolved by using other substrates such as 2,6-dimethoxyphenol. Pyrogallol (Alvarez and King, 1969) and O-dianisidine (Kay and Basile, 1987) have also been used as substrates. Gove and Hoyle (1975) have described that peroxidase isozymes could be stained with benzidine, scopoletin and guaiacol. The reason for differential staining of peroxidase activity may be due to differing degrees of sensitivity between hydrogen donors and certain isozymes. Separation and visualization of peroxidase isozymes involves four major steps.

(1) Preparation of gels, (2) application of sample, (3) electrophoresis and (4) visualization.

Preparation of gels

Generally polyacrylamide gel electrophoresis (PAGE) (Davis, 1964) and vertical slab gel apparatus as described by Studier (1973) is used to separate peroxidase isozymes.

REAGENTS FOR PAGE

Running (Small pore) gel

Stock A	30 g acrylamide and 0.8 g N,N'-methylene bisacrylamide in 100 ml water.
Stock B	36.6, tris, 0.23 ml N, N, N,'N'- tetramethyl ethylene diamine (TEMED) and 48 ml 1N HCl in 100 ml distilled water (pH 8.9)
Stock C	4 mg riboflavin in 100 ml distilled water
Stock D	140 mg ammonium persulfate in 10 ml distilled water. (This stock should be prepared fresh).

Stacking (Large pore) gel

Stock E	**10g acrylamide and 2.5g, N,N'-methylene bisacrylamide in 100ml water.**
Stock F	5.98g tris, 0.46ml TEMED and 48 ml 1N HCl to 100 ml solution, pH 6.7
Stock G	560 mg ammonium per sulfate in 100 ml water (freshly prepared)
Stock H	34.2 g sucrose in 100 ml water.
	All stocks are prepared in double distilled water and stored at 4°C in brown glass bottles.

Electrode buffer is prepared by adding 6.0 of tris and 28.8 glycine in 1.0 litre of distilled water and diluted ten times before use. The gel is prepared by mixing the stocks as per requirement (Table 1 and 2).

TABLE 1: Ratio of stock solutions for running gel

Stock solutions	Percent Gel		
	10%	7.5%	5%
A	9	2	4.5
B	3	1	3.0
C	6	2	7.5
D	6	2	6.0
Water	3	1	6.0

TABLE 2: Ratio of stock solutions for stacking gel

Stock solution	Percent Gel	
	2.5%	5%
E	1	1
F	0.5	0.25
G	0.5	0.25
H	2.0	0.5

The mixture of running gel is degassed immediately and poured between the two glass plates which are separated by 1.5mm spacer and sealed across the bottom with 1 per cent agar prepared in appropriately diluted stock B (x8), the gel surface is over layered with isobutanol or water and left under fluorescent/sunlight for 30-60 minutes. After the polymerization, isobutanol/water is removed and the stocks E,F,G and H are mixed, degassed and poured over running gel. A comb is also inserted between the plates. The gel is again left for polymerisation for 30-60 minutes (preferably overnight). The gel plates are then fixed onto electrophoretic apparatus and the compartments are filled with electrode buffer. The comb is gradually removed by sliding it vertically upwards and sample is loaded in wells.

Application of Sample

Sample is prepared by mixing equal volumes of enzyme extract and 1.0 M sucrose solution and a few drops of 0.01 per cent bromophenol blue (BPB). Samples may then be loaded into the gel slots formed in the gel using microsyringe.

Electrophoresis

The gels are subjected to an electric current of 25 mA in a refrigerated cabinet, until the tracking dye (BPB) reaches almost to the lower end of the gel. The gel is then freed from glass plates and transferred to an appropriate staining solution for visualization of bands.

Visualization of Isoperoxidases

The gels are immersed in a freshly prepared solution containing O-dianisdine in 95% ethanol (1 g/l), diluted 1:1 (v/v) with acetate buffer (pH 4.5) (0.88 M sodium acetate plus 0.62 N acetic acid). After 30 minutes the above solution is replaced with 0.3% hydrogen peroxide solution until the peroxidase bands are visible. Other staining techniques comprise of guaiacol- H_2O_2 (Siegel and Galston, 1967), aminoethyl carbazole and orthodianisidine (Siegel and Galston, 1967) and diaminobenzene tetrahydrochloride with H_2O_2 (Benvenuto *et al.*, 1983).

PRESENT STATUS

It is still unresolved whether cell cultures, derived from different parts of a parent plant or seedling show persistent differences in properties *in vitro*. Variation in isozyme pattern has been reported during growth of the callus cultures of root, hypocotyl and cotyledon derived from seedlings of *Phaseolus vulgaris* cv. Contender (Arnison and Boll, 1974). Appearance of new isoperoxidases in cultures of stem segments of *Populus nigra* showing rooting confirm the role of peroxidase in rooting (Bhattacharya *et al.*, 1975). Chandra *et al.* (1971) also reported the development of some isoperoxidases in parts producing roots and showed that isozymes did not develop in cultures treated with streptomycin which inhibits the production of roots. The *in vitro* derived roots from explants of tobacco transformed by *Agrobacterium rhizogenes* show a distinctive isozyme pattern of peroxidase (Benvenuto *et al.*, 1983). The transformed explant showed the presence of isoperoxidase groups GII and GIV in addition to the GI and GIII present in normal explant. The increase in overall activity and changes in isozyme pattern have been observed during rhizogenesis in different plant species including tobacco (Simola, 1973; Quoirin *et al.*, 1974; van Hoff and Gasper, 1976; Mader and Bopp, 1976; Nanda *et al.*, 1973 and Gurumurti and Nanda, 1974). Appearance of transient isoperoxidases and persistent differences during part of the growth cycle has been reported by several investigators (Khavkin and Sukhorzhevskaia, 1979; Sharma, 1992; Swarnkar *et al.*, 1986a, b, 1987 a, c; Balasimha and Subramonian, 1983). Rooting capacities on different rooting media of *Sequoiadendron giganteum* have been compared taking peroxidase activity as marker (Berthon *et al.*, 1987). Increased peroxidase activity during callus initiation was also observed in the stem segments of *Solanum surattense* (Swarnkar *et al.*, 1986b) and in cultured pith of *Pelargonium* (Lavee and Galston, 1968). Initiation of callus from the stem segments of *Sinapis alba* had additional peroxidase bands (Bajaj *et al.*, 1973). Shoot bud differentiation in tobacco pith cultures was characterized by the appearance of three anodic and two cathodic isoperoxidases (Rawal and Mehta, 1978), Androgenesis in *Datura innoxia* observed appearance of four new isoperoxidases (Swarnkar *et al.*, 1987b). Isoperoxidases have been used as a biochemical marker in leaf and root morphogenesis in Barley (Coppens and Dewitte, 1990) and in somatic embryogenesis in carrot (Joresbo *et al.*, 1989). In the studies on *Nicotiana tabacum* L. cv. Wisc. 38, it has been observed that various isoperoxidases were associated with different developmental events (Kay and Basile, 1987). Peroxidase activity was found to be lowest at the stage of maximum differentiation in the development of Vanda seedlings (Alvarez and King, 1969). Isoperoxidases can also be used as biochemical markers in differentiating embryogenic and non-embryogenic calli (Rao *et al.*, 1990) and in normal and habituated tissues (Guar *et al.*, 1993).

In tissue cultures of tobacco the GI group of peroxidase isozyme was found to be involved in H_2O_2 formation which accelerates the polymerization of cinamic alcohols- the precursors of lignin (Mader *et al.*, 1980). Isoperoxidases have also been found to be indicative of heavy metal toxicity in plants (Nashikkar and Chakrabarti, 1994). Some investigators have found increase in peroxidase activity as well as qualitative changes in the isozyme pattern after infection of a hypersensitive plant with TMV (Simons and Ross, 1970 and Weststeijn, 1976). Isoperoxidases have often been used in studies on molecular aspects of host pathogen interaction. Increase in two extracellular peroxidase (P 7.7 and P 8.5) was observed following inoculation of barley with the powdery mildew (Kerby and Somerville, 1992). A cationic peroxidase (Pl 8.6; MW 43KD) was also observed in rice during an incompatible interaction with *Xanthomonas oryzae* pv *oryzae* (Reimers *et al.*, 1992). The anionic peroxidases of tomato are encoded by two genes tap1 and tap2 (Roberts and Kolattukudy, 1989). Absence of expression of these genes in transgenic tomato plant having chimeric antisense transcript corresponding to 5 region of tap 1 and tap 2 prevented would- ABA- and pathogen- induced synthesis of anionic peroxidase (Sherf *et al.*, 1993). These results have given a better understanding of functional ontogeny of isoperoxidases in plant growth and development. In tobacco callus tissue growth hormones like cytokinin, auxin and gibberellin were shown to have effects on peroxidase activity (Lee, 1972). In the studies on *Spirodela polyrrhiza* L. abscisic acid was seen to induce a novel peroxidase but that too was antagonized by cytokinin (Katerina and Smart, 1994). Peroxidase activity has been shown to be altered by light (Sharma *et al.*, 1976), kinetin and light (Sharma and Biswal, 1976) and calcium (Sticher *et al.*, 1982). Thus isoperoxidases, being one of the major metabolic catalysts, show greater diversity in their forms and functions as compared to other enzyme groups.

Isozyme patterns of plant organs and tissues may be a direct reflection of particular genes that are operating at a given stage of growth and differentiation. Since peroxidase is one of the enzymes among those for which loci are the most represented on chromosomes (Scandalios, 1969, 1974), it was compared in parental species of *Nicotiana sylvestris* and *N. tomentosiformis* and their interspecific hybrid (Trinh *et al.*, 1981). The hybrids show an additive pattern of parental peroxidases (Sheen, 1972; Yen and Sadanga, 1977). This apparently demonstrates the Mendelian inheritance of peroxidase isozymes as suggested by Hoess *et al.* (1974). Another advantage of the use of isozymes as marker systems for parasexually produced hybrids is that this technique allows one to estimate the relative contribution of each parental genome to the hybrid pattern. In triploid maize endosperm for example, where the maternal parent contributes twice the genetic information of the paternal parent, the gene dosage effect is readily apparent on the zymogram pattern. They are thus helpful in several plant breeding programmes where variants with selective characters are of prime importance. Somatic hybrids are detected by the help of isozymes, since isozymes are direct gene products and are less susceptible to phenocopy responses and modification by environmental variables.

FUTURE PROSPECTS

Isozyme analysis has come up as genetic and biochemical tool in the study of plant physiology. Since peroxidase isozymes are broadly represented among higher plant species, and are easily detectable, they can be used as powerful markers to study the developmental changes occurring during morphogenesis. By studying the isoperoxidases one can elucidate various processes at genetic level and thereby desirable traits can be introduced by regulating the control mechanisms. In many physiological processes of the cell like cell wall formation, lignification, H_2O_2 formation, detoxification, wound-healing, disease-resistance etc., peroxidase enzyme plays a key role, therefore the technique of isoperoxidase analysis can be a great help in manipulating these processes. Rooting remains always one of the critical steps of the micropropagation technique (Jordan *et al.*, 1980; Welander, 1981 and Zimmermann and Broome, 1981). Variation in peroxidase activity can be used as a marker of inductive and initiative phases of root formation. Peroxidase isozymes find their applications in monitoring the stability of cells in culture, detection of hybrid cells, assessment of possible developmental stages in cultured cells, and studies of gene regulation. Isozyme variations are not related to growth conditions (Noh and Minocha, 1990) therefore for selection of useful traits from somaclonal variants obtained from cell and tissue culture, isozymes provide a useful marker. Thus, characterization of the cell cultures and the study of transmissible characters must be directed towards biochemical and physiological properties rather than morphological characters.

CONCLUSION

Plant peroxidase has multiple molecular forms which vary in contents with the species, the tissue, the stage of growth and the application of growth substances (Ahannon, 1968). They play a significant role in the regulation of cell growth and differentiation (Galston and Dalberg, 1954). It has often served as a parameter of metabolic activity during growth alterations. They show greater diversity in their forms and functions as compared to other enzyme groups. The results on peroxidase pattern in different plant species are difficult to compare for several reasons. Several of the extraction procedures hitherto used are known to change mobilities of isoperoxidases on gel (Scandalios, 1974; van Huystee and Cairns, 1980). The common procedures for isozyme separation, visualization and characterization included chromatography, electrophoresis, enzyme kinetics, precipitation etc. Even the most commonly used method electrophoresis is also a highly versatile technique. Different types of electrophoresis is also a highly versatile technique. Different types of electrophoretic methods like zone electrophoresis, disc electrophoresis, i.e.f. and each with different support media (paper/ starch/ acrylamide/ agarose) have been employed to separate isoperoxidases. Furthermore, the report that different substrate (guaiacol/benzidine/scopoletin) differ in the degree of their sensitivity to isoperoxidases (Gove and Hoyle, 1975) has further aggravated the situation. Different extraction procedures also have great bearing on isozyme analysis in plant tissues. Most of the workers have isolated and characterized the soluble isoperoxidases (Arnison and Boll, 1974; Bajaj *et al.*, 1973; Lee, 1972 and Swarnkar *et al.*, 1985, 1986a,b,c). Some workers have used soluble as well as ionically bound isoperoxidases (Adatthody and Racusen, 1967; Kevers *et al.*, 1981 and Mader *et al.*, 1980) while a few have used three fractions including the third covalently bound fraction separated by hydrolysing tissues overnight with cellulase and pectinase (Kay and basile, 1987). Therefore, there is a need to derive consensus over the methodology and substrates used for identifying the peroxidase isozymes in plants. If necessary, the isozymes be established by combination of as many methods as possible.

Looking at the diversity of forms and functions exhibited by isoperoxidases, these enzymes are likely to emerge as a powerful developmental probe for plant physiologists, biochemists and molecular biologists.

REFERENCES

Abeles, F.B., Hershberger, W.L. and Dunn, L.J. 1989. Hormonal regulation and intracellular localization of a 33-Kd of cationic peroxidase in excised cucumber cotyledons. Plant Physiol. 89: 664-668.

Adatthody, K.K. and Racusen, D. 1967. On the extraction of peroxidase isozymes from bean leaves. Can. J. Bot. 45: 2237-2242.

Alvarez, M.R. and King, D.D. 1969. Peroxidase localization, activity and isozyme patterns in the developing seedlings of *Vanda* (Orchidaceae). Amer. J. Bot. 56: 180-186.

Andreev, L.N. and Shaw, M. 1965. A note on the effect of rust infection on peroxidase isozymes in flax. Can. J. Bot. 43: 1479-1483.

Arnison, P.G. and Boll, W.G. 1974. Isozymes in cell cultures of bush bean (*Phaseolus vulgaris* cv. Contender).: Isozymatic changes during the callus culture cycle and differences between stock cultures. Can. J. Bot. 52: 2621-2629.

Asins, M.J., Benito, C. and Perez de la Vega, M. 1982. Changes and chromosomal location of peroxidase isozymes during hexaploid wheat kernel maturation. Z. Pflanzenzuchtg. 89: 121-129.

Bajaj, Y.P.S., Bopp, M. and Bajaj, S. 1973. Patterns of peroxidases and differentiation in *Sinapis alba*. Phytomorphology 23: 43-52.

Balasimha, D. and Subramonian, N. 1983. Role of phenolics in auxin induced rhizogenesis and isoperoxidases in Cacao (*Theobroma cacao* L.) stem cuttings. Indian J. Exp. Biol 21: 65-68.

Bassiri, A. and Carlson, A.S. 1979. Isozyme pattern in tobacco plant parts and their derived calli. Crop Sci. 19: 909-914.

Benvenuto, E., Ancora, G., Spano, L. and Constantino, P. 1983. Morphogenesis and isoperoxidase characterization in tobacco "Hairy root" regenerants. Z. Pflanzenphysiol. 110: 239-245.

Berthon, J.Y., Boyer, N. and Gasper, Th. 1987. Sequential rooting media and rooting capacity of *Sequoiadendron giganteum in vitro*. Peroxidase activity as marker. Plant Cell Reports, 6: 341-344.

Bhattacharya, N.C., Kaur, N.P. and Nanda, K.K. 1975. Transients in isoperoxidases during rooting of etiolated stem segments of *Populus nigra*. Biochem. Physiol. Pflanzen. 167: 159-164.

Birecka, H.A. and Miller, A. 1974. Cell wall and protoplast isoperoxidases in relation to injury, indole acetic acid and ethylene effects. Plant Physiol. 53: 569-574.

Chandra, G.R., Gregory, L.E. and Worley, J.M. 1971. Studies on initiation of adventitious roots on mung bean hypocotyl. Plant and Cell Physiol. 12: 317-324.

Conroy, J.M., Borzelleca, D.C. and McDonnel, L.A. 1982. Homology jof plant peroxidases. Plant Physiol. 69: 28-31.

Coppens, L. and Dewitte, D. 1990. Esterase and peoroxidase zymograms from Barley (*Hordeum vulgare* L.) callus as a biochemical marker system of embryogenesis and morphogenesis. Plant Science 67: 97-105.

Coumans-Gilles, M.F., Kevers, C., Coumans, M. and Gasper, T. 1982. Auxin content and metabolism in auxin requiring and non-requiring calluses. Plant Tissue Culture. Proc. 5th Int. Cong. Plant Tissue & Cell Culture, Tokyo (Ed. Fujiwara, C.A.) pp. 197-198.

Davis, B.J. 1964. Disc electrophoresis. II. Methods and application to human serum proteins. Ann. N.Y. Acad. Sci. 121: 404-427.

Dougall, D.K. 1965. The biosynthesis of protein amino-acids in plant tissue cultures. I. Isotope competition experiments using glucose 4-^{14}C and the protein amino acids. Plant Physiol. 40: 891-897.

Espelie, K.E., Franceschi, V.R. and Kolattukudy, P.E. 1986. Immunocytochemical localization and time course of appearance of an anionic peroxidase associated with suberization in wound healing potato tissue. Plant Physiol. 81: 487-492.

Evans, J.J. and Alldridge, N.A. 1965. The distribution of peroxidase in extreme dwarf and normal tomato (*Lycopersicon esculentum*). Phytochemistry, 4: 499.

Fry, S.C. 1986. Cross-linking of matrix polymers in the growing cell walls of angiosperms. Ann. Rev. Plant Physiol. 37: 165-186.

Galston, A.W. and Dalberg. L.Y. 1954. The adaptive formation and physiological significance of Indole-acetic acid oxidase. Amer. J. Bot. 41: 373-380.

Gaur, A., Bhardwaj, L., Merilon, J.M. and Ramawat, K.G. 1993. changes in phospho-lipids, fatty acids, oxidative enzymes, phenolics and protein levels during growth of normal and habituated tissues of *Cocculus pendulus*. Indian J. Exp. Biol., 31: 987-990.

Goldberg, R., Imberty, A., Liberman, M. and Prat, R. 1986. Relationships between peroxidatic activities and cell wall plasticity. In: Molecular and physiological aspects of plant peroxidases. Greppin, H., Penel, C., Gasper, T. (Eds.), Univ. of Geneva, Geneva, Switzerland, pp. 208-220.

Gove, J.P. and Hoyle, M.C. 1975. The isozymic similarity of Indole-acetic acid oxidase to peroxidase in Birch and Horseradish. Plant Physiol. 56: 684-687.

Gurumurti, K. and Nanda, K.K. 1974. Peroxidase isozymes of mung bean hypocotyl cuttings during rooting. Phytochemistry, 13: 1089-1093.

Hamill, A.E. and Brewbaker, J.L. 1969. Isozyme polymorphism in flowering plants. IV. The peroxidase isozymes of maize (*Zea mays*). Physiol. Plant. 22: 945-958.

Harkin, J.M. and Obst. J.R. 1973. Lignification in trees: indication of exclusive peroxidase participation. Science 180: 296-298.

Heinz, D.J. and Mee, G.W.P. 1971. Morphologic, cytogenetic and enzymatic variation in Saccharum species hybrid clones derived from callus tissue. Amer. J. Bot., 58: 257-262.

Hinnman, R.L. and Lang, J. 1965. Peroxidase catalyzed oxidation of indole-3-acetic acid. Biochemistry 4: 144-158.

Hoess, R.H., Smith, H.H. and Stowell, C.P. 1974. A genetic analysis of peroxidase isozymes in two species of *Nicotiana*. Biochem. Genet. 11: 319-323.

IUPAC-IUB. Commission on biochemical nomenclature 1977. J. Biol. Chem. 252: 5939-5941.

Jordan, M., Itturriaga, L. and Feucht, W. 1980. Inhibition of root formation in *Prunus avium* hypocotyls by chorogenic acid *in vitro*. Gartenbauwissenschaft., 45: 15-17.

Joresbo, M., Andersen, J.M., Okkels, F. and Rajagopal, R. 1989. Isoperoxidase as markers of somatic embryogenesis in carrot cell suspension cultures. Physiol. Plant. 76: 10-16.

Katerina, C. and Smart, C.C. 1994. The abscisic acid induction of a noval peroxidase is antagonized by cytokinin in *Spirodela polyrrhiza* L. Plant Physiol., 105: 497-507.

Kay, L.E. and Basile, D. 1987. Specific peroxidase isozymes are correlated with organogenesis. Plant Physiol. 84: 99-105.

Kerby, K. and Somerville, S.C. 1992. Purification of an infection related, extracellular peroxidase from barley. Plant Physiol., 110: 397-402.

Kevers, C., Coumans, M., DeGreef, W., Hofinger, M. and Gasper, T. 1981. Habituation in sugarbeet callus: auxin content, auxin protectors, peroxidase pattern and inhibitors. Physiol. Plant. 51: 281-286.

Khavkin, E.E. and Sukhorzhervskaia, T.B. 1979. Maintenance of isozymes spectra in callus and suspension cultures derived from internodes of maize (*Zea mays* L). Biochem. Physiol. Pflan. 174: 431-437.

Lagrimini, L.M. 1991. Wound-induced depositionof polyphenols in transgenic plants over expressing peroxidase. Plant Physiol. 96: 577-583.

Lagrimini, L.M., Burkhart, W., Moyer, M. and Rothstein, S. 1987. Molecular cloning of complementary DNA encoding the lignin-forming peroxidase from tobacco: Molecular analysis and tissue specific expression. Proc. Natl. Acad. Sci. USA. 84: 7542-7546.

Lavee, S. and Galston, A.W. 1968. Hormonal control of peroxidase activity in cultured *Pelargonium* pith. Amer. J. Bot. 55: 890-893.

Lee, T.T. 1972. Interaction of cytokinin, auxin and gibberellins on peroxidase isozymes in tobacco tissue culture *in vitro* Can. J. Bot. 50: 2471-2477.

Mader, M. and Bopp, M. 1976. Neue Vorstellungen Zum Problem der isoperoxidasen anhand der Trennung durch Disk. Electrophorese und isoelektrische Fokussierung. Planta. 128: 247-253.

Mader, M., Ungemach, J. and Schlo, B.P. 1980. The role of peroxidase isozyme groups of *Nicotiana tabacum* in hydrogen peroxide formation. Planta. 147: 467-470.

Mako, V. and Novacky, A. 1969. Contribution to the study of plant peroxidase isozymes by means of disc electrophoresis on acrylamide gel. Biologia. 21: 128-132.

Markert, C.L. and Moller, F. 1959. Multiple forms of enzymes: tissue, ontogenic and species specific pattern. Proc. Natl. Acad. Sci. USA. 45: 753-763.

McAdam, J.W., Nelson, C.J. and Sharp, R.E. 1992. Peroxidase activity in the leaf elongation zone of Tall Fescue. Plant Physiol. 99: 872-878.

Nanda, K.K., Bhattacharya, N.C. and Kaur, N.P. 1973. Effect of morphactin on peroxidases and its relationship to rooting hypocotyl cuttings of *Impatiens balsamina*. Plant & Cell Physiol. 14: 207-211.

Nashikkar, V.J. and Chakrabarti, T. 1994. Catalase and peroxidase activity in plants. An indiction of heavy metal toxicity. Indian J. Exp. Biol. 32: 520-521.

Nessel, A. and Mader, M. 1977. On the physiological significance of the isozyme groups of peroxidase from tobacco demonstrated by biochemical properties. I. Separation, purification, chemical and physical properties. Z. Pflanzenphysiol. 82: 235-246.

Noh, Eun-Woon and Minocha, S.C. 1990. Pigment and isozyme variation in aspen shoots regenerated from callus culture. Plant Cell Tissue and organ culture. 23: 39-44.

Novacky, A. and Hampton, R.E. 1968. The effect of substrate concentration on the visualization of isoperoxidases in disc electrophoresis. Phytochemistry. 7: 1143-1145.

Quoirin, M., Boxus, P. and Gasper, T. 1974. Root initiation and isoperoxidases of stem tip cutting from mature *Prunus* plants. Physiol. Veg. 12: 165-174.

Rao, K.V., Suprasanna, P. and Reddy, G.M. 1990. Biochemical changes in embryogenic and non-embryogenic calli of *Zea mays* L., Plant Science. 66: 127-130.

Rawal, S.K. and Mehta, A.R. 1978. Changes in peroxidase isoenzyme pattern during organogenesis in tobacco callus cultures. In: 4th Intl. Cong. Plant Tissue and Cell Culture, Calgaryl, Canada, p. 130.

Reimers, P.J., Guo, A., and Leach, J.E. 1992. Increased activity of a cationic peroxidase associated with an incompatible interaction between *Xanthomonas oryzae* pv. *oryzae* and rice (*Oryza sativa*). Plant Physiol. 97: 528-536.

Roberts, E. and Kolattukudy, P.E. 1989. Molecular cloning, nucleotide sequence and abscisic acid induction of a suberization associated highly anionic peroxidase. Mol. Gen Genet. 217: 223-232.

Saunders, B.C., Siedle, A.G.H. and Stark, B.P. 1964. Peroxidase. Butterworth, London.

Scandalios, J.G. 1969. Genetic control of multiple molecular forms of enzymes in plants: A review. Biochem. Genet. 3: 37-79.

Scandalios, J.G. 1974. lozymes in development and differentiation. Ann. Rev. Plant Physiol. 25: 38-78.

Shannon, L.M. 1968. Plant isozymes. Ann. Rev. Plant Physiol. 19: 187-210.

Sharma, P. 1992. Biochemical studies on tissue culture of some legumes. Ph.D. Thesis, Univ. of Rajasthan, Jaipur.

Sharma, R. and Biswal, U.C. 1976. Effect of kinetina dn light on peroxidase activity of detached barley leaves. Z. Pflanzenphysiol. 78: 169-172.

Sharma, R., Sopory, S.K. and Guha-Mukherjee, S. 1976. Phytochrome regulation of peroxidase activity in maize. Plant Sci. Let. 6: 69-75.

Sheen, S.J. 1972. Isozymic evidence bearing on the origin of *Nicotiana tabacum* L. Evolution. 26: 143-154.

Sherf, B.A., Bajar, A.M. and Kolattukudy, P.E. 1993. Abolition of an inducible highly anionic peroxidase activity in transgenic tomato. Plant Physiol. 101: 201-208.

Siegel, B.J. and Galston, A.W. 1967. The isoperoxidases of *Pisum sativum*. Plant Physiol. 42: 221-226.

Simola, L.K. 1973. Changes in the activity of several enzymes during root differentiation in cultured cells of Atropa belladonna. Z. Pflanzenphysiol. 68: 373-378.

Simons, T.J. and Ross, A.F. 1970. Enhanced peroxidase activity associated with induction of resistance to tobacco mosaic virus in hypersensitive tobacco. Phytopathology. 60: 383-384.

Srivastava, O.P. and van Huystee, R.B. 1977. Interaction among phenolics and peroxidase isozymes. Bot. Gaz. (Chicago) 138: 457-464.

Sticher, L., Penel, C. and Greppin, H. 1982. Calcium requirement for the secretion of peroxidases by plant cell suspensions. J. Cell Sci. 48: 345-353.

Studier, F.W. 1973. Analysis of bacteriophage T early RNAs and proteins in slab gels, J. Mol. Biol. 79: 237-249.

Swarnkar, P.L. 1985. Biochemical studies on morphogenesis and differentiation in tissue cultures. Ph.D. Thesis, Univ. of Rajasthan, Jaipur, India.

Swarnkar, P.L., Bohra, S.P. and Chandra, N. 1986a. Biochemical changes during growth and differentiation of the callus of *Solanum surattense*. J. Plant Physiol. 126: 75-81.

Swarnkar, P.L., Bohra, S.P. and Chandra, N. 1986b. Biochemical studies on initiation of callus in *Solanum surattense*. J. Plant Physiol. 126: 293-296.

Swarnkar, P.L., Bohra, S.P. and Chandra, N. 1987a. Organogenesis in callus cultures of *Solanum surattense*: Some physiological and biochemical aspects In: Expanding horizons in Botanical research, pp. 138-143. Arihant Publishers, Jaipur.

Swarnkar, P.L., Bohra, S.P. and Chandra, N. 1987b. Biochemical changes during androgenesis in Datura innoxia. Curr. Sci. 56: 730-731.

Swarnkar, P.L., Bohra, S.P. and Chandra, N. 1987c. Biochemical changes during rooting in cultured stem segments of *Petunia hybrida*. Ann. Biol. 3: 32-35.

Thorpe, T.A. 1980. Organogenesis *in vitro*: Structural, physiological and biochemical aspects. Int. Rev. Cytol. IIA (Ed. I.K. Vasil), Academic Press, London, New York, pp. 71-111.

Thorpe, T.A., Tran-Thanh-Van, M. and Gasper, T. 1978. Isoperoxidases in epidermal layers of tobacco and changes during organ formation *in vitro*. Physiol. Plant. 44: 388-394.

Trinh, T.H., Gasper, T., Tran Thanh Van K. and Marcotte, J.L. 1981. Genotype, ploidy and physiological state in relation to isoperoxidases in *Nicotiana*. Physiol. Plant. 53: 153-157.

van Hoff, P. and Gasper, T. 1976. Peroxidase and isoperoxidase changes in relation to root initiation of *Asparagus* cultured *in vitro*. Sci. Hortic. 4: 27-31.

van Huystee, R.B. and Cairns, W.L. 1980. Appraisal of studies on induction of peroxidase and associated porphyrin metabolism. Bot. Rev. 46: 429-446.

Welander, M. 1981. The rooting ability of shoots raised *in vitro* from the apple rootstocks A2 and M26 in adult growth phase. Symposium on Clonal Forestry, Uppsala, Sweden, pp. 117-129.

Weststeijn, E.A. 1976. Peroxidase activity in leaves of *Nicotiana tabacum* var. Xanthi nc. before and after infection with tobacco mosaic virus. Physiol. Plant Pathol. 8: 63-71.

Yamauchi, N. and Watada, A.F. 1991. Regulated chlorophyll degradation in spinach leaves during storage. J. Am. Soc. Hortic. Sci. 116: 58-62.

Yen, S.T. and Sadanga, K. 1977. Inheritance of leaf peroxidases in oats. Can. J. Genet. Cytol. 19: 303-312.

Zimmermann, R.H. and Broome, O.K. 1981. Phloroglucinol and *in vitro* rooting of apple cultivar cuttings. J. Amer. Soc. Hort. Sci., 106: 648-659.

Regeneration and Phenolics in Explants and Callus of *Syzygium cumini* (L.) Skeels

Shashi Mathur, L.K. Pareek and N. Chandra

INTRODUCTION

Micropropagation of elite plants of tree species is of obvious advantage (Ahuja, 1991) and attempts have been made in this direction in several tree species (Das and Mitra, 1990). A very common problem in culturing tree explants is browning of cultured tissue and leaching of brown purplish chemicals into the medium within a few hours of implantation of the explant. Even very young explants taken from trees e.g., young shoots, floral parts etc. show the same behaviour which has been well documented in the published literature (Zaprometov, 1989; Yadav *et al.*, 1990; Compton and Preece, 1986). In the present study also very young vegetative explants of mature tree of *Syzygium cumini* showed excessive amount of exudation in vitro and did not survive. A detailed investigation on the qualitative nature of chemicals involved in browning of chemicals and their comparative levels in different types of explants of tree and regenerants as well as different calluses were made in jamun (*Syzygium cumini* (Linn.) Skeels, Myrtaceae) which is quite an important tree species in our country and has delicious fruits and medicinal properties. The effect of growth regulators was also seen on the regeneration/ proliferation of callus/ explants.

MATERIALS AND METHODS

Various explants such as very young leaves, nodal and internodal segments, shoot-tips were taken from the tree growing in the University garden. After surface sterilisation with 0.1% mercuric chloride they were cultured on modified MS (Morashige and Skoog, 1962) medium containing half strength major salts and chelated iron, 6% (w/v) sucrose, 400.0 mg/l glutamine, 100.0 mg/l ascorbic acid and supplemented with various concentrations of auxins/ cytokinins. Immature embryos were excised aseptically from sterilized young green fruitlets (1.2-1.6 cm long). All the cultures were initially kept in dark for 2 days and then shifted to a growth chamber maintained at 26 ± 2°C with 16 hr photoperiod provided with white fluorescent light (1000 lux). Different calluses were separated and subcultured on various levels of 2,4-D and BAP. Phenolic compounds such as orthodihydric phenols, flavonols and anthocyanins were quantitatively estimated from alcohol extracts of the tissue by the method of Swain and Hillis (1959) using Arnow's reagent, Vanillin reagent and Peroxide reagent respectively.

RESULTS AND DISCUSSION

Explants from mature trees

Nodal and internodal segments and leaf explants taken from mature tree showed excessive amount of exudation into the medium. Various methods were tried to overcome the problem of browning viz. presoaking of explants in water or in antioxidant solution (100.0 mg/l polyvinylpyrrolidone + 100.0-500.0 mg/l ascorbic acid + 50.0-100.0 mg/l citric acid), addition of antioxidants into the medium either singly or in combination, keeping the cultures in dark initially for 2-3 days, frequent transfer of explants onto a fresh medium, initial culture of explants in the liquid medium containing different antioxidants and then their subsequent transfer to solidified MS medium supplemented with same concentration of growth hormone but devoid of antioxidants. Presoaking of explants in anti-oxidant solution and addition of antioxidants like polyvinylpyrrolidone have been reported to be effective in controlling the browning of the cut ends of explants in *Eucalyptus* sps (Arya and Shekhawat, 1986) and *Tectona grandis* (Gupta *et al.*, 1980) respectively. But in the present investigations it did not help. All the explants turned black and did not show any response. Quantitative estimation of different phenolic compounds revealed that all these explants contained high levels of orthodihydric phenols flavonols and anthocyanins (Table 1).

TABLE 1: Showing morphogenetic response on different media and phenolic contents (mg/g Fresh Wt.) in explants of mature tree.

Explants	Morphogenetic response		Phenolic contents		
	2,4-D 2.0mg/l	BAP 2.0mg/l	Orthodihydric phenols (mg/g)	Flavonols (mg/g)	Anthocyanins (mg/g)
Shoot tips	-	-	13.300	19.350	1.750
Leaf	-	-	12.100	19.800	1.500
Nodal segment	-	-	13.200	20.450	1.410
Internodal segment	-	-	12.400	20.050	1.330
Immature	Callus	Germination	5.800	9.560	0.950

Immature Embryo

Immature embryos excised from young green fruitlets showed comparatively less browning and survived in vitro on reducing the major salts to half strength, increase of sucrose and incorporation of ascorbic acid in the medium and keeping the cultures in dark initially for 2 days followed by regular transfer of explants onto a fresh medium. Reduction in the concentration of nitrogen containing compounds in the MS medium by half has been reported to be an effective means of reducing browning of the medium, in *Rhododendron* (Anderson, 1975). Increase of sucrose in the medium, addition of ascorbic acid in the medium and regular transfer of explants to a fresh medium for 2-3 days has also been reported to be effective in controlling browning of the explant and medium (Litz, 1984a, Broome and Zimmermann, 1978). It is known that oxidation products of phenols

(Quinones) form brown or black oxidation polymers which form complex with plant proteins and kill the cells eventually. Antioxidants prevent the polymerisation of quinones and reduce the probability of reacting with protein (Loomis and Battaile, 1966). Callus formation was observed from both cotyledons as well as embryonal axis of the immature embryo (Fig. I-1). Compact calluses of three different colours (green, cream and white) formed from the cotyledon but only white friable callus formed from the axis of immature embryo. Quantitative estimations of phenolic compounds showed that they contained approximately half of O.D. phenols, flavonols and anthocyanins than that of explants of mature trees (Table 1).

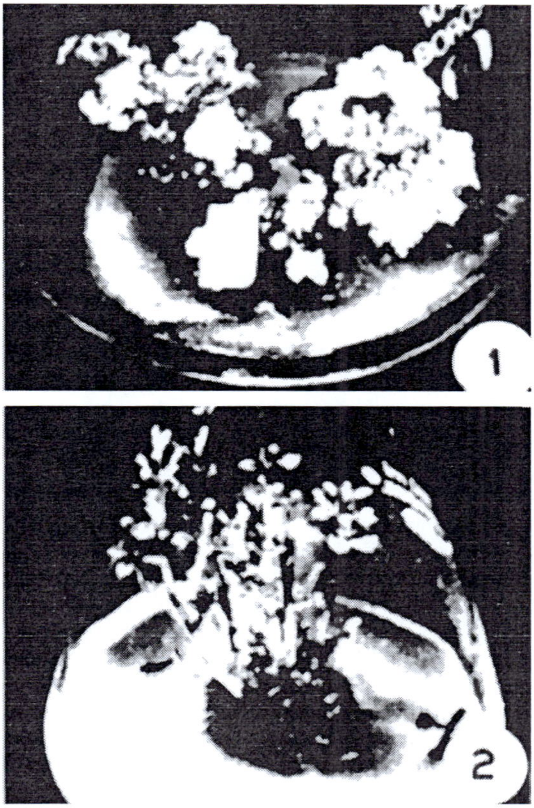

FIG. I. (1). Callus formation from immature embryo on modified MS medium supplemented with 2,4-D. (2). Shoot differentiation from callus derived from cotyledon of immature embryo.

Callus Derived from Embryo

Different calluses showed different behaviour in leaching as well as in their morphogenetic response when subcultured on the same medium. Friable callus did not show any browning of the medium and proliferated, but did not differentiate shoot buds whereas green cotyledonary callus exuded black exudates but survived and showed shoot bud formation on different concentrations of BAP. Cream cotyledonary callus did not show any exudation but regenerated shoots on different concentrations of BAP either alone or in combination with auxins (Fig.I-2). White cotyledonary callus showed excessive leaching and formed rose coloured root primordia but could not be maintained. On quantitative estimation of phenolic compounds it was found that white cotyledonary callus contained maximum amount of O.D. phenols, flavonols and anthocyanin (Table 2).

TABLE 2: Showing morphogenetic response on different media and phenolic contents (mg/g Fresh Wt.) in different types of calli derived from immature embryo.

Different types of Calli	Morphogenetic Response		Phenolic contents		
	2,4-D 2.0mg/l	BAP 2.0mg/l	Orthodihydric phenols (mg/g)	Flavonols (mg/g)	Anthocyanins (mg/g)
Cream cotyledonary callus	Callus	Shoot 6.80 ± 0.37	2.090	3.550	0.110
Green cotyledonary callus	Callus	Shoot 2.60 ± 0.04	2.410	17.600	0.530
White cotyledonary callus	Rose coloured root primordia	Callus	3.390	19.500	1.480
White friable callus	Callus	1.340	3.450	0.060	

Explants from Regenerated Shoots

Explants such as nodal and internodal segments, shoot tips and leaf explants taken from shoots regenerated in vitro did not show browning of the medium. Nodal and shoot-tip explants showed proliferation of shoots on BAP containing medium. Leaves and internodal segments showed callus formation (Table 3). Quantitative estimation of different phenolic compounds revealed that these explants contained very small amounts of O.D. phenols and flavonols and completely lacked the anthocyanins. In chestnut also the amount of phenolic compounds present in the seedling explants has been reported to be less than in the buds from twigs of trees (Cherve *et al.*, 1983).

Formation of phenolic compounds in cultured tissues is reported to be affected by several factors such as light, growth regulators, antioxidants etc. Among different auxins, 2,4-D has been reported to be the most efficient auxin, for induction and maintenance of callus and for decreasing the synthesis of phenolics in cultures (Cherve *et al.*, 1983). In the present study also 2,4-D proved to be the best auxin for the induction and growth of the callus. Addition of antioxidants into the medium is reported to decrease the polyphenol oxidase activity in the cultured tissues thereby reducing browning of the tissue. In the present study also addition of ascorbic acid in the medium was found necessary for checking the browning and the healthy growth of tissues.

TABLE 3: Showing morphogenetic response on different media and phenolic contents (mg/g fresh wt.) in explants of regenerated shoots in vitro.

Explants of regenerated shoots	Morphogenetic Response		Phenolic contents		
	2,4-D 2.0mg/l	BAP 2.0mg/l	Orthodihydric phenols (mg/g)	Flavonols (mg/g)	Anthocyanins (mg/g)
Shoot-tips	Callus	Shoots 19.20 ± 1.30	0.0190	0.140	-
Leaf	Callus		0.090	0.140	-
Nodal segment	Callus	Shoots 10.0 ± 1.20	0.080	0.120	
Internodal segment	Callus		0.080	0.110	-

Sukrasno and Yeoman (Sahai and Shuler, 1984) have reported the amount of phenolic esters and glycosides to be much higher in the slow growing compact callus grown on MS medium than in the rapidly growing friable callus grown on SH medium of *Capsicum frutescens*. In the present investigation also it was observed that among different calluses white cotyledonary callus contained maximum amount of phenolics such as O.D. phenols, flavonols and anthocyanins which was very slow growing and did not survive.

The present study shows that the callus proliferation or morphogenic response is related to the lower levels of phenolic substances specially orthodihydric phenols and lower levels/ total absence of anthocyanins.

REFERENCES

Ahuja, M.R. 1991. Plant Research and Development, 33: 106-120.

Anderson, W.C. 1975. Propagation of Rhododendrons by tissue culture. Part I. Development of a culture medium for multiplication of shoots. Proc. Int. Plant Prop. Soc. 25: 129-135.

Arya, H.C. and Shekhawat, N.S. 1986. Clonal multiplication of tree species in the Thar Desert through Tissue Culture. Forest Ecology and Management. 16: 201-208.

Broome, O.C. and Zimmermann, R.H. 1978. In vitro propagation of blackberty. Hortscience. 13: 151-153.

Chevre, A.M., Gill, S.S., Mouras, A. and Saleoses, G. 1983. *In vitro* vegetative multiplication of chestnut. J. Hortic. Sci. 58: 23-29.

Compton, M.E. and Preece, J.E. 1986. Newsletter, 50: 9-18.

Das, T. and Mitra, G.C. 1990. Plant Cell Tissue and Organ Culture, 22: 95-103.

Gupta, P.K., Nagdir, A.L., Mascarenhas, A.F. and Jagannathan, V. 1980. Tissue Culture of forest trees: clonal multiplication of Tectona grandis L. (Teak) by tissue culture. Plant Sci. Lett. 17: 259-268.

Litz, R.E. 1984a. In vitro responses of adventitious embryos of two polyembryonic Eugenia species. Hort. Sci. 19: 720-722.

Loomis, W.D. and Battaile, J. 1966. Plant phenolic compounds and the isolation of plant enzymes. Phytochemistry 5: 423-438.

Morashige, T. and Skoog, F. 1962. A revised medium for rapid growth and bioassays with tobacco tissue cultures. Physiol. Plant. 15: 473-497.

Sahai, O.P. and Shuler, M.L. 1984. Biotech. Bioeng, 26: 111.

Sukrasno, N. and Yeman, M.M. in VII Intl. Congr. on plant Tissue and Cell Culture, Amsterdam, 320, Abstracts.

Swin, T. and Hillis, W.E. 1959. The phenolic constituents of Prunus domestica I. The quantitative analysis of phenolic constituents. J. Sci. Food Agric. 10: 63-68.

Yadav, U., Lal, N. and Jaiswal, V.S. 1990. Plant Cell Tissue and Organ Culture. 21: 87-92.

Zaprometov, M.N. 1989. The formation of Phenolic compounds in plant cell and Tissue Cultures and the possibility of its regulation. In: Advances in Cell culture. Vol. 7. (Ed. K. Marmorosch and G.H. Sato), pp. 201-225. Academic Press, INC. California.

Cellular Approaches to Salt Tolerance in Wheat

E. J. Joseph and I. D. Nikiforova

INTRODUCTION

Soil salinity is one of the most important environmental stresses limiting the productivity of agricultural systems around the world. It is considered as the largest soil toxicity problem in tropical Asia (Greenland, 1984), the severity of which is gradually being aggravated by the build-up of salts in soil through irrigated agriculture. In India alone about 30-50% of irrigated land is salt-affected (Flowers *et al.*, 1986).

Plant cell and tissue culture system is now being widely used to select cell lines with increased tolerance to salt stress. However, the beneficial application of cellular selection techniques in crop improvement programme calls for not only the selection of stable tolerant cell lines but also their regeneration into whole plants that express and in turn transmit tolerance to the progenies. Although putative salt-tolerant cell lines were isolated from a variety of crop plants, regeneration of plants from selected cell lines was successful only in some few cases like tobacco (Nabors *et al.*, 1980), rice (Narayanan and Sree Rangasamy, 1989; Subhashini and Reddy, 1989), Sorghum (Bhaskaran *et al.*, 1986), wheat (Butenko *et al.*, 1988), tomato (Garcia-Reina, *et al.*, 1988), citrus (Ben-Hayyim and Goffer, 1989) and flax (Mc Hughen and Rowland, 1988). The objectives of the present study were to select salt-tolerant cell lines, to study their growth characteristics and to regenerate whole plants from the selected cell lines in wheat.

MATERIALS AND METHODS

Experimental material

In vitro selection for salt tolerance was conducted in 4 varieties of spring wheat (*Triticum aestivum* L.) viz., Tzelinnnaya-21, Tzelinnaya-Yubileinnaya, Saratovskaya- 29 and Moscovskaya-35, chosen from different eco-geographic regions of the erstwhile Soviet Union.

Callus initation and culture

The immature grains were collected from the middle 8-10 florets of the inflorescence at 14-16 days after anthesis. The immature embryos were aseptically isolated from surface-sterilized [sodium hypochlorite (0.2-0.3% available chlorine)] immature seeds and placed on artificial growth medium (14-15 per Petri dish) with the scutellum facing upwards for callus initiation.

The growth medium contained inorganic salts and vitamins of MS-medium (Murashige and Skoog, 1962), supplemented with 2mg/L 2,4-D, 100mg/L asparagine, 10mg/L $AgNO_3$, 20g/L sucrose and 7g/L agar. The morphogenic calli, indicted by the presence of green spots, alone were sub-cultured to fresh media at the end of each passage (28-30 days). The cultures were grown continuously under cool white fluorescent light (15W/m^2) at 27°C and 70% relative humidity.

Measurement of growth

The weight of callus tissues was determined aseptically and their growth measured in terms of their relative growth (RG), calculated by the formula:

$$\text{Relative growth (RG)} = W1\text{-}W0/W0 \times 100$$

where, W0 and W1 are the initial fresh weight and final fresh weight of the callus at the beginning and end of a passage, respectively (Bhaskaran *et al.*, 1983).

Cellular selection

The selection for salt tolerance was started at the 4th passage. For this weighed pieces (20-35 mg) of calli were placed on callus selection media (14-15 per Petri dish) containing graded concentrations of the stress inducing agents, NaCl (85mM, 171mM and 256mM) or Na_2SO_4 (35mM, 70mM and 105mM). Between 28th and 30th day of culture the percentage survival of tissues was recorded, the calli weighed aseptically and the growing non-necrotic parts of tissues alone sub-divided and moved to fresh selection media. At every sub-culture the callus tissues that derived from each of the initial calli subjected to the selection course were separately ear-marked and the descendants of a particular callus grouped under a common somaclonal number (a cell line). Long-term culture and selection (10-15 passages) of tissues on the stress-media was the means for the isolation of stress-tolerant cell lines. The stability of the acquired character in the selected cell lines was tested by growing them for two passages on medium devoid of the stress factor and then measuring the growth after reverting them to the selection media. Cross-tolerance of the selected cell lines to NaCl and Na_2SO_4 was tested.

Regeneration from selected cell lines

The regeneration from the selected cell lines was obtained on regeneration medium containing 0.5mg/L IAA and 1mg/L BAP in the presence of the stress factor (171mM NaCl). The regenerants having 4-5 small leaflets were transferred to harmone-free medium containing half the strength of inorganic salts and 120mM NaCl in test tubes. The surviving plantlets (R1) were transplanted to sterile sand in small organic cups and later to soil in pots. The plants were grown to maturity in the greenhouse.

RESULTS AND DISCUSSION

Isolation of salt-tolerant cell lines

The immature embryos of all the 4 varieties of wheat readily produced morphogenic calli on callus growth medium. In order to get a sufficient stock of callus tissues through profuse multiplication and also to generate cellular variability the young calli were initially cultured on normal growth medium for 3 passages prior to the commencement of selection process. At each sub-culture during the selection phase the callus tissues were sub-divided into micro clumps weighing 20-35 mg for ensuring a better contact between the cells and the medium to prevent the escape of non-tolerant forms during selection.

Through long-term selection for 10-15 passages on the selection media 5 cell lines of the variety Tzelinnaya-21 and 3 cell lines of the variety Tzelinnaya-Yubileinnaya, were isolated as tolerant lines. Out of the 8 cell lines, 7 were selected on NaCl medium and one on Na_2SO_4 medium (Table-1). The comparatively lower inhibitory effect of NaCl to Na_2SO_4 in the case of wheat was responsible for the isolation of more number of salt-tolerant cell lines on NaCl media (Joseph and Nikiforova, in preparation). The innate stress-tolerant character of Tzelinnaya varieties contributed to the *in vitro* selection of salt-tolerant cell lines in those varieties (Joseph and Nikiforova, in prep.).

TABLE 1: Stress tolerant cell lines and regenerants of wheat obtained after long-term selection *in vitro*

Name of cell line	Stress inducing factor	Duration of selection (months)	No. of regenerants obtained
A-SSC-1	NaCl	10	17
A-SSC-3	NaCl	10	19
A-SSC-6	NaCl	13	18
A-SSC-11	NaCl	17	14
A-SSC-12	NaCl	14	4
B-SSC-4	NaCl	10	5
B-SSC-5	Na_2SO_4	14	1
B-SSC-6	NaCl	14	28

A- Tzelinnaya-21, B- Tzelinnaya- Yubileinnaya, SSC-Selected somaclone.

Stability of tolerance

The growth response of the selected cell lines when they were again subjected to stress conditions after maintaining them on stress-free media for two passages revealed that the cell lines highly varied in the degree of stability of the acquired tolerance (Table-2). In general, the selected cell lines exhibited a slower growth rate than non-selected calli of the same age on salt-free medium. All the selected cell lines, except A-SSC-11, were superior to the non-selected calli in terms of their relative growth when they were transferred back to the stress-media from salt-free media. The line A-SSC-11 almost lost the salt-tolerant trait upon the release of the selection pressure. This was due to the occurrence of a temporary physiological adaptation to salt stress, instead of genetic adaptation, in the line A-SSC-11. Such physiological adaptations during *in vitro* selection process were noticed earlier also in tobacco (Hasegawa *et al.*, 1984) and sugar beet (Chandler *et al.*, 1989). The line A-SSC-6 demonstrated equal/better growth on selection media than on salt-free media. This lower relative growth of the line A-SSC-6 on salt-free media is an indication of a possible shift towards halophytic physiology in that line.

TABLE 2. Growth characteristics of the selected cell lines after two passages of culture on stress-free media.

Name of cell line	Control of $NaCl/Na_2SO_4$	171mM NaCl		70mM Na_2SO_4	
	relative growth* (mg)	relative growth* (mg)	% to control	relative growth* (mg)	% to control
A-NSC	580 ± 88	94 ± 23	16.2	116 ± 18	20.0
A-SSC-1	293 ± 43	137 ± 14	46.8	221 ± 27	75.4
A-SSC-3	369 ± 16	147 ± 19	39.8	226 ± 23	61.2
A-SSC-6	164 ± 33	152 ± 18	92.7	192 ± 35	117.1
A-SSC-11	492 ± 127	103 ± 25	20.9	185 ± 30	37.6
B-NSC	269 ± 21	33 ± 12	12.3	94 ± 6	36.1
B-SSC-	536 ± 77	190 ± 25	35.5	306 ± 41	57.0
B-SSC-	220 ± 38	89 ± 19	40.5	178 ± 14	80.9

A- Tzelinnaya-21, B- Tzelinnaya- Yubileinnaya, SSC-Selected somaclone, NSC - Non selected callus.
* Mean ± SE

The cell lines which were selected for salt tolerance on NaCl-media were found to be cross-tolerant to Na_2SO_4 and vice versa (Table-2). This was significant while taking into consideration the existence of varietal specificities in cellular growth response to NaCl and Na_2SO_4 in wheat (Joseph and Nikiforova, in prep.). Furthermore, contrary to the present finding, the NaCl selected cell lines of alfalfa (Stavarek and Rains, 1983) citrus (Kochba *et al.*, 1982) and Chinese cabbage (Paek *et al.*, 1988) were not found to be tolerant to the same extent on Na_2SO_4 media.

Regeneration into whole plants

The salt stress (171mM NaCl) was imposed during the regeneration process also so as to prevent the emergence of non-tolerant forms. The regeneration medium containing NaCl initially induced green zones and leafy structures in the case of non-selected calli too. But, in due course of time they turned yellow and died. The regenerants obtained from selected cell lines were further screened during the developmental stage on medium containing 120mM NaCl inside test tubes (data not presented). The selected regenerants exhibited differential growth response, but finally majority of them developed into normal plantlets. The regenerants were obtained from all the selected cell lines (Table-1). The plantlets were transplanted first to sterile sand and then to soil in pots. The plantlets grew and developed to maturity as normal plants in the green house.

The plants obtained by *in vitro* selection for salt tolerance, after passing through the *in vitro* tests at three stages (callus, regeneration process and plantlet), may be considered as stable forms. This study threw more light on the feasibility of selecting stress-tolerant strains from cultured plant cells exploiting the somaclonal variations found in tissue cultures. However, it is apparent that the variant cells tolerant to salt, resulting from either genetic modification or physiological adaptation can co-exist in culture. Hence, the real proof for genetic basis of tolerance in the selected regenerants should come from the sexual progenies of the regenerants.

ACKNOWLEDGMENT

The first author is highly thankful to the Russian Government for providing the fellowship and facilities for doctoral studies. The author is also thankful to the Centre for Water Resources Development and Management, Kozhikode for sanctioning study leave and to the Government of India for giving the opportunity to go abroad for higher studies.

REFERENCES

Ben-Hayyim, G. and Goffer, J. 1989. Plant regeneration from a NaCl-selected salt-tolerant callus culture of Shamouti orange (*Citrus sinesis* L. Osbeck). Plant Cell Rep. 7: 680-683.

Bhaskaran, S., Smith, R.H. and Schertz, K.F. 1983. Sodium chloride-tolerant callus of *Sorghum bicolor* (L) Moench. Z. Pflanzenphysiol. 112: 459-463.

Bhaskaran, S., Smith, R.H. and Schertz, K.F. 1986. Progeny screening of sorghum plants regenerated from sodium chloride-selected callus for tolerance. J. Plant Physiol. 122: 205-210.

Butenko, R.G., Nikiforova, I.D. and Chernov, V.A. 1988. Growth and morphogenesis in cell cultures of spring wheat under stress conditions and selection of tolerant cell lines. Proceedings of the International Conference. Postdam, Germany. 9-19.

Chandler, S.F., Paek, K.J., Hachy, J.E. and Thorpe, T.A. 1989. Sodium sulphate tolerance in *Beta vulgaris* (sugar beet). Bot. Gaz. 150: 247-250.

Flowers, T.J., Troke, P.F. and Yeo, A.R. 1977. The mechanisms of salt tolerance in halophytes. Ann. Rev. Plant Physiol. 28: 29-121.

Garcia-Reina, G., Moreno, V. and Luque, A. 1988. Selection for NaCl tolerance in cell culture of three Canary Island tomato land races. 1. Recovery of tolerant plantlets from NaCl-tolerant cell strains. J. Plant Physiol. 133: 1-6.

Greenland, D.J. 1984. Exploited plants: rice. Biologist, 31: 291-295.

Hasegawa, P.M., Breassan, R.A., Handa, S. and Handa, A.K. 11984. Cellular mechanisms of tolerance to water stress. Hort. Sci. 19-371-377.

Kochba, J., Ben-Hayyim, G., Spiegal-Roy, P., Saad, S. and Neumann, H. 1982. Selection of stable salt-tolerant callus cell lines and embryos in *Citrus sinensis* (L) Osbeck and *C. aurantium* L. Z. Pflanzenphysiol. 106: 111-118.

Mc Hughen, A. and Rowland, G. 1988. Canada's first biotic crop variety: an early maturing, rust-resistant flax from cellular selection for salt tolerance. Genome. 30: 469.

Murashige, T. and Skoog, F. 1962. A revised medium for rapid growth and bio-assay with tobacco tissue cultures. Plant Physiol. 15: 473-479.

Nabors, M.W., Gibbs, S.B., Bernstein, C.S. and Meis, M.E. 1980. NaCl-tolerant tobacco plants from cultured cells. Z. Pflanzenphysiol. 97: 13-17.

Narayanan, K.K. and Sree Rangasamy, S.R. 1989. Inheritance of salt tolerance in progenies of tissue culture selected variants of rice. Curr. Sci. 58: 1204-1205.

Paek, K.Y., Chandler, S.F. and Thorpe, T.A. 1988. Physiological effects of Na2SO4 and NaCl on callus cultures of *Brassica compestris* (Chinese cabbage). Physiol. Plant. 72: 160-166.

Stavarek, S.J. and Rains, D.W. 1984. The development of tolerance to mineral salts. Hort. Sci. 19: 377-382.

Subhashini, K. and Reddy, G.M. 1989. Evaluation of the progeny under stress of regenerated salt-tolerant rice. J. Genet. Breed. 43: 125-130.

Manipulation of Wheat Protein Quality

P. R. Shewry, A. S. Tatham, N. G. Halford, F. Barro, P. Barcelo and P. Lazzeri

INTRODUCTION

Wheat is the most important crop in the world, with over 560 million tonnes harvested in 1992. It is used to prepare a wide range of foods, including various types of bread, pasta and noodles, as well as for livestock feed and as a source of industrial raw material. This widespread use in reflected in the existence of a large number of different cultivars, which have been selected by plant breeders for their adaptedness to specific environments, agronomic performance, resistance to pests and pathogens and end use properties. This traditional breeding approach has been immensely successful, but is now being complemented by a new approach: the use of genetic engineering. This allows the transfer of specific genes, which may derive from different types of wheat, from related cereals or even from animals or microbes. It therefore provides an opportunity to introduce novel variation which is not present within the gene pool available to plant breeders.

We are therefore developing procedures for the genetic engineering of wheat and using them to manipulate the protein composition in order to improve the quality for bread making and other food uses.

TRANSFORMATION OF WHEAT AND RELATED CEREALS

A number of methods have been developed for plant transformation, but only one of these has achieved wide success for transformation of wheat and related small gain cereals. This is particle bombardment, in which the DNA is coated onto gold particles which are then accelerated into the cells, using either a gunpowder charge, compressed gas or electrical discharge. This has been used to develop highly efficient transformation systems for wheat (Vasil *et al.*, 1992, 1993; Weeks *et al.*, 1993), barley (Wan and Lemaux, 1993; Ritala *et al.*, 1994) and tritordeum (Barcelo *et al.*, 1994). Tritordeum is a novel cereal crop, produced as a hybrid between tetraploid pasta wheat (*T. durum*) and a wild barley species (*Hordeum chilense*). It therefore lacks the *D genome* associated with bread making quality in *T. aestivum* (bread wheat), and has the genome composition AABBHchHch. It is well adapted to the hot mediterranean climate, for example Southern Spain, and has considerable potential as a novel food crop (Alvarez *et al.*, 1992).

We have established reliable transformation systems for wheat and tritordeum, based on the bombardment of immature embryos and immature influrescences, respectively. In this work the gold particles are coated with two separate DNAs, one carrying marker genes and the second the gene of interest. Two marker genes are used, to allow selection and confirmation and transformation, respectively. Our most recent work has used the bar gene which confers resistance to the herbicide phosphinothricine as the selectable marker and the Uid A (ß-galacturonidase) gene as the scoreable marker. The transformed plants can therefore be selected by growth on phosphinothricine and confirmed as transformed by assaying leaf tissue for gus (ß-galacturonidase) activity. The presence of the gene of interest can be demonstrated by polymerase chain reaction (PCR) using appropriate primers, and its expression by analysis of mRNA and protein fractions using specific probes and antibodies.

TARGETS FOR WHEAT TRANSFORMATION

The availability of robust transformation systems for wheat provides an opportunity to manipulate many aspects of the growth, development, performance and quality of the crop. Of particular interest are resistances to abiotic stresses (e.g. cold, heat, drought, waterlogging, salinity, high levels of minerals) and to pests and pathogens, the latter allowing reduced inputs of crop protection chemicals with economic and environmental benefits. It should also be possible to improve the efficiency of uptake and utilization of nitrogen and other mineral nutrients, by manipulating roof architecture or the structures and properties of specific transporter proteins and enzymes. This will reduce the need for expensive fertilizers and the levels of nitrates in natural and potable waters. These opportunities have been discussed in detail in a number of recent review articles (for example Lazzeri and Shewry, 1993) and will not be considered further here. Instead, we will briefly discuss work in progress in our own laboratories on the improvement of bread making quality of wheat and tritordeum.

IMPROVEMENT OF BREAD MAKING QUALITY

The ability to form leavened bread from flours is determined largely by the grain storage proteins, which form the gluten fraction. gluten is present as a network in doughs, which allows the entrapment of the carbon dioxide which is released by the yeast during proving, resulting in a light porous crumb structure which is fixed by baking. Similarly the gluten proteins confer the strength which is required for making other types of bread (e.g. flat and pocket breads), for pasta and noodles and for other uses in the food industry. Gluten confers a combination of two physical properties to dough, elasticity and extensibility (viscous flow), and a precise balance between these is required for each specific end use. Poor bread making quality of UK and European wheat's often results from the presence of gluten which is insufficiently elastic, so much of our work has focused on understanding and manipulating this property (Tatham *et al.*, 1990; Shewry *et al.*, 1994a,b).

Although gluten is a complex mixture of over 50 individual proteins, it appears that one group of components is particularly important in determining the elastic properties (Payne, 1987). These are called the high molecular weight (HMW) subunits and form part of the polymeric glutenin fraction. Individual cultivars of bread wheat contain either 3,4 or 5 individual HMW subunit proteins, each accounting on average, for about 2% of the total extractable proteins (Halford *et al.*, 1992). This difference in HMW subunit number may result in a quantitative effect on quality, with more proteins resulting in a larger amount of glutenin polymers and hence more elastic gluten and dough. It also indicates that it should be possible to improve the elasticity of gluten by inserting more HMW subunit genes to increase the gene dosage. In addition to this quantitative effect on quality, there also appear to be qualitative differences associated with the presence of specific allelic forms of HMW subunits (Payne, 1987). This is presumably due to intrinsic differences between the structures and properties of the individual proteins.

We are transferring genes for additional quality associated HMW subunits to wheat and tritordeum, under the control of their own endosperm-specific promoters. Preliminary results have shown a range of expression levels, varying from very low to above the levels of expression of the endogenous HMW subunit genes. Selected seeds are currently being multiplied in order to provide sufficient grain to determine the effects of the additional proteins on the amount and composition of gluten and the rheological and functional properties of doughs. This will provide valuable new information on the relationship between the HMW subunits and end use quality, and should ultimately lead to the production of novel wheat's with improved properties.

ACKNOWLEDGMENTS

IACR receives grant-aided support from the Biotechnology and Biological Sciences Research Council of the United Kingdom.

REFERENCES

Alvarez, J.B., Ballesteros, J., Sillero, J.A. and Martin, L.M. 1992. Tritordeum: a new crop of potential importance in the food industry. Hereditas, 116: 193-197.

Barcelo, P., Hagel, C., Becker, D., Martin, A. and Lorz, H. 1994. Transgenic cereal (tritordeum) plants obtained at high efficiency by microprojectile bombardment of inflorescence tissue. The Plant Journal 5, 583-592.

Halford, N.G., Field, J.M., Blair, H., Urwin, P., Moore, K., Robert, L., Thompson, R., Flavell, R.B., Tatham, A.S. and Shewry, P.R. 1992. Analysis of HMW glutenin subunits encoded by chromosome 1A of bread wheat (Triticum Barcelo, P., Hagel, C., Becker, D., Martin, A. and Lorz, H. 1994. Transgenic cereal (tritordeum) plants obtained at high efficiency by microprojectile bombardment of inflorescence tissue. The Plant Journal 5, 583-592.

Halford, N.G., Field, J.M., Blair, H., Urwin, P., Moore, K., Robert, L., Thompson, R., Flavell, R.B., Tatham, A.S. and Shewry, P.R. 1992. Analysis of HMW glutenin subunits encoded by chromosome 1A of bread wheat (Triticum aestivum L.) indicates quantitative effects on grain quality. Theoretical and Applied Genetics, 83: 373-378.

Lazzeri, P. and Shewry, P.R. 1993. Biotechnology of cereals. Biotechnology and Genetic Engineering Reviews, Ed. M. Tombs. Intercept Ltd., U.K. pp. 79-146.

Payne, P.I. 1987. Genetics of wheat storage proteins and the effect of allelic variation on breakmaking quality. Annual Review of Plant Physiology, 38: 141-153.

Ritala, A., Aspergren, K., Kurten, U., Salmenkallio-Marttila, M., Mannonen, L., Hannus, R., Kauppinen, V., Teeri, T.H. and Enari, T.M. 1994. Fertile transgenic barley by particle bombardment of immature embryos. Plant Molecular Biology, 24: 317-325.

Shewry, P.R., Tatham, A.S., Halford, N.G., Davies, J., Harris, N. and Kreis, M. 1994a. Improvement of barley and wheat quality by genetic engineering. In: Improvement of Cereal Quality by Genetic Engineering, Eds. R.J. Henry and J.A. Ronalds, Plenum Press, London and New York, pp. 79-86.

Shewry, P.R., Tatham, A.S., Halford, N.G., Barker, J.H.A., Hannappel, U., Gallois, P., Thomas, M. and Kreis, M. 1994b. Opportunities for manipulating the seed protein composition of wheat and barley in order to improve quality. Transgenic Research 3: 3-12.

Tatham, A.S., Shewry, P.R. and Belton, P.S. 1990. Structural studies of cereal prolamins, including wheat gluten. In: Advances in Cereal Science and Technology, Ed. Y. Pomeranz. American Association of Cereal Chemists, St. Paul, Minnesota, U.S.A. Vol. 10: 1-78.

Vasil, V., Castillo, A.M., Fromm, M.E. and Vasil, I.K. 1992. Herbicide resistant fertile transgenic wheat plants obtained by microprojectile bombardment of regenerable embryogenic callus. Bio/Technology, 10: 667-675.

Vasil, V., Srivastava, V., Castillo, A.M., Fromm, M.E. and Vasil, I.K. 1993. Rapit production of transgenic wheat plants by direct bombardment of cultured immature embryos. Bio/Technology, 11: 1553-1558.

Wan, Y. and Lemaus, P.G. 1994. Generation of large numbers of independently transformed fertile barley plants. Plant Physiology, 104: 37-48.

Weeks, J.T., Anderson, O.D. and Blechl, A.E. 1993. Rapid production of multiple independent lines of fertile transgenic wheat. Plant Physiology, 102: 1077-1084.

Cellular Selection and Characterization of Pea Lines Resistant to Culture Filtrate of *Fusarium oxysporum*

Ashok Kumar, D. K. Srivastava, V. K. Gupta and U. K. Kohli

INTRODUCTION

Fusarium wilt caused by *Fusarium oxysporum* f. sp. *pisi* is a serious disease of autumn pea in mid-hills of Himachal Pradesh. Recent agro-technique of growing pea in this season has attracted great attention (i.e. September to November). It is estimated that this disease causes losses upto 60 per cent (Kumar, 1993).

Wilt in plant is characterized by the loss of turgidity. It may be due to partial plugging of vascular bundles by the pathogen or due to toxic substances secreted by the pathogen. The browning or brick red colour of the infected roots are the typical symptoms of the disease. The disease is caused by soil borne pathogen the chemical control of the disease is not only costly, but also not feasible.

Use of conventional breeding methods i.e. selection and hybridization for developing resistant lines/cultivars takes a considerable time. Therefore, the present investigations were aimed at to develop resistant clones/lines through the use of *in vitro* cell selection technique. The cellular selection by various workers (Acrioni *et al.*, 1987; Behnke, 1979; Chawla and Wenzel, 1987 and Vidhyasekaran *et al.*, 1990) in alfalfa, potato, barley and rice have been adopted to develop resistant cell lines through *in vitro* cell selection technique.

MATERIALS AND METHODS

Plant Material

The certified seeds of cultivar Arkel were procured from the Department of Vegetable Crops, Dr. Y.S. Parmar, University of Horticulture and Forestry, Nauni-Solan. The seeds were surface sterilized in mercuric chloride ($HgCl_2$) at 0.1% for 2 minutes and were inoculated on Murashige and Skoog (1962) half strength medium containing 0.5% sucrose and devoid of plant growth regulators. Leaf explants of 15 days old aseptically grown seedlings were used.

Establishment and maintenance of callus

To find out the optimum growth of the callus, the calli were further subcultured on MS medium supplemented with various concentration and combination of growth regulators viz. NAA ranging from 0.2-1.00 mg/l and BAP at 0.5-1.5 mg/l for 45 days. The growth of the callus was scored at intervals of 15 days.

Isolation and assessment of culture filtrate

The axenic cultures of *Fusarium oxysporum* f. sp. *pisi* were obtained from the Department of Plant Pathology, Dr. Y.S. Parmar University of Horticulture and Forestry, Nauni-Solan. The pure culture of pathogen was incubated in liquid Richard's medium (Hendrix and Nielson, 1958). From ten days old culture of pathogen, the mycelial mats, microconidia, macroconidia and chalamydospores were removed through strainer and filtered through Whattman paper 42, after centrifuging the spore suspension at 30,000 rpm for 15 minutes. Finally, the filter sterilized culture filtrate was obtained by passing through 0.22 mm millipore filter. The toxicity of culture filtrate was assessed by the use of radical inhibition techniques in moong bean (Awasthi, 1983).

Cell plating on selective medium

The selective media were prepared by mixing the filter sterilized culture filtrate with sterilized molten MS medium (callus maintenance medium) to achieve volume by volume (v/v) concentration of 0%, 5%, 7.5%, 10%, 12.5%, 15% and 20%. The medium was thoroughly mixed with culture filtrate and poured in sterilized petridishes under laminar flow. Cell clumps of about one millimeter in diameter were cultured on the selective medium. They were sealed with the parafilm and incubated at 25 ± 2°C with 16 hrs. photoperiod. The readings were scored after 45 days of culture (Table 1).

TABLE 1: Selection of cell lines resistant to culture filtrate of *Fusarium oxysporum* f.sp. *pisi*.

Sr. No.	% Concentration of culture filtrate	Growth behaviour of cell lines after 45 days		Percentage survival of cell lines
		Callus Colour	Growth of Callus	
1.	0.00	Golden	+++	100% cell alive
2.	5.00	Greenish	++	80% cell alive
3.	7.50	Greenish Brown	+	60% cell alive
4.	10.00	Brownish	-	Only 3 clones Survive
5.	12.50	Brownish	-	Only 2 clones survive
6.	15.00	Dark Brown	-	100% cell killed
7.	20.00	Dark Brown	-	100% cell killed

Index: +++ = Normal growth of cells, + = Inhibition of cell lines, ++ = Satisfactory Growth, - = Complete inhibition (except two/three clones)

Determination of soluble proteins

Free tissue (calli) were homogenised in phosphate buffer (0.05 M, pH 7.00) and the homogenate were centrifuged at 10,000 g for 30 minutes. The proteins in the resulting supernatant were precipitated with TCA (final concentration 10%) and estimated according to Lowery *et al.*, 1951).

Determination of phenols

Phenol content of the selected and control cell lines was analyzed by using the method of Farkas and Kiraley (1962).

Assay of acid phosphatase (EC 3.1.3.2)

Acid phosphatase was assayed using p-nitrophenylphosphate as the substrate following the method of Shinshi and Kato (1979) with the exception that $MgCl_2$ (10 mM) was included in the reaction mixture. The enzyme activity was expressed as mmol p-nitrophenol liberated per hour per gram fresh weight.

RESULTS AND DISCUSSION

Establishment of callus culture

Leaf explants from 15 days old aseptically grown seedlings were cultured on the callus induction medium. After callus initiation, they were subsequently subcultured on the callus maintenance medium, which was standardized by using various concentrations of BAP and NAA (data not shown). The colour of the callus after subculturing turned to peculiar golden yellow. There was no change in the colour of the medium, which was used for initiation and maintenance of callus.

Isolation of cell lines

During the cellular selection procedure continuous selection method was used to isolate the cell line tolerant to the culture filtrate. Toxicity of the culture filtrate of *Fusarium oxysporum* f.sp. *pisi* was tested by standard radical inhibition technique. Cell clumps of about 1 mm in diameter were exposed to the increasing concentration of the culture filtrate (i.e. 0%, 5%, 7.5%, 10%, 12.5%, 15% and 20%) in the medium. Upon incubation for 45 days, the cells which could tolerate these concentrations of the culture filtrate grew to form cell clones (Table 1). It was observed that the growth of the cells on the selective medium was slow as compared to the control. Increasing the concentration of culture filtrate upto 7.5%, approximately 40% of the cell clumps growth was inhibited Incorporation of culture filtrate beyond 7.5% inhibited the cells growth adversely. A complete growth inhibition was observed on the medium containing 15% and 20% of the culture filtrate. All the cells turned dark brown and died on these media. Except 2 or 3 cell clumps, most of the cell clumps showed no further growth on the selective medium containing 10% and 12.5% of the culture filtrate. They became brown and died after about 3-4 weeks. In the present study, 12.5% of the culture filtrate was found optimum for selection of tolerant cell lines. At this concentration only 2 clones were selected. Although selection of cell lines was also done at 10% of the culture filtrate 3 clones were selected) but it is quite pertinent to select and regenerate the plants at higher concentration of culture filtrate. Later on, the selected clones (at 12.5% level) were subcultured on the callus maintenance medium for proliferation and biochemical characterization.

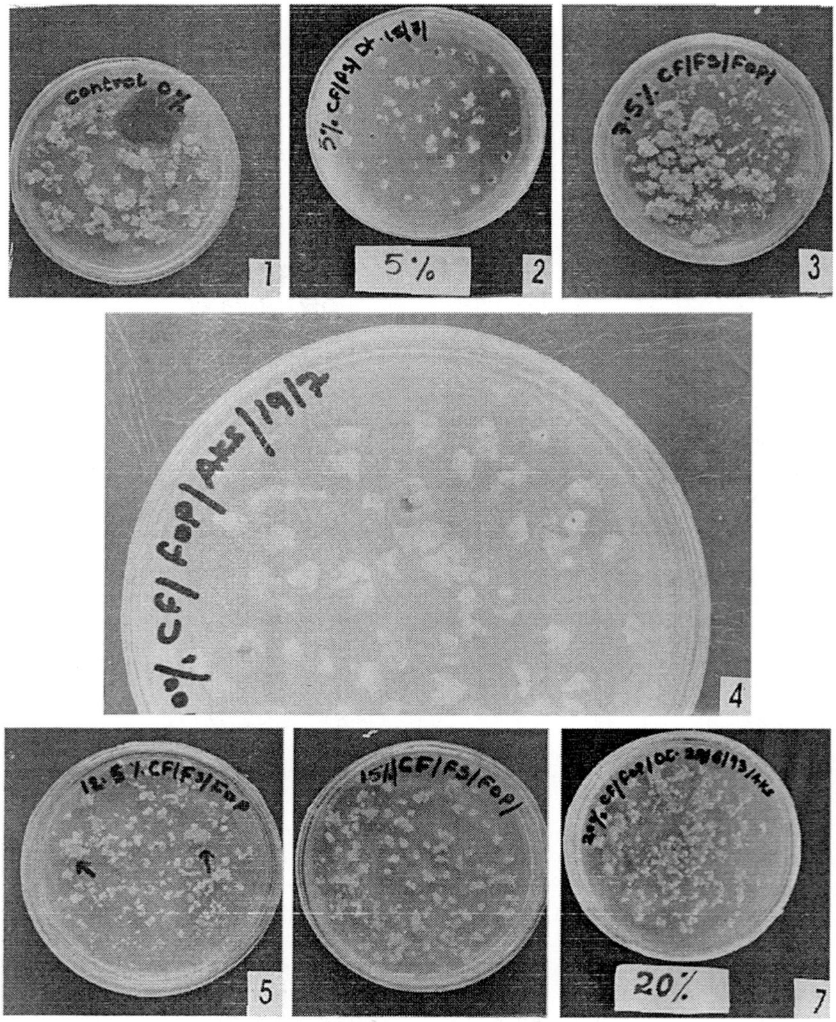

FIG. 1-7: *In vitro* selection for obtaining cell lines resistant to culture filtrate of pathogen *Fusarium oxysporum* f.sp. *pisi.* cell clumps were cultured on the selective medium containing different concentration of culture filtrate(0-20%). Fig. 1-7 shows the grow of cell colonies after 45 days of incubation (Except fig. 2). In Fig. 5 arrow marks show the resistant clones developed on the selective medium containing 12.5 per cent culture filtrate.

Although a purified toxin may have been a more appropriate selective agent, our partially fractionated culture filtrate proved effective. A number of other investigators have also obtained successful selections using culture filtrates to select resistant cell lines/plant

(Acrioni *et al.*, 1987; Carlson, 1980; Hammerschlag, 1988; Hartman *et al.*, 1984; Malepszy and El Kazzaz, 1990; Selvapandiyan, *et al.*, 1987).

Biochemical characterization

Compared to the control the selected clones revealed higher level of soluble proteins, phenols (except two clones in case of phenols) and the activity of acid phosphatase (Table 2).

TABLE 2: Comparison of the levels of soluble proteins, phenols and activity of acid phosphatase in the control and selected clones of pea resistant to culture filtrate of *Fusarium oxysporum* f.sp. *pisi*.

Sr. No.	Clones	Soluble Proteins mg/g fresh weight	Phenols mg/g fresh weight	Acid Phosphatase activity μmol p-nitro phenol liberated/hr./gram fresh weight.
1.	Control	0.991	163.25	22.83
*Selected clones				
2.	I-10	2.09	232.65	25.36
3.	II-10	1.83	164.83	23.20
4.	III-10	1.92	148.62	30.96
5.	IV-12.5	2.87	167.43	36.72
6.	V-12.5	3.97	157-89	33.92

*Selected clones: I-10, II-10, III-10 are the clones selected on the medium containing 10% of culture filtrate and IV-12.5 and V-12.5 are the clones selected on the medium containing 12.5% of culture filtrate.

It has been reported that the environmental stresses are known to enhance the activity of acid phosphatase in selected cell lines and plants (Pan, 1987; Thakur, 1991; Srivastava *et al.*, 1992). Acid phosphatase activity has also been used as an index of salt stress tolerance (Pan, 1987).

The synthesis of phenols, phytoalexins and lignins are three key factors responsible for disease resistance in higher plants (Friend *et al.*, 1981 and Hahlrock and Scheel, 1989). After infection by pathogen, an important host enzyme ß-1,3 glucanase is induced which coverts nontoxic glycosides to toxic phenols which are inhibitory to pathogen (Hilderbrand and Schroth, 1964). The higher levels of phenols in some of the selected clones (Table 2) contribute towards its resistance to *Fusarium* wilt. Accumulation of phenols in resistant variety/cell clones than the susceptible one has also been reported (Sharma *et al.*, 1983; Hahlbrock and Scheel, 1989; Prakash *et al.*, 1994). Wisniewski and Bogle (1984) reported that levels of the phenols in *Prunus persica* increase in the advance of the fungus *Cytospora lencostoma* and this increase may function as a mechanism that show pathogen advance.

The potential impact of *in vitro* cell selection to obtain resistant plant material is great, for many horticulturally important pea cultivars, which are otherwise susceptible to *Fusarium* wilt. *Fusarium* wilt tolerant cell clones in combination with advances in pea regeneration (Kathan and Jacobsen, 1990 and Ozcan *et al.*, 1992) may eventually lead to the recovery of *Fusarium* wilt tolerant pea plants.

REFERENCES

Acrioni, S., Pezzolli, M. and Damiani, F. 1987. *In vitro* selection of alfalfa plants resistant to *Fusarium oxysporum* f.sp. *medicagenesis*. Theor. and Appl. Genet. 74: 700-705.

Awasthi, J. 1983. Toxin production and physiology of *Trichothecium roseum* caused by pink mold rot of apple. Thesis, M.Sc. HPKVV, Palampur.

Behnke, Monika. 1979. Selection of potato callus for resistance to culture filtrate of *Phytophthora infestans* and regeneration of resistant plants. Theor. and Appl. Genet. 55: 69-71.

Carlson, P.S. 1973. Methionine sulfoximine resistant mutants of tobacco. Science. 180: 1366-1368.

Chawla, H.S. and Wenzel, G. 1987. *In vitro* selection of barley and wheat for resistance against *Helminthosporium sativum*. Theor. and Appl. Genet. 74: 841-845.

Farkas, G.L. and Kirley, Z. 1962. Role of phenolic compounds in the physiology of plant disease resistance. Phytopath. Z. 44: 105-150.

Friend, J. 1981. Plant phenols, lignification and plant disease. Progr. Phytochem. 7: 197-261.

Hahlbrock, K. and Scheel, D. 1989. Physiology and molecular biology of phenylpropanoid metabolism. Ann. Rev. Plant Physiol, Plant Mol. Bio. 40: 347-369.

Hammerschlag, F.A. 1988. Selection of peach cells for insensitivity to culture filtrate of *Xanthomonas compestris* pv. *pruni* and regeneration of resistant plant. Theor. and Appl. Genet. 76: 865-869.

Hartman, C.L., McCoy, T.J. and Knous, T.R. 1984. Selection of alfalfa (*Medicago sativa*) cell lines and regeneration of plant resistant to toxin(s) produced by *Fusarium oxysporum* f.sp. *medicagenesis*. Plant Sci. Lett. 34: 183-194.

Hendrix, F.F. and Nielson, L.W. 1958. Invasion and infection of crops other than form a suscepts of *Fusarium oxysporum* f.sp. *batatas*. Phytopath. 55: 148-155.

Hilderbrand, D.C. and Schroth, M.N. 1964. Arbutin-hydroquinone complex in pear as a factor in fire blight development. Phytopath. 54: 640-645.

Kathan, A.D. and Jacobsen, H. 1990. *Agrobacterium tumefaciens* mediated transformation of *Pisum sativum* using binary and cointegrate vector. Plant Cell Rep. 9: 276-279.

Kumar, A. 1993. Studies on resistance to *Fusarium* wilt in some cultivars of garden pea (*Pisum sativum* L.). Thesis M.Sc. Dr. Y.S. Parmar, University of Horticulture and forestry, Solan, H.P.

Lowery, O.H., Rosebrough, N.J., Farr, A.L. and Randall, R.J. 1951. Proteins measurement with folin reagent. J. Bio. Chem. 193: 265-275.

Malepszy, S. and El-Kazzaz, A. 1990. *In vitro* culture of *cucumis sativus* XI selection of resistance to *Fusarium oxysporum* f.sp. *cucumerinum*. Acta Hort. 280: 455-458.

Murashighe, T. and Skoog, F. 1962. A revised medium for rapid growth and bioassay with tobacco tissue culture. Physiol. Plant. 15: 473-497.

Ozcan, S., Barghchi, M, Firak, S. and Draper, J. 1992. High frequency adventitious shoot regeneration from immature cotyledons of pea (*Pisum sativum* L.). Plant Cell Rep. 11: 44-47.

Pan, Shu-mei. 1987. characterization of multiple acid phosphatase an salt stressed spinach leaves. Aus. J. Plant Physiol. 14: 117-124.

Prakash, S., Sehgal, A., Singh, R. and Choudhary, J.B. 1994. Isolation and characterization of cell lines resistant to crude culture filtrate of *Fusarium* in chickpea. J. Plant Biochem. and Biotech. 3: 63-65.

Sacristan, M.D. 1983. Resistance response to *Phoma lingum* of plant regenerated from selected cell and embryogenic culture of haploid *Brassica napus*. Theor. and Appl. Genet. 54: 129-132.

Selvapandian, A., Mehta, A.r. and Bhat, P.N. 1988. Cellular breeding approach for development of *Fusarium* wilt resistant tobacco. Proc. Ind. National Acad. B54: 391-394.

Sharma, S.G., Narayan, R., Lal, S. and Chaturvedi, C. 1983. Role of phenolic compounds in resistance of maize to leaf blight caused by drechslera state of *Cochliobolus helterostrophus*. Indian Phytopath. 36: 43-46.

Shinshi, H.a nd Kato, K. 1979. Purification and properties of acid phosphatase from cultured tobacco cells. Phytochem. 18: 243-245.

Srivastava, D.K., Gupta, V.K. and Sharma, D.R. 1992. Isolation and characterization of water stress tolerant callus cultures in tomato (*Lycopersicon esculentum* cv. Kt-1). In: Proceedings of national seminar on plant physiology- Physiological approaches for crop and plant production under environmental stresses. Abstract No. 1.29. pp 15b.

Thakur, P.S. 1991. Effect of tricontanol and mixtalol on amino acids, chlorophyll contents and acid phosphatase activity during water deficit *Dedonaea viscosa*. Ind. J. Exp. Biol. 29: 985-987.

Vidhyasekaran, P., Long, D.H., Borromeo, E.S., Dapata, F.J. and Mew, T.W. 1990. Selection of brown spots resistant rice plants from *Helminthosporium oryzae* toxins resistant callus. Ann. Appl. Biol. 117: 515-523.

Wisniewski, M.a nd Bogle, A.L. 1984. Interaction between *Cytospora lenscotoma* and host phenolic compounds in dormant peach trees. J. Amer. Soc. Hort. Sci. 109: 563-566.

Somatic Embryogenesis and Plant Regeneration from Immature Cotyledons of Chickpea (*Cicer arietinum* L.)

A. P. Sagare, K. Suhasini and K. V. Krishnamurthy

INTRODUCTION

A major pre-requisite for the production of transgenic plants is an efficient *in vitro* plant regeneration system. In legumes, immature explants have been found to be most responsive for induction of somatic embryos. Somatic embryogenesis and plant regeneration from immature zygotic embryos of soybean (Barwale *et al.*, 1986, Ranch *et al.*, 1985, Li *et al.*, 1985, pea (Kysely *et al.*, 1987, Tetu *et al.*, 1990, Kysely and Jacobsen, 1990), peanut (Ozias-Akins, 1989) has been reported.

In chickpea, a foremost cultivated legume, plant regeneration via cotyledon like structures from immature cotyledons has been reported (Shri and Davis, 1992). Based on earlier reports in legumes where immature explants were found to be most responsive, immature cotyledonary segments and embryo axes were selected as explants for our studies.

In this paper we communicate our results in chickpea which demonstrates the selection of immature explants at the correct developmental stage, effect of stress for an early response for induction and development of somatic embryos in immature embryo axes explants, effect of culture vessel and ontogeny of embryogenesis.

MATERIALS AND METHODS

Plant material

Seeds of cultivar PG12 were collected from Mahatma Phule Agricultural University, Rahuri, India. Chickpea plants were grown in field and pods were collected 22-25 d post anthesis. Pods were surface sterilized according to the protocol described earlier (Suhasini *et al.*, 1994).

Culture of immature embryo axes and cotyledonary segments

Seven immature embryo axes and cotyledonary segments (5-10 mm^2) were cultured per petri-plate (55 x 15 mm). The induction medium comprised of Murashige and Skoog's (MS) basal medium (Murashige and Skoog, 1962) supplemented with 3.0% sucrose, 0.6% agar and 3.0 mg/l 2,4,5-trichlorophenoxyacetic acid (2,4,5-T) (solid T3 medium) (Sagare *et al.*, 1993).

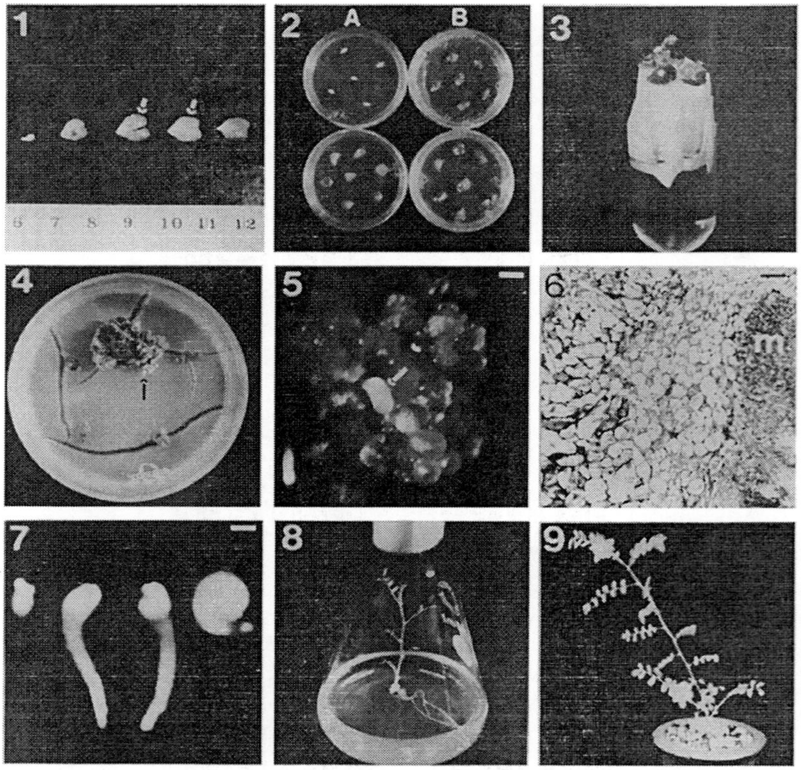

FIG. 1-9. Induction and development of somatic embryos from immature cotyledons of chickpea. (1) Immature seeds without seed coat at different developmental stages. Arrows indicate the correct developmental stages used as an explant. (2) Immature embryo axes and cotyledonary segments at the time of culture (A) and after 8 weeks of culture on induction medium (B). (3) Immature cotyledonary segments cultured on liquid T3 medium after 15 d of incubation. (4) Immature embryo axes after 8 months of incubation on induction medium. Arrow shows cotyledonary-stage somatic embryo. (5) Cotyledonary-stage somatic embryo (arrow) developed on immature embryo axes after subculture on the same induction medium in test tubes. Bar = 1.1 mm. (6) Section of cotyledonary segment after 30 d of culture on solid T3 medium showing meristematic cells (m) at cut end. Bar = 125 mm. (7) Germinated somatic embryos. Bar = 1.38 mm. (8) Converted somatic embryo. (9) Somatic embryo regenerated plant in soil.

In addition, immature cotyledonary segments were also cultured in liquid induction medium (liquid T3 medium) in test-tubes supported with Whatman No. 1 filter paper bridges. All cultures were incubated at 25 ± 2°C under cool white fluorescent light (38 $mmol.m^{-2}.s^{-1}$) for eight weeks.

Development, maturation and conversion of somatic embryos

Globular-stage somatic embryos formed on both the explants after eight weeks of culture were transferred to solid T3 medium in test-tubes for their further development. Immature embryo axes explants with globular-stage somatic embryos were left as such without a subculture, for 8-9 months to evaluate the effect of stress. These cultures were incubated under conditions described above. For maturation of cotyledonary-stage somatic embryos, they were transferred to half-strength MS medium with 3.0% sucrose, 0.1 mg/l abscisic acid (ABA) and 0.6% agar as gelling agent in test-tubes for 10 d and then transferred to conversion medium which consisted of half-strength MS medium with 3.0% sucrose, 1.0 mg/l zeatin riboside and 0.6% agar. Fully converted embryos were transferred to pots containing peat: vermiculite: solrite (1:1:1 v/v) hardened and transferred to greenhouse.

Histology

Immature cotyledonary segments after 30 d of incubation on solid T3 medium were fixed in formalin: acetic acid: alcohol (FAA) (5:5:90 by volume) for 48 h and embedded in paraffin after passing through t-butanol series. Sections of 8-10 mm were cut and stained with hematoxylin-eosin and mounted with DPX and observed microscopically.

RESULTS AND DISCUSSION

Immature embryo axes and cotyledonary segments from 12-15 mm size seeds (Fig. 1) harvested 22-25 d post-anthesis gave the optimum response for the induction of somatic embryogenesis. Both the explants produced embryogenic callus after 8 weeks of culture (Fig. 2). globular-stage somatic embryos were formed on the surface of the callus. These embryos were subcultured onto the same induction medium in test-tubes for their further development into cotyledonary-stage somatic embryos. To reduce the time required for the formation of embryogenic callus, immature cotyledonary segments were cultured in liquid T3 medium in test tubes. Callus formation was observed within 15 d of culture (Fig. 3).

Immature embryo axes when left on the same induction medium in test-tubes without a subculture for 8-9 months, cotyledonary-stage somatic embryos developed (Fig. 4).

In rare instances globular-stage somatic embryos derived from immature embryo axes developed into cotyledonary-stage somatic embryos when subcultured in test-tubes in the same induction medium (Fig. 5) which was not observed earlier (Sagare *et al.*, 1993). Cotyledonary-stage somatic embryos had two well developed cotyledons with a prominent shoot and root meristem.

Cotyledonary-stage somatic embryos were cultured on half-strength MS medium supplemented with 0.1 mg/l ABA for maturation. Germination (Fig., 6) and conversion (Fig. 7) of these embryos were achieved on half-strength MS medium supplemented with 1.0 mg/l zeatin riboside. Converted plantlets were transferred to soil (Fig. 8) and grown to maturity.

After 30 d of culture on induction medium, immature cotyledonary segments were fixed to study the ontogeny of somatic embryos. A section of cotyledonary segment revealed the depletion of food reserves, intact epidermal cells and meristematic activity at the cut end of the explant (Fig. 9) as against an earlier report of Shri & Davis, 1992, where no cotyledon like structures were formed at the wound site of the explant. On the contrary, epidermal cells showed meristematic activity in soybean (Hepher *et al.*, 1988).

Our findings with immature explants of chickpea could be useful for studying origin and development of somatic embryos.

ACKNOWLEDGMENTS

The authors are grateful to Dr. R.B. Deshmukh for the supply of seeds. The Research Fellowships awarded by C.S.I.R., Government of India, New Delhi to A.P.S. and K.S. are acknowledged. Financial support from Department of Biotechnology, Government of India is also gratefully acknowledged.

REFERENCES

Barwale, U.B., Kerns, H.R. and Widholm, J.M. 1986. Plant regeneration from callus cultures of several soybean genotypes via embryogenesis and organogenesis, Planta 167: 473-481.

Christou, P. and Yang, N.S. 1989. Developmental aspects of Soybean (*Glycine max*) somatic embryogenesis, Ann Bot. 64: 225-234.

Kysely, W., Myers, J.R., Lazzeri, P.A., Collins, G.B. and Jacobsen, H.J. 1987. Plant regeneration via somatic embryogenesis in pea (*Pisum sativum* L.), Plant Cell Rep. 6: 305-308.

Kysely, W. and Jacobsen, H.J. 1990. Somatic embryogenesis from pea embryos and shoot apices, Plant Cell Tiss Org. Cult. 20: 7-14.

Li, B., Langridge, J. and Szlay, A.A. 1985. Somatic embryogenesis and plantlet regeneration in the soybean *Glycine max*, Plant Cell Rep: 4: 344-347.

Malik, K.A. and Saxena, P.K. 1992. Somatic embryogenesis and shoot regeneration from intact seedlings of *Phaseolus acutifolium* A., *P. aureus* (L.) Wilczek, *P. coccineus* L., and *P. wrightii* L., Plant Cell Rep. 11: 163-168.

Murashige, T. and Skoog, F. 1962. A revised medium for rapid growth and bioassays with tobacco tissue cultures, Physiol Plant 15: 473-497.

Ozias-Akins, P. 1989. Plant regeneration from immature embryos of peanut, Plant Cell Rep. 8: 217-218.

Ranch, J.P., Oglesby, L. and Zielinski, A.C. 1985. Plant regeneration from embryo-derived tissue cultures of soybeans, *In vitro* Cell Dev Biol. 21: 653-658.

Sagare, A.P., Suhasini, K. and Krishnamurthy, K.V. 1993. Plant regeneration via somatic embryogenesis in chick pea (*Cicer arietinum* L.), Plant Cell Rep, 12: 652-655.

Shri, P.V. and Davis, T.M. 1992. Zeatin-induced shoot regeneration from immature chickpea (*Cicer arietinum* L.) cotyledons, Plant Cell Tiss. Org. Cult. 28: 45-51.

Suhasini, K., Sagare, A.P. and Krishnamurthy, K.V. 1994. Direct somatic embryogenesis from mature embryo axes in chickpea (*Cicer arietinum* L.), Plant Sci. 102: 189-194.

Tetu, T., Sangwan, R.S. and Sangwan-Norreel, B.S. 1990. Direct somatic embryogenesis and organogenesis in immature zygotic embryos of *Pisum sativum* L., J. Plant Physiol, 137: 102-109.

Direct Somatic Embryogenesis in Chickpea (*Cicer arietinum* L.)

K. Suhasini, A. P. Sagare and K. V. Krishnamurthy

INTRODUCTION

Chickpea, a major grain legume is susceptible to a number of diseases like wilt caused by *Fusarium oxysporum*, blight caused by *Ascochyta rabiei*, and yields are affected by the pot borer *Helicoverpa armigera*. Although disease resistance is available in wild species of *Cicer*, the cross incompatibility restricts its transfer to cultivated species (Singh, 1991). Biotechnological approach using gene transfer technology is a probable answer for which an efficient plant regeneration system is essential. Recently, plant regeneration via somatic embryogenesis from immature cotyledonary segments (Sagare *et al.*, 1993) young leaflets (Barna and Wakhlu, 1993; Kumar *et al.*, 1994) and mature embryo axes (Suhasini *et al.*, 1994) have been reported. But the low frequency of embryogenesis restricts the use of these protocols for transformation studies.

Mature embryo axes which is available throughout the year was selected as an explant and various cultivars of chickpea were screened for their embryogenic potential in an attempt to increase the frequency of embryogenesis. In the present communication we have outlined the response of mature embryo axes on solid and liquid induction medium and histological study of different development stages of somatic embryos.

MATERIALS AND METHODS

Plant material

Seeds of cultivars PG1, PG5, PG12 and Chaffa were collected from Mahatma Phule Agricultural University, Rahuri, and cultivar C235 from National Bureau of Plant Genetic Resources (NBPGR), New Delhi. Seeds were surface sterilized by washing first with commercial detergent (Teepol), for 5 min., 70% ethanol for 1 min., 0.1% mercuric chloride for 5 min., rinsed with sterile distilled water 5 or 6 times and incubated in sterile distilled water on gyratory shaker at 50 rpm for 16 h at 25 ± 2°C.

Culture of mature embryo axes

Mature embryo axes of cultivars PG1, PG5, PG12, Chaffa and C235 were dissected aseptically from presoaked seeds and cultured for induction of embryogenesis. About seven explants were cultured per petri-plate containing Murashige and Skoog's (MS) medium (Murashige and Skoog, 1962) supplemented with varying concentrations of 2,4,5-trichlorophenoxyacetic acid (2,4,5-T) (0.1, 1.0, 3.0, 5.0 and 10.0 mg/l), 3.0% sucrose and 0.6% agar. Two mature embryo axes of cultivar PG12 were cultured per test tube containing MS medium supplemented with varying concentrations of 2,4,5-T (as mentioned supplemented with varying concentrations of 2,4,5-T (as mentioned above) supported with Whatman No. 1 filter paper bridges. All cultures were incubated at 25 ± 2°C under cool white fluorescent light (23 $\mu mol.m^{-2}.s^{-1}$).

Maturation and conversion of somatic embryos

For maturation and conversion of somatic embryos, cotyledonary-stage somatic embryos were separated from the explant and incubated on half-strength MS major salts, full strength minor salts and vitamins (half-strength MS medium) with 3.0% sucrose, 0.1 mg/l abscisic acid and 0.6% agar as gelling agent in test tubes for 10 d and then transferred to conversion medium which consisted of half-strength MS medium with 3.0% sucrose, 1.0 mg/l zeatin and 0.6% agar in Magenta boxes (Sigma).

Histology

Somatic embryos at different developmental stages were fixed in formalin: acetic acid: alcohol (FAA) (5:5:90 by volume) for 48 h and embedded in paraffin after passing through t-butanol series. Sections of 8-10 mm were cut and stained with hematoxylin-eosin and mounted with DPX and observed microscopically.

RESULTS AND DISCUSSION

In our earlier report we have demonstrated somatic embryogenesis and plant regeneration from immature cotyledonary segments (Sagare *et al.*, 1993). Since the availability of immature explants of chickpea is restricted only to the season (January-March), mature embryo axes which is available throughout the year was selected as an explant. Our earlier report on immature cotyledonary segments showed 2,4,5-T to be an effective auxin for induction of embryogenesis. Mature embryo axes of various cultivars of chickpea (i.e.) PG1, PG5, PG12, Chaffa and C235 were cultured on varying concentrations of 2,4,5-T to evaluate their embryogenic response.

Somatic embryogenesis from mature embryo axes has been reported in other important legumes like peanut (McKently, 1991), soybean (Christou & Yang, 1989) and *Phaseolus* (Malik and Saxena, 1992). After 10-12 jewels of culture on varying concentrations of 2,4,5-T, embryogenic response of the different cultivars was evaluated. All cultivars showed formation of globular structures except cultivar PG5. Among the different concentrations of 2,4,5-T tested, only 3.0 mg/l gave maximum response in all the genotypes (Table 1). Since cultivar PG12 gave maximum response (65.0%) in formation of globular-stage somatic embryos, further studies were carried out with PG12 alone.

TABLE 1: Effect of different concentrations of 2,4,5-T on embryogenic response of mature embryo axes of different cultivars of chickpea.

2,4,5-T (mg/l)	Cultivar				
	PG1	PG5	PG12	C235	Chaffa
0.1	NR	NR	NR	NR	NR
1.0	++ (8.33%)	NR	++ (2.0%)	NR	+++ (8.33%)
3.0	++ (16.66%)	NR	+++ (65.0%)	+++ (41.0%)	+++ (41.66%)
5.0	+++ (16.66%)	NR	++ (42.0%)	+++ (33.33%)	NR
10.0	NR	NR	++ (20.0%)	+++ (8.33%)	NR

++ = 3-8 globular-stage embryos per explant. +++ = 8-many globular stage embryos per explant.
NR = No response, (Figures in parentheses indicate percentage embryogenic response).

After 10-12 weeks of culture on solid T3 medium (MS basal medium supplemented with 3.0 mg/l 2,4,5-T with gelling agent), explants of cultivar PG12 which showed formation of globular stage somatic embryos had to be subcultured onto the same medium in test tubes for their further development into cotyledonary stage somatic embryos. Cotyledonary-stage somatic embryos were usually observed after 4-6 weeks of culture. However, on prolonged incubation on the induction medium in petri-plates globular-stage somatic embryos dedifferentiated to form callus. In an attempt to reduce the time required for the formation of cotyledonary-stage embryos, explants of cultivar PG12 were cultured on MS liquid medium supplemented with varying concentrations of 2,4,5-T (0.1-10.0 mg/l) in test tubes. After 30 d of culture, explants were scored for the presence of cotyledonary-stage somatic embryos. About 5% and 25% of the explants showed the presence of cotyledonary-stage somatic embryos on 3.0 and 5.0 mg/l 2,4,5-T respectively (Table 2) (Fig. 1A). Though the percentage embryogenic response was high in 5.0 mg/l 2,4,5-T, the number of cotyledonary-stage somatic embryos per explant was only 1-2. The number of cotyledonary-stage somatic embryos per explant in 3.0 mg/l 2,4,5-T was 2-12 (Fig. 1B). Embryos at different developmental stages i.e., globular, heart, torpedo and cotyledonary were observed (Fig. 1C).

TABLE 2: Effect of 2,4,5-T on somatic embryogenesis from mature embryo axes of chickpea cultivar PG12 cultured on liquid medium*

2,4,5-T mg/l	% response to form cotyledonary-stage somatic embryos	No. of cotyledonary- stage somatic embryos per explant.
0.1	0	-
1.0	0	-
3.0	5.0	2-12
5.0	25.0	1-2
10.0	0	-

*Explants were dissected from seeds soaked in sterile distilled water and agitated at 50 rpm on gyratory shaker.

On scrutiny, somatic embryos were found to arise from shoot tip and cotyledonary node region of the explant in chickpea (Suhasini *et al.*, 1994) while in peanut they arose from epicotyl region of zygotic embryo (McKently, 1991) and from hypocotyl region of zygotic embryo in *Phaseolus* (Malik and Saxena), 1992).

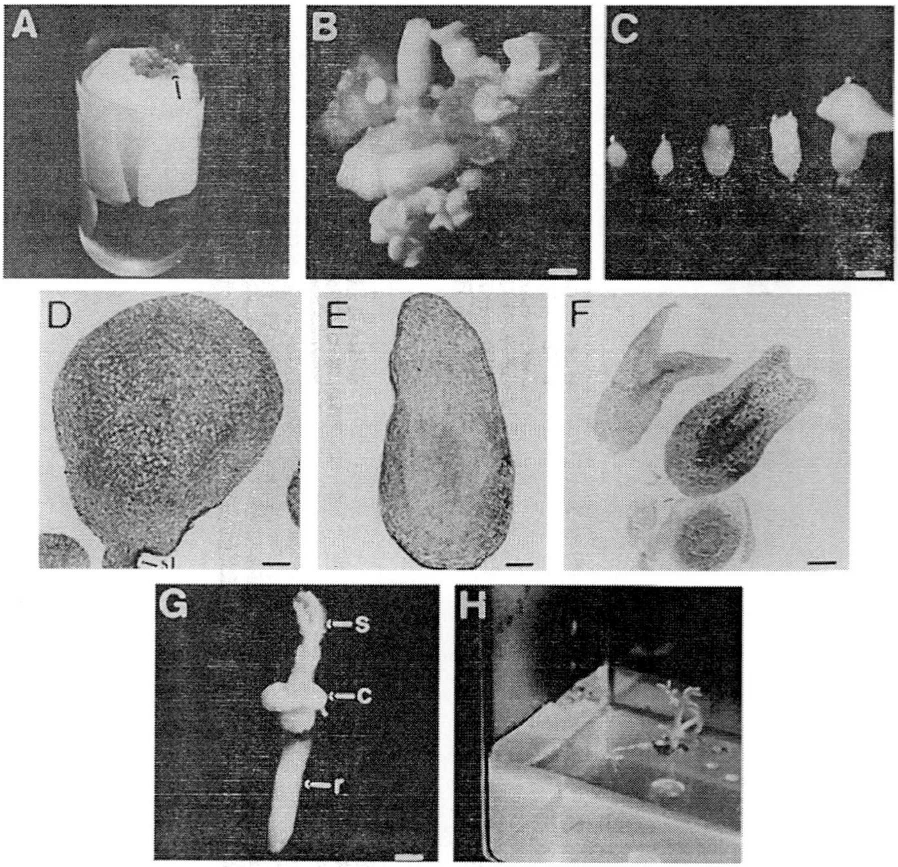

FIG. 1 Induction and development of somatic embryos from mature embryo axes of chickpea. (A) Somatic embryos (arrow) developed on mature embryo axes after 30 d of culture on liquid T3 medium. (B) Magnified view of shoot apex region of explant showing cluster of cotyledonary-stage somatic embryos. Bar = 925 mm. (C) Developmental stages of chickpea somatic embryogenesis. Bar = 625 mm. (D) Light micrograph of globular-stage somatic embryo with suspensor-like stalk (sl). Bar = 65 mm. (E) Light micrograph of torpedo-stage somatic embryo. Bar = 130 mm. (F) Section passing through cotyledonary node region of mature embryo axes showing cluster of somatic embryos at different developmental stages. Bar = 140 mm. (G) Converted somatic embryo showing shoot pole (s), cotyledons (c) and root pole (r). Bar = 895 mm. (H) Somatic embryo derived plantlet in Magenta Box.

Globular and torpedo-stage somatic embryos obtained from mature embryo axes cultured on solid T3 medium and cluster of somatic embryos obtained from the cotyleconary node region of mature embryo axes cultured on liquid medium with 3.0 mg/l 2,4,5-T were studied histologically. Globular-stage somatic embryo showed the presence of suspensor-like stalk (sl) (Fig. 1D) and torpedo-stage somatic embryo revealed the presence of vascular initials (Fig. 1E). A section passing through a cluster of embryos revealed embryos at different developmental stages (Fig. 1F) which confirmed their non synchronous development.

Converted embryos showed root and shoot pole development (Fig. 1G). For further development, these converted embryos were cultured on conversion medium in Magenta boxes (Fig. 1H) until they attained a height of 6-8 cm. Plantlets were then transferred to pots, hardened and grown to maturity.

Since direct somatic embryogenesis without an intervening callus phase leads to production of plants without somaclonal variation, this protocol would be useful in genetic transformation studies.

ACKNOWLEDGMENTS

The authors are grateful to Dr. R.B. Deshmukh for the supply of seeds. The Research Fellowships awarded by C.S.I.R., Government of India, New Delhi to K.S. and A.P.S. are acknowledged. Financial support from Department of Biotechnology, Government of India is also gratefully acknowledged.

REFERENCES

Barna, K.S. and Wakhlu, A.K. 1993. Somatic embryogenesis and plant regeneration from callus cultures of chickpea (*Cicer arietinum* L.), Plant Cell Rep, 12: 521-524.

Christou, P. and Yang, N.S. j1989. Developmental aspects of Soybean (*Glycine max*) somatic embryogenesis, Ann Bot. 64: 225-234.

Kumar. D.V., Kirti, P.B., Sachan, J.K.S. and Chopra, V.L. 1994. Plant regeneration via somatic embryogenesis in chickpea (*Cicer arietinum* L.), Plant Cell. Rep. 13: 468-472.

Malik, K.A. and Saxena, P.K. 1992. Somatic embryogenesis and shoot regeneration from intact seedlings of *Phaeolus acutifolium* A., *P. aureus* (L.) Wilczek, *P. coccineus* L., and *P. wrightii* L. Plant Cell Rep. 11: 163-168.

McKently, A.H. 1991. Direct somatic embryogenesis from axes of mature peanut embryos, In Vitro Cell Dev. Biol. 27: 197-200.

Murashige, T. and Skoog, F. 1962. A revised medium for rapid growth and bioassays with tobacco tissue cultures, Physiol Plant 15: 473-497.

Sagare, A.P., Suhasini, K. and Krishnamurthy, K.V. 1993. Plant regeneration via somatic embryogenesis in chick pea (*Cicer arietinum* L.), Plant Cell Rep. 12: 652-655.

Singh, D.P. 1991. Chickpea in Genetics and Breeding of Pulse Crops, Kalyani Publishers, pp. 6-61.

Suhasini, K., Sagare, A.P. and Krishnamurthy, K.V. 1994. Direct somatic embryogenesis from mature embryo axes in chickpea (*Cicer arietinum* L.), Plant Sci, 102: 189-194.

Regeneration of Water Stress Tolerant Callus Culture of Tomato (*Lycopersicon esculentum*)

D. K. Srivastava, V. K. Gupta and D. R. Sharma

INTRODUCTION

The potential of *in vitro* cell selection for the production of novel genotypes with valuable attributes for agriculture has attracted considerable attention and resulted in a number of reports on the selection of abiotic and biotic stress tolerant cell lines/plants (Nabors *et al.*, 1984; McHugen and Swartz, 1984; Hasegawa *et al.*, 1984; Handa *et al.*, 1986; Hasegawa *et al.*, 1986; McCoy, 1987; Kumar and Sharma, 1989; Siddeswar and Kavikishore, 1989 and Sabbah and Tal, 1990). Plant cell culture methods allow for the selection of variants such as those resistant to water stress which would have practical application in agriculture. A potential application of such studies is the subsequent regeneration of tolerant plants which has been achieved in rice (Siddeswar and Kavikishore, 1989).

The present study deals with the *in vitro* selection of water stress tolerant cell line of tomato, characterization of selected cell line and its subsequent regeneration into complete plantlets.

MATERIAL AND METHODS

Establishment of callus culture

Callus culture of tomato (*Lycopersicon esculentum* L. cv. *Solan gola*) were initiated from leaf segments of aseptically grown seedlings. Leaf segments of about 0.5-1.0 cm. were cultured on MS medium (Murashige and Skoog, 1962) supplemented with BAP (3mg/l) and NAA (1mg/l). All the cultures were incubated under 5000 lux at 25 ± 2° with 16 hours photoperiod. The MS medium supplemented with 3% sucrose, 100 mg/l meso-inositol, gelled with 0.8% agar and pH adjusted 5.8 was used as the basal medium. For subculturing, concentration of the hormones was reduced to half. The calli were subcultured at intervals of 4 weeks.

Selection of water stress tolerant cell line

Polyethylene glycol (PEG) induces water stress in the medium by acting as a non-penetrating osmotic agent that lowers water potential of the medium. Cell clumps of about 1 mm. in diameter were exposed to the increasing concentration of polyethylene glycol (PEG, M.W. 6000) ranging from 10 g/l to 100 g/l. The cells were selected at the concentration of 100 g/l PEG. Selected callus (on the selective medium) and non-selected callus (on the normal medium) were biochemically characterized for soluble proteins, free proline and free amino acids.

Determination of soluble proteins

Free tissue (calli) were homogenized in phosphate buffer (0.05 M, pH 7.0) and homogenate centrifuged at 10,000 g for 30 minutes. The proteins in the resulting supernatant were precipitated with TCA (final concentration, 10%) and were estimated according to Lowery *et al.,* (1951).

Determination of free proline

Free proline was estimated according to Bates *et al.* (1973).

Extraction and estimation of free amino acids

Free amino acids were extracted with hot 90% ethanol on the boiling water bath. The extraction was repeated thrice and the supernatants were pooled. After evaporation to dryness on a water bath, the residue was dissolved in water to a known volume. The aqueous extract was clarified with basic lead acetate removing excess lead with sodium oxalate crystals. The extracts were filtered and used for estimation of free amino acids. Total free amino acids were determined using ninhydrin according to Lee and Takahashi (1966).

Regeneration of plants from the selected cell clones

Shoots were regenerated from water stress tolerant callus cultures of tomato using the regeneration medium of McCormick *et al.* (1986). For shoot elongation the cultures were transferred to MS medium supplemented with kinetin (1 mg/l) and IAA (0.1 mg/l). About 3-5 cms. long shoots were then transferred to the root regeneration medium (MS medium supplemented with 0.1 mg/l of NAA).

RESULTS AND DISCUSSION

Selection of PEG tolerant cell line

The cells were selected at 100 g/l polyethylene glycol (PEG M.W. 6000). Upon incubation for 40 days, most of the cells died but only a few cells which could tolerate this concentration of PEG grew to form cell clones. Selected clones were successfully subcultured on the selective medium (100 g/l PEG) for 8 weeks and then transferred to the regeneration medium. The selected clones were characterized while they were growing on the selective medium itself (Fig. 1).

Soluble proteins

There was an accumulation of soluble proteins in the selected clones. The selected cells contained twice the soluble protein content as compared to the control (Table-1). In contrast, Handa *et al.* (1983) reported a decline in soluble protein levels in the adapted cell line.

Free amino acids and free proline

Water stress resulted in accumulation of free amino acids including proline. The levels of free proline and free amino acids were respectively about four fold and two fold higher in the selected cells as compared to the control (Table -1). This suggested that the mechanism of synthesis of free proline in the selected cells may be independent of that regulating the levels of other amino acids. Enhanced levels of free amino acids and free proline have also been reported in the selected cells of other plant species (Handa *et al.*, 1983; Handa *et al.*, 1986). Increase in the free amino acids content did not occur at the expense of soluble proteins which were also higher in the selected cells.

TABLE 1: Comparison of the level of solutes in control and PEG induced water stress tolerant tomato cell line.

Solute	**Cell lines**	
	Control	PEG induced water stress tolerant
Soluble proteins mg/g fresh weight	2.25 (32.61)	5.42 (40.15)
Free proline µmoles/g fresh weight	3.35 (48.55)	13.51 (100.07)
Free amino acids µmoles/g fresh weight	3.80 (55.07)	7.84 (58.07)

Values within the parentheses are on dry weight basis.

Increase in the proline could occur either due to fresh synthesis or from breakdown of proline rich proteins during stress. In the present case, *de novo* synthesis may account for an increase in the free proline, as the soluble proteins did not decrease in the water stress tolerant cells. The mechanism of tolerance has been shown to be involved in protection of plant tissues against stress by acting as N-storage compounds, osmosolute and hydrophobic protectant for enzymes and subcellular organelles (Handa *et al.*, 1986; Stewart and Lee, 1974; LeRudulier *et al.*, 1984).

Regeneration of plants from the selected cell clones

The presence of high concentration of BAP in the nutrient medium promoted the regeneration of shoots from water stress tolerant callus cultures of tomato (Fig. 2). The shoots did not elongate on this medium, but when transferred to another medium containing low concentration of kinetin and IAA, the shoots elongated (Fig. 3-4). On transfer to the MS medium supplemented with 0.1 mg/l of NAA, the regenerated shoots developed roots within 7-8 days. The plants are being multiplied and would be tested for water stress tolerance.

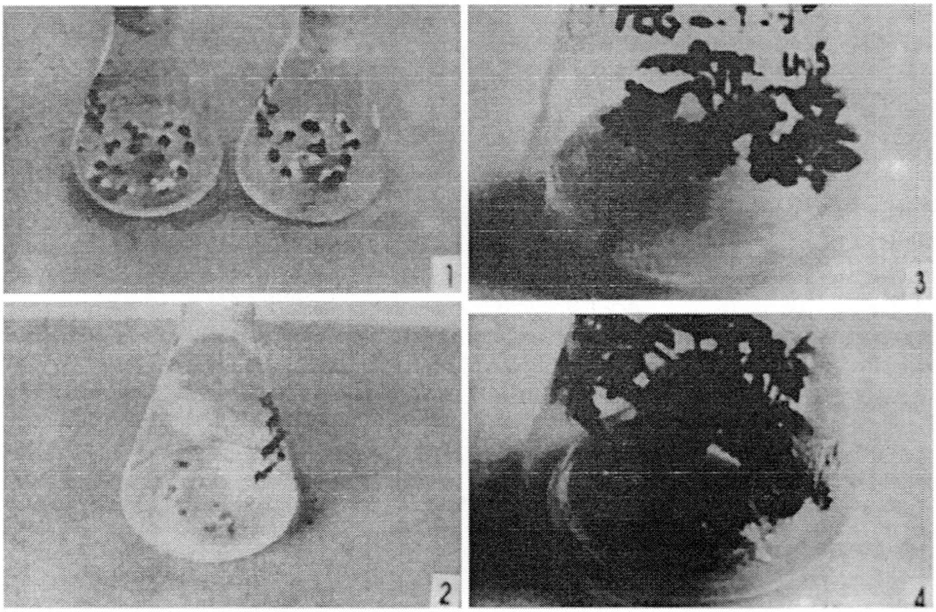

FIG. 1-4. Regeneration in water stress tolerant cell lines of tomato. (1) Growth of selected clones on the stress medium. (2) Regeneration of shoot from the selected clone. (3-4) Shoot proliferation from the selected clone.

REFERENCES

Bates, L.S., Waldren, R.P. and Teare, I.D. 1973. Rapid determination of free proline for water stress studies. Plant and Soil. 39: 205-207.

Handa, S., Bressan, R.A., Handa, A.K., Carpita, N.C. and Hasegawa, P.M. 1983. Solutes contributing to osmotic adjustment in cultured plant cells adapted to water stress. Plant Physiol. 73: 834-843.

Handa, S., Handa, A.K., Hasegawa, P.M. and Bressan, R.A. 1986. Proline accumulation and adaptation of cultured plant cells to water stress. Plant Physiol. 80: 938-945.

Hasegawa, P.M>, Bressan, R.A., Handa, S. and handa, A.K. 1984. Cellular mechanism of tolerance of water stress. Hort. Science. 19: 7-13.

Hasegawa, P.M., Bressan, R.A. and Handa, A.K. 1986. Cellular mechanism of salinity tolerance. Hort. Science. 21: 1317-1324.

Kumar, V. and Sharma, D.R. 1989. Isolation and characterization of sodium chloride resistant callus culture of *Vigna radiata* (L) Wilczek var. radiata. J. Expt. Bot. 40: 143-147.

Le Rudulier, P., Strom, A.M., Dandekar, A.M., Smith, L.T. and Valentine, R.C. 1984. Molecular biology of osmoregulation. Science. 224: 1064-1068.

Lee, Y.P. and Takahashi, T. 1966. An improved colorimetric determination of amino acids with the use of ninhydrin. Annl. Bioch. 14: 71-77.

Lowery, O.H., Rosebrough, N.J., Farr, A.L. and Randall, R.J. 1951. Protein measurement with folin phenol reagent. J. Bio. Chem. 193: 265-275.

McCormick, S., Niedermeyer, J., Fry, J., Barnason, A., Horsch, R. and Fraley, R. 1986. Leaf disc transformation of cultivated tomato (*L. esculentum*) using *Agrobacterium tumefcciens*. Plant cell Reports. 5: 81-84.

McCoy, T.J. 1987. Characterization of alfalfa (*Medicago sativa* L.) plants regenerated from selected NaCl-tolerant cell lines. Tissue Culture Assoc. Meeting, Washington, DC. pp. 92-96.

McHughen, A. and Swartz, M. 1984. A tissue culture derived salt-tolerant line of flax (*Linum usitatissimum*). J. Plant Physiology. 117: 109- 117.

Murashighe, T. and Skoog, F. 1962. A revised medium for rapid growth and bioassay with tobacco tissue culture. Physiol. Plant. 15: 473-497.

Nabors, M.W., Gibbes, G.E., Bernstein, C.S. and Meis, D.E. 1980. NaCl tolerant tobacco plants from cultured cells. Z. Pflanzenphysiol; 97: 13-17.

Sabbah, S. and Tal, M. 1990. Development of callus and suspension culture of potato resistant to NaCl and mannitol and their response to stress. Plant Cell, Tissue and Organ Culture, 21: 119-128.

Siddeswar, G. and Kavikishore, P.B. 1989. Plant regeneration form polyethylene glycol adapted callus of rice. Current Science. 58: 926-928.

Stewart, G.R. and Lee, J.A. 1974. The role of proline accumulation in halophytes. Planta. 120: 279-289.

Biochemical Mechanism of Salt Tolerance in Tomato Cultivars

G. Jansi Rani and G. Balakrishnamurthy

INTRODUCTION

Arid or coastal regions are often unarable due to high salinity. As public demands for water and land increase, the prospect of adopting these areas for agricultural production seems imminent. A wide range of tolerance to salinity is exhibited by plants (Berstein and Hayward, 1957) and several general strategies have evolved: avoidance, exclusion, compartmentation and true tolerance.

When a plant is exposed to high salinity, osmotic adjustment within the cells is required to avoid loss of water and dehydration. A class of small organic molecules like proline and protein were found to accumulate in higher plants. Salt stress affects plant growth by water deficit, high internal concentration of toxic ions and impaired uptake of essential nutrients. In this present study, the effect of NaCl on four different tomato cultivars at different levels of NaCl on proline, protein and ionic accumulation were assessed.

MATERIALS AND METHODS

For the study of nutrient ion accumulation and the biochemical mechanism of salt tolerance, leaf bit calli of tomato cultivars namely Co.2, Co.3, PKM.1 and UC.82 stressed at four different levels of NaCl (0.5, 1.0, 1.5 and 2.0 per cent) were assessed at two stages namely (1) after 35 days in stress condition, (2) after 105 days in stress. The proline content of the stressed calli was estimated by following the method of Bates (1973). The protein content of the calli was estimated by adopting the method of Lowry *et al.* (1951). To estimate the calcium and magnesium content, Jackson's (1973) method was used. Sodium and potassium contents were estimated by following the procedure of Toth and Prince (1949).

RESULTS AND DISCUSSION

Proline and Protein

The influence of different concentrations of NaCl on the proline content of leaf bit calli of four tomato cultivars at 35 and 105 days after NaCl stress are given in Fig. 1. It was seen that an increase in salt levels and duration of stress, showed a significant increase in proline content. Significant difference in proline content was noted between cultivars, callus lines of the salt tolerant cultivar UC.82 (28.18 mg/g) showed higher proline content compared to salt sensitive cultivars Co.2 and Co.3 (22.64 and 24.04 mg/g respectively) indicating that proline may act as a metabolic strategem in maintaining the osmotic balance. This result is in line with the reports of Subhashini and Reddy (1987), Dracup (1991). Simultaneously in all the cultivars, a concomitant increase in the proline content was observed with increased levels (from 0.5 to 2.0%) of NaCl suggesting that the proline accumulates under salt stress conditions and this result corroborates with the earlier findings of Rai *et al.* (1983). During salt stress at both stages, sensitive cultivars Co.2 and Co.3 accumulated lesser proline than the tolerant cultivar PKM.1 and UC.82 which is in line with the reports of Chae *et al.* (1989) who also proved that proline accumulation was high in tolerant cultivar.

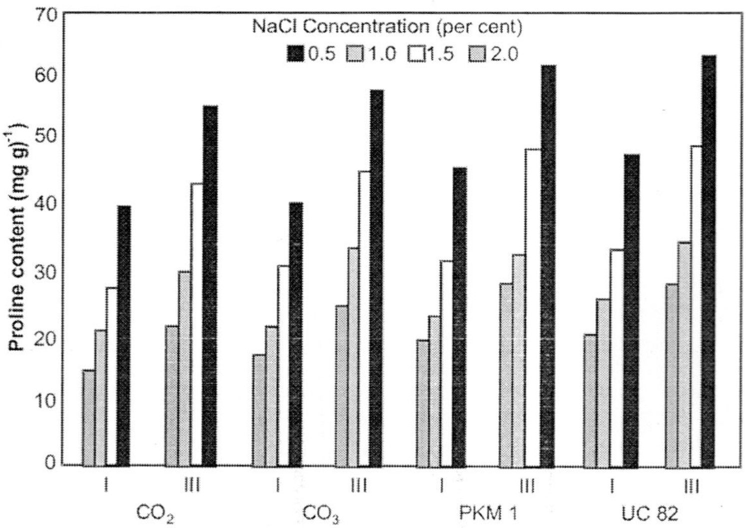

FIG. 1: Influence of NaCl on the proline content (mg g^{-1}) of tomato callus.

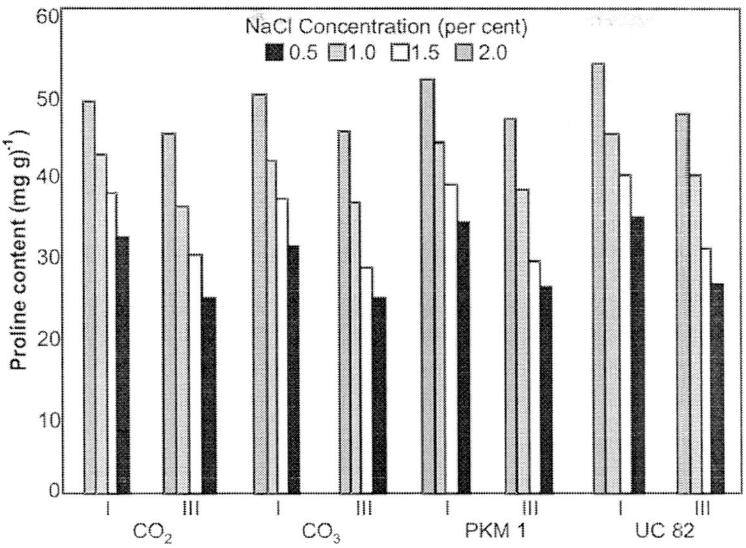

FIG. 2. Influence of NaCl on the protein content (mg g^{-1}) of tomato callus.

The protein content exhibited a different picture with regard to salt tolerant tomato cultivars (Fig. 2). Protein content declines with increasing levels of salinity. The sensitive cultivar. Co.2 (32.62 mg/g) showed more decline in protein content than the other cultivars. This salt induced reduction in protein content may be attributed to the decreased protein synthesis, accelerated proteolysis and denaturation of enzymes (Billard *et al.*, 1982). The salt stress has been found to suppress the protein content in the stressed calli of four cultivars. This result is similar to report of Rajasekaran and Shanmugavelu (1983). Hence it is observed that with increase in duration of stress, a reduction in protein content was observed. At 0.5 per cent NaCl, the reduction was from 51.6 to 46.6 mg at 35 and 105 days respectively. Likewise at 2.0 per cent level, the protein content decreased from 33.2 to 25.9 mg at 35 and 105 days after stress respectively. Irrespective of the cultivars, response for salt tolerance, the protein content of leaf bit calli of tomato reduced with increasing concentrations and duration of stress in the medium. This is concomitant with the report of Bhattacharya (1991) who observed that the protein content of embryo derived calli of rice reduced with increasing levels of NaCl in the medium.

Sodium, potassium, calcium and magnesium

The tolerant cultivar UC.82 (1.67%) recorded higher Na^+ content followed by PKM.1 (1.62%) than the sensitive cultivars Co.3 (1.60\%) and Co.2 (1.43%). In all the four cultivars, increase in salt concentration and duration of stress showed an increase in sodium content (Fig. 3).

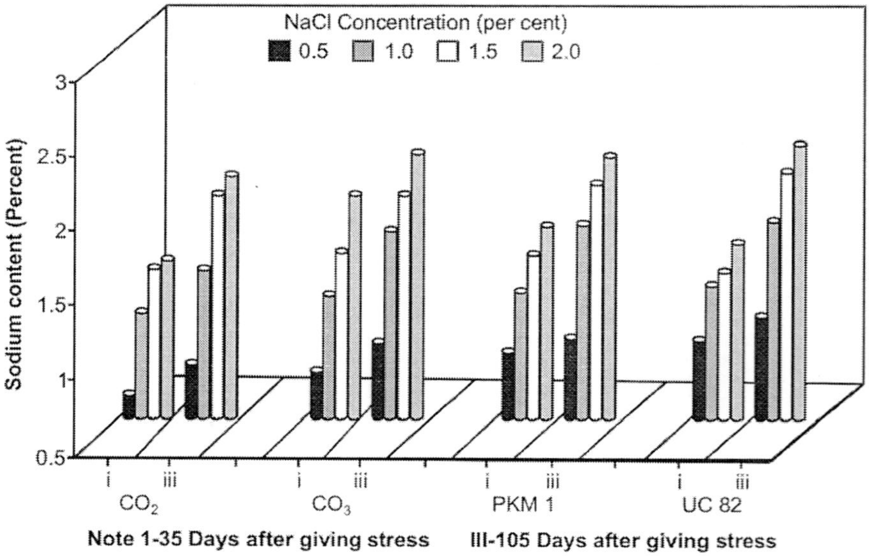

FIG. 3. Influence of NaCl on the sodium content (per cent) of leaf bit calli of tomato cultivars.

Increased NaCl levels resulted in parallel decrease of K^+, Ca^{++} and Mg^+ ions in the stressed calli of all the cultivars (Fig. 4, 5, 6). The decrease of K^+ ion may probably be due to a lower efficiency of K^+ ion may probably be due to a lower efficiency of K^+ displacement from the membrane by Na^+ (Cramer *et al.*, 1986). The present study showed the decline of K^+ content from 2.73% at 0.5 per cent level to 1.84% at 2.0 per cent NaCl level after 35 days in stress indicating that the accumulation of Na^+ in the callus was accompanied by loss of K^+ at all levels of NaCl. Similar results were also reported by Gulati and Jaiwal (1992). The Mg^+ content also declined with fall of Ca^{++}. The per cent calcium content reduced from 1.08% (0.5% NaCl) to 0.85% (2.0% NaCl) at 35 days and 1.06 to 0.82% at 105 days after stress.

The tolerant cultivars UC.82 (1.15%) and PKM.1 (1.00%) recorded higher Ca content than the sensitive cultivars Co.2 (0.77%) and Co.3 (0.82%) indicating that the better ability of UC.82 in permitting the influx of Ca in spite of higher intake of Na might have played a role in its salt resistance. Similarly Arjunan and Chandrasekaran (1988) also reported high amount of Ca in salt tolerant cultivars. Pathmanabhan (1973) suggested that Mg content was higher in tolerant cultivars than the susceptible cultivars. In this study also the tolerant cultivars UC.82 and PKM.1 showed higher Mg content (0.63 and 0.57%) than Co.3 (0.37%) and Co.2 (0.33%). The increase in duration of stress from 35 to 105 days also slightly reduced the Ca^{++}, Mg^+ and K^+ contents in all the cultivars.

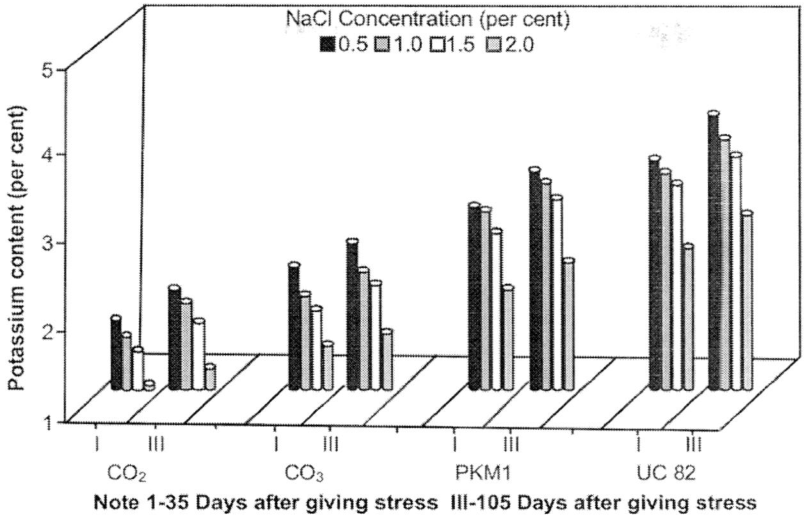

FIG. 4. Influence of NaCl on the potassium content (per cent) of leaf bit calli of tomato cultivars.

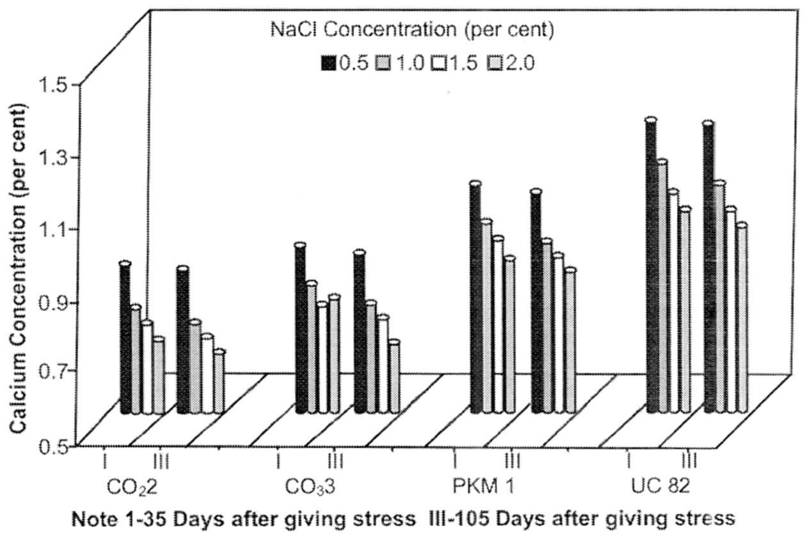

FIG. 5. Influence of NaCl on the calcium content (per cent) of leaf bit calli of tomato cultivars.

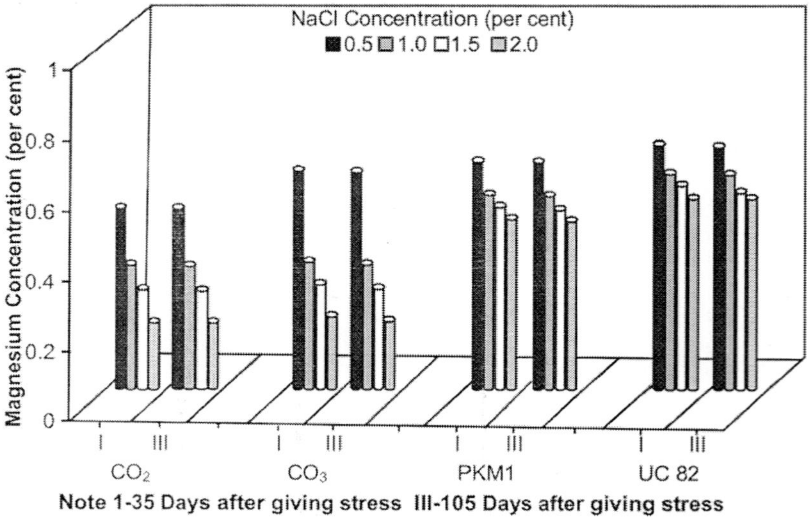

FIG. 6. Influence of NaCl on the magnesium content (per cent) of leaf bit calli of tomato cultivars.

The biochemical and nutritional analyses of the present study indicated that the overall changes that have taken place on the endogenous constituents upon stressing with NaCl were found to certain extent related to salt tolerance on tomato.

There was a positive relationship between salt levels and proline, sodium contents. A negative relationship was observed between salt levels and protein, calcium, magnesium and potassium contents. The proline, calcium and magnesium contents may be considered for screening the genotypes against salt stress.

REFERENCES

Arjunan, A. and Chandrasekaran, S. 1988. Tolerance in rice (*Oryza sativa* L.) in relation to salt uptake and yield. Indian J. Plant Physiol., 31: 403-406.

Bates, L.S. 1973. Rapid determination of tree proline for water stress studies. Plant and Soil, 39: 205-207.

Bernstein, L. and Hayward, H.S. 1957. Physiology of salt tolerance. Ann. Rev. Plant Physiol. 9: 25-46.

Bhattacharya, S. 1991. Differential responses of rice (*Oryza sativa* L.) callus derived from three cultivars in saline environment. Phytomorph., 41: 133-137.

Billard, J.P., Blont, P. and Boucand, J. 1982. Electrophoretic modifications of the soluble leaf proteins of *Gaueda maritima* var. macrocarpa, *Atriplex hortensis* and *Phaseolus vulgaris* in relation to the concentration of NaCl in the culture medium. Soil Fertilizers, 48: 9379.

Chae, Y.A., Kim, Y.H. and Hen, Y.G. 1989. *In vitro* breeding for salt tolerant rice. 3. Detection of reliable selection criterion for salt-tolerance evaluation. Korean J. Breed., 21: 16-21.

Cramer, G.R., Lauchli, A. and Polito, V.S. 1985. Displacement of Ca^{++}, Na^{+} from the plasmalemma of root cells. A primary response to salt stress? Plant Physiol., 79: 207-211.

Dracup, M. 1991. Increasing salt tolerance of plants through cell culture requires greater understanding of tolerance mechanisms. Aust. J. Plant Physiol., 18: 1-15.

Gulati, A. and Jaiwal, P.K. 1992. Comparative salt responses of callus cultures of *Vigna radiata* (L.) Wilczek to various osmotic and ionic stresses. J. Plant Physiol., 141: 120-124.

Jackson, M.L. 1973. Soil chemical analysis. Constable and Co. Ltd., London.

Lowry, O.H., Rosenbrough, N.J., Farr, L.A. and Randall, R.J. 1951. Protein measurement with Folin Phenol reagent. J. Biol. Chem. 193: 265-295.

Pathmanabhan, G. 1973. Evaluation of salt tolerance in sorghum of seedling stage. M Sc. (Ag.) Thesis, TNAU, Coimbatore.

Rai, V.K., Singh, G., Thakur, P.S. and Banyal, S. 1983. Protein amino acid relationship during water stress in relation to drought resistance. Plant Physiol. Biochem. 10: 161-167.

Rajasekaran, L.R. and Shanmugavelu, K.G. 1983. Effect of different types of soil and quality of water on certain physiological factors in tomato. In Proc. Nat. Seminar on Prod. Tech. Tomato and Chillies, Coimbatore, TNAU, p. 64.

Subhashini, K. and Reddy, G.M. 1987. Selection, characterization and regeneration of salt tolerant calli of scented Rice. Proc. Symp. Plant, Cell and Tissue culture of Economically Important Plants (Ed.) G.M. Reddy, 147-151.

Toth, S.J. and Prince, A.L. 1949. Estimation of cation exchange capacity and exchangeable calcium, potassium and sodium contents of soils by Flame photometer techniques. Soil Sci., 67: 439-445.

In Vitro Evaluation and Regeneration of Tomato Under NaCl Stress

G. Jansi Rani and G. Balakrishnamurthy

INTRODUCTION

High salt concentration in soils of arid zone lands present a major problem to agricultural and horticultural productivity. This is especially serious since the area of salt affected land is increasing annually. One of the solutions to this widespread problem is to select and breed lines with enhanced salt tolerance. A recent technique that may assist the traditional plant breeding approach is Plant Tissue Culture (Chandler *et al.*, 1986). In this approach tissue selection can be carried out on a large number of individual cells which is much more economical of time and space.

Stavarek and Rains (1984) proposed identification, selection and regeneration of salt tolerant genotypes from cell culture as a way to breed for salt tolerance. In the present study, attempts were made to identify and regenerate the salt tolerant mutant lines of cv. Co.3 under NaCl stress.

MATERIALS AND METHODS

Tomato seeds of the salt sensitive but high yielding locally preferred cultivar Co.3 were grown in pots under glass house conditions. Young leaves at the fourth node were excised from 20-25 days old seedlings. Leaf segments were isolated aseptically (Balakrishnamurthy and Irulappan, 1992) and inoculated in MS medium with growth regulator combination of BAP 2.5 mg l^{-1} + IAA 0.5 mg l^{-1} for callus induction. After 35 days, calli with uniform size were collected and inoculated in fresh medium supplemented with NaCl at varying concentrations (0.5, 0.6, 0.7, 0.8, 0.9 and 1.0 per cent) which were found to be optimum for callus survival (Jansi Rani and Balakrishnamurthy, 1994). The screening of calli under NaCl stress consisted of five passages each with 35 days. The surviving calli at the end of each passage were taken to their respective NaCl medium in the five passages. After the fifth passage, surviving calli in all the treatments were cultured in MS medium supplemented with BAP (3.0 mg l^{-1}) and IAA (0.5 mg l^{-1}) for differentiation of shoots. The shoots thus obtained were cultured in MS medium supplemented with GA (1.0 mg l^{-1}) (T_1) and IAA (0.5 mg l^{-1}) + IBA (1.0 mg l^{-1}) (T_2) for induction of roots. The regenerated plantlets were screened for their survival and also under field condition for salt tolerance.

RESULTS AND DISCUSSION

In the present study, to generate somaclonal variants and to identify salt tolerant lines in the cultivar Co.3, leaf bit calli were screened in culture medium containing NaCl from 0.5 to 1.0 per cent. Results indicated that the survival of calli was the highest (58.35%) in the first passage and was the lowest in the fifth passage (14.24%) indicating that the increase in duration of stress caused decline in survival (Table 1).

TABLE 1: Influence of NaCl at different concentrations on the survival rate (per cent) of Co.3 leaf bits through 5 passages

	NaCl concentration (per cent)						
Passages	0.5	0.6	0.7	0.8	0.9	1.0	Mean
1	67.12 (8.14)	64.11 (8.03)	61.72 (7.88)	59.22 (7.72)	57.54 (7.61)	41.62 (6.48)	58.35 (7.66)
2	50.32 (7.12)	47.51 (6.92)	45.87 (6.80)	44.25 (6.68)	41.62 (6.48)	37.45 (6.15)	44.99 (6.74)
3	43.64 (6.63)	38.31 (6.22)	29.88 (5.50)	27.50 (5.28)	24.13 (4.95)	22.00 (4.73)	30.73 (5.58)
4	37.41 (6.15)	28.25 (5.35)	25.23 (5.06)	23.24 (4.86)	16.52 (4.11)	0.00 (0.61)	21.64 (4.69)
5	34.12 (5.87)	24.13 (4.95)	14.91 (3.91)	9.11 (3.08)	1.42 (1.34)	0.00 (0.61)	14.24 (3.82)
Mean	46.82 (6.87)	40.46 (6.39)	35.52 (5.99)	32.66 (5.75)	28.25 (5.35)	20.21 (4.54)	33.99 (5.86)

In the first passage, the survival rate of calli (67.12%) was higher at 0.5 per cent NaCl than at 1.0 per cent NaCl level (41.2%) whereas in the fifth passage a low rate of survival (34.12%) at 0.5 per cent and no survival at 1.0 per cent NaCl was observed. Similarly a gradual decline in the rate of survival of calli from 0.5 to 1.0% NaCl was reported by Li *et al.* (1988) and Mallek and Richez (1988).

The surviving calli after passing through five passages were taken for shoot differentiation and regeneration. The response for shoot differentiation of the NaCl adapted calli (0.5 to 0.9%) ranged from 86.7 per cent to 8.3 per cent in 20 and 36.3 days respectively. No regeneration was observed in 1.0 per cent NaCl adapted calli (Table 2). Similar trend was observed by Bajaj and Gupta (1986) in *Pennisetum purpureum* where the maximum response for regeneration of plantlets were obtained from cells tolerant to 0.5 per cent NaCl. However, the regeneration was severely depressed in NaCl tolerant calli which were subjected to higher levels of NaCl (Li, 1990 and Collins *et al.*, 1990).

TABLE 2: Efficiency of NaCl adapted leaf bit calli of Co.3 tomato to shoot regeneration in salt free medium with BAP 3.0 mg l^{-1} + IAA 0.5 mg l^{-1}

NaCl adapted callus	Per cent response for shoot differentiation	Days taken for shoot differentiation	Mear no. of shoots	Mean length of shoots (cm)
0.0	98.3	15.0	5.3	7.9
0.5	86.7	20.0	5.0	5.7
0.6	81.7	21.7	3.3	5.0
0.7	75.0	24.7	2.7	3.7

NaCl adapted callus	Per cent response for shoot differentiation	Days taken for shoot differentiation	Mear no. of shoots	Mean length of shoots (cm)
0.8	66.7	25.3	2.3	3.3
0.9	8.3	36.3	0.7	0.7
1.0	0.0	0.0	0.0	0.0

The regenerated shootlets were taken to the root differentiation medium containing GA 1.0 mg l^{-1} (T_1) and IAA 0.5 mg l^{-1} + IBA 1.0 mg l^{-1} (T_2). Lengthy roots were produced at T_2 than T_1 (Table 3).

The calli adapted to 0.5 per cent NaCl showed high response (93.3 per cent) for rooting in T_1 and 20.6 days for root initiation whereas in T_2, the response for rooting was low (80.7 per cent) and root initiation was earlier (13 days). Similarly the 0.8 per cent NaCl adapted calli gave 70.7 and 65.0 per cent response and showed root initiation in 25 and 20 days in T_1 and T_2 respectively.

TABLE 3: Efficiency of NaCl adapted leaf bit calli of Co.3 tomato root regeneration in soil free medium

Treatments + NaCl adapted calli		Per cent response for root initiation	Days taken for root initiation	Mean no. of roots	Mean length of roots (cm)
T1 +	0.0	98.3	18.7	14.7	5.4
T2 +	0.0	83.3	8.7	8.0	6.9
T1 +	0.5	93.3	20.6	12.0	4.3
T2 +	0.5	80.7	13.0	5.3	6.2
T1 +	0.6	85.7	21.5	11.0	3.4
T2 +	0.6	75.3	15.0	5.0	5.9
T1 +	0.7	78.3	23.0	9.0	2.9
T2 +	0.7	72.5	78.0	4.5	5.5
T1 +	0.8	70.7	25.0	4.9	2.6
T2 +	0.8	65.0	20.0	3.5	4.7

T1: MS + GA 1.0 mg l^{-1}; T2: MS + IAA 0.5 mg l^{-1} + IBA 1.0 mg l^{-1}

These observations lend evidence to the earlier findings (Coleman and Greyson, 1977) which showed GA_3 stimulated root formation in a number of regeneration systems. This can be explained in the light of findings by varga and Humphries (1974) who reported enhancement of rooting by GA_3 by stimulating endogenous auxin synthesis in a number of non regenerating systems.

The plantlets regenerated from calli adapted to 0.5 per cent NaCl showed higher survival than the plantlets regenerated from calli adapted to higher levels of NaCl indicating the toxic effects of salt or osmotic stress. This result is in line with the findings of Kishor and Reddy (1989) which showed that the regeneration of salt adapted calli diminished rapidly with time and concentration of NaCl.

Similarly, loss of ability to regenerate or less frequency of regeneration of tissues which have been exposed to high concentration of NaCl was found in carrot (Komizenko and Khretonova, 1973), alfalfa (Smith and McComb, 1981) and Pennisetum (Bajaj and Gupta, 1986).

The regenerated plantlets were hardened under mist chamber and are being screened under field conditions for further study.

REFERENCES

Bajaj, Y.P.S. and Gupta, R.K. 1986. Different tolerance of callus culture of *Pennisetum americanum* (L.) and *P. purpureum* Schum to sodium chloride. J. Plant Physiol., 125: 491-495.

Balakrishnamurthy, G. and Irulappan, I. 1992. *In vitro* Propagation in Tomato (*Lycopersicon esculentum* Mill.). New Trends in Biotechnology eds. N.S. Subba Rao, C. Balagopalan and S.V. Ramakrishna. Oxford and IBH Pub. Co. Pvt. Ltd.

Chandler, S.F., Mandal, B.B. and Thorpe, T.A. 1986. Effect of sodium sulphate on tissue culture of *Brassica napus* cv. Westar and *Brassica campestris* L. cv. Toibin. J. Plant Physiol., 126: 105-117.

Coleman, W.K. and Greyson, R.I. 1977. Analysis of root formation in leaf discs of Lycopersicon esculentum Mill. cultured *in vitro*. Ann. Bot., 41: 307-320.

Collins, J.C., Ibrahim, K.M. and Collin, H.A. 1990. *In vitro* selection for salt tolerance in *Coleus blumei*. Acta. Hort., 280: 431-435.

Flippone, E. 1985. *In vitro* growth and differentiation of tomato (*Lycopersicon esculentum*) tissue on high level of NaCl. Genetica Agraria, 39: 323.

Jansi Rani, G. and Balakrishnamurthy, G. 1994. *In vitro* studies on salt tolerance in tomato (*Lycopersicon esculentum* Mill). M.Sc. Thesis Tamil Nadu Agricultural University, Coimbatore.

Kishor, P.B.K. and Reddy, G.M. 1989. Salt stress in callus cultures of rice. Second International Rice Genetics. Symposium, Manila, Philippines, Poster, 26.

Komizenko, E.I. and Khretonova, T.I. 1973. Effect of NaCl on the process of somatic embryogenesis and plant regeneration in carrot tissue cultures. Soviet Pl. Physiol., 20: 220-227.

Li, C. 1990. Induction, selection and preliminary identification of variants for tolerance to NaCl of *Medicago sativa* L. Genetica Agraria, 42: 75.

Li, N.J., Filippone, E., Cardi, T. and Console, R. 1988. *In vitro* and *in vivo* evaluation of tolerance to salt in tomato. Genetics Agraria, 42: 79.

Mallek, E. and Richez, M. 1988. Effect of salinity on tomato: response of the F_2 hybrid (*L. esculentum* x *S. pennellii*) x *S. pennellii*. Plant Physiol. Biochem., 26: 209.

Smith, M.K. and McComb, J.A. 1981. Effect of NaCl on the growth of whole plants and their corresponding callus cultures. Aust. J. Plant Physiol., 8: 267-275.

Stavarek, S.J. and Rains, D.W. 1984. Cell culture technique: Selection and physiological studies of salt tolerance. In: R.C. Staples and G.H. Toenniessen (Eds.) Salinity tolerance in plants. pp. 321-334. John Wiley, New York.

Varga, M. and Humphries, E.C. 1974. Root formation on petioles of detached primary leaves of dwarf bean (*Phaseolus vulgaris*) pretreated with gibberellic acid, triiodobenzoic acid and cytokinins. Ann. Bot. N.S., 39: 803.

In Vitro Selection of Atrazine Tolerant Cell Lines and Regeneration Studies in *Brassica campestris*

G. S. Virk and Jaswinder Kaur

INTRODUCTION

Herbicides have become an indispensable tool of modern agriculture, since they act as economically superior weed control. Plant tissue culture plays a significant role in the production of many herbicide-resistant varieties with better yield and it has met with more success than through conventional plant breeding. The progress achieved in the genetic manipulation of plants and the ability to transfer foreign genetic material from a variety of sources to plants have opened an exciting and productive area of research. A great area of pioneering research remains to be done to develop a range of viable herbicide-resistant crop plants and it is likely to be a very complicated process. The essential field testing of transgenic and regenerated plants is time consuming, tedious and has to be implemented under the conditions in which the risk of introducing and spreading new plants in agriculture ecosystem is minimized.

Oilseed crops are among important cash crops and good source of storage lipids in plant kingdom. *Brassica campestris* having enormous economic value is being extensively employed in tissue culture studies. The present investigation was undertaken to isolate the atrazine tolerant cell lines in *Brassica campestris* employing tissue culture techniques.

MATERIAL AND METHODS

Seeds of Brassica campestris were procured from Oilseed Section of Punjab Agricultural University, Ludhiana and were used for callus induction, preparation of suspension cultures and selection of herbicide tolerant cell lines.

The herbicide atrazine employed to raise tolerant cell lines is a wettable powder containing 50% (w/w) atrazine which is 2-chloro-4-ethylamino-6-isopropylamino-5-triazine ($C_8H_{14}ClN_5$) having molecular weight 215.7 g. It was procured from Northern Minerals Ltd., Gurgaon.

In vitro grown seedlings under aseptic conditions served as the source of inoculum for different experiments. MS medium was employed for raising callus and preparation of suspension cultures. MS medium was notified by supplementing it with various combinations and concentrations of different growth regulators viz. 2,4-D, IAA, KN, NAA and GA_3 (Table 1). Explants viz. shoot apex, root apex, hypocotyl, young leaves and cotyledons were surface sterilized with 0.02% $HgCl_2$ and thoroughly washed before inoculation. The cultures were incubated at 26 ± 1 °C in well illuminated growth chamber having 80% humidity. Callus initiation was noticed after 5-6 days of inoculation and it was allowed to grow for 4-5 weeks before subculturing.

TABLE 1: Growth regulators, their concentration, solvent, storage temperature and sterilization conditions.

Growth regulator	Concentration (mg/l)	Solvent	Storage	Sterilization
2,4-D	2	50% ethyl alcohol	RT	Co-autoclave
IAA	2	1 N NaOH	-0 °C	Co-autoclave
KN	4	1 N HCl	-0 °C	Co-autoclave
NAA	2	Absolute alcohol	-0 °C	Co-autoclave
GA_3	2	Absolute alcohol	-0 °C	Sietz Filter

RT = Room temperature

For raising suspension cultures, 1.5 g of fast growing calli were inoculated aseptically into 100 ml flask containing 20 ml of liquid medium and a few glass beads. These were kept on rotary shaker at a speed of 100-200 rpm at 25 °C. The suspension cultures were multiplied by pipetting out 5 ml fine suspension into the flask having 20 ml liquid culture medium and 2,4-D. The growth of the cells in these flasks was measured weekly with the help of haemocytometer. Bergman's cell plating technique was followed for raising single cell suspension cultures.

TABLE 2: Regeneration media, concentration of growth regulators and kind of medium.

Supplementation of MS medium	Concentration of growth regulators (mg/l)	Kind of medium
MS (Control)	-	-
MS + KN	MS + 4	Shoot inducing
MS + 2,4-D + KN	MS + 2 + 2	Shoot inducing
MS + NAA	MS + 2	Root inducing
MS + 2,4-D + NAA	MS + 2 + 2	Root inducing
MS + IAA	MS + 2	Root inducing
MS + 2,4-D + IAA	MS + 2 + 2	Root inducing

Different concentrations (0.5, 1, 1.5 and 2%) of herbicide (atrazine) were prepared in distilled water at pH 5.8. The required amount of atrazine at a particular concentration was added to the suspension culture which was mixed with previously autoclaved media in 1:1 ratio. This mixture was poured into sterilized petri plates. These were inoculated, sealed with Johanson's tape and incubated in dark at 26 ± 1 °C. The cultures were examined weekly for the growth of resistant colonies. The observed colonies at lower doses were picked up, their suspension was prepared and spread onto the medium having higher concentrations and lethal concentration. The resistant colonies observed in lethal concentration were transferred to MS medium supplemented with growth regulators in order to develop callus. Attempts were made to regenerate callus on different regeneration media (Table 2).

RESULTS

Among various explants inoculated, only cotyledons and injured roots attached to seeds resulted in well proliferated calli whereas root, shoot and leaf apices did not lead to callusing (Table 3).

TABLE 3: Percentage of callus initiation with different explants.

Explant	Medium	Number of replicates	Number of tubes with callus initiation	Percentage of callus initiation
Injured root attached to seed	MS + 2,4-D	32	26	81.25
Cotyledon	MS + 2,4-D	32	8	25.00
Shoot tips	MS + 2,4-D	32	-	-
Root tips	MS + 2,4-D	32	-	-
Leaf	MS + 2,4-D	32	-	-

Callus initiation was found to be maximum in the case of injured root attached to seed.

Callus initiation

MS medium was found to be the most suitable for callus induction when its efficiency was compared with LS and B-5 media (Table 4). MS medium was supplemented with 2,4-D, IAA and NAA and the best results were obtained with 2,4-D (2 mg/l). It resulted in direct callusing from injured roots attached to seed within a week. After this, the callus grew well. Calli were also obtained when LS and B-5 media were supplemented individually with 2,4-D (2 mg/l), but the explant inoculated was cotyledons. The supplementation of MS medium with 2,4-D (2 mg/l) and NAA (2 mg/l) resulted in feeble callus along with the development of small roots (Table 5).

TABLE 4: Comparison of different culture media for callus initiation.

Medium	Number of tubes inoculated	Number of tubes with callus initiation	Percentage
MS	32	26	81.25
LS	32	10	31.25
B_5	32	7	21.75

TABLE 5: Effect of different auxins and culture media on callus induction.

Callus source	Medium	Auxin	Concentration (mg/l)	Response towards callus initiation
Injured root attached to seed	MS	2,4-D	2	Callus started initiation after 6-7 days of inoculation.
Injured root attached to seed	MS	NAA	2	Callus started initiation after 2 weeks of inoculation.
Injured root attached to seed	MS	IAA	2	Callus started initiation after 4 weeks of inoculation.
Cotyledons	LS	2,4-D	2	Callus initiated after 10 days of inoculation but grew very slowly.

Callus source	Medium	Auxin	Concentration (mg/l)	Response towards callus initiation
Cotyledons	LS	NAA	2	Callus initiated after 10 days of inoculation but grew very slowly.
Cotyledons	B_5	2,4-D	2	Callus initiated after 15 days of inoculation with feeble development.
Cotyledons	B_5	NAA	2	Callus initiated after 15 days of inoculation with feeble development.

Maintenance and subculturing of callus

Calli were maintained at 26 ± 1°C under fluorescent tube lights and multiplied by periodical subculturing on MS medium supplemented with 2,4-D. Only well proliferated calli with sufficient growth were chosen for subculturing. The increase in the growth of callus was determined on the basis of increase in final weight of callus after 5 weeks (Table 6). The results indicated that the increase in weight was more on MS medium supplemented with 2,4-D.

Table 6: Growth rate of callus

Callus source	Medium	Initial weight after 15 days of inoculation (mg)	Final weight after 35 days of inoculation (mg)	Increase in weight of callus in 20 days (mg)
Injured root attached to seed	MS + 2,4-D	545	905	360
Cotyledon	B_5 + 2,4-D	520	753	233

Regeneration

The effect of NAA, KN, IAA and GA3 was studied individually and in combination with each other on the induction of differentiation and regeneration of callus.

Effect of individual treatments of growth regulators

NAA (2 mg/l) supplemented MS medium resulted first in vitorous growth of callus and then there was formation of thick roots. KN (4 mg/l) resulted in enhanced growth of shoots and roots and it also played a regulatory role for regeneration in liquid MS medium. GA_3 (2 mg/l) first resulted in enhancing the growth of callus which was followed by feeble differentiation of roots and shoots (Table 7).

TABLE 7: Effect of individual and combined treatments of growth regulators on regeneration.

Medium	Growth regulator	Concentration of growth regulator (mg/l)	Effect of growth regulator
MS	NAA	2	Initially the callus grew vigorously followed by the formation of short and thick roots.
MS	KN	4	Callus started differentiating into shoots and roots.
MS	GA_3	2	Initially the callus grew well which was

Medium	Growth regulator	Concentration of growth regulator (mg/l)	Effect of growth regulator
			followed by feeble formation of roots and shoots.
MS	2,4-D + IAA	2 + 2	Long and thin roots differentiated.
MS	2,4-D + NAA	2 + 2	Initially the callus growth was enhanced followed by the formation of thick roots.
MS	2,4-D + KN	2 + 4	Enhanced growth of callus followed by differentiation of shoots only.
MS	NAA + KN	2 + 4	Callus started directly differentiating into roots and shoots followed by the formation of plantlets.
MS	NAA + GA_3	2 + 2	Callus did not differentiate into roots or shoots even after 3 months nor the growth of callus was enhanced.
MS	NAA + IAA + KN	2 + 2 + 4	Formation of green coloured embryoid-like structures after 35 days which did not develop further.

Effect of combined treatment of growth regulators

The results are presented in Table 7.

2,4-D + IAA: 2,4-D (2 mg/l) + IAA (2 mg/l) supplementation of MS medium resulted in differentiation of roots only. The roots formed were thin and long when compared to thick and short roots formed by individual treatment with IAA.

2,4-D + NAA: 2,4-D (2 mg/l) + NAA (2 mg/l) supplementation of MS medium resulted in the formation of thick roots. It also enhanced the growth of callus.

2,4-D + KN: 2,4-D (2 mg/l) + KN (4 mg/l) supplementation of MS medium resulted in increased growth of callus and only shoot differentiation was noticed.

NAA + KN: NAA (2 mg/l) + KN (4 mg/l) supplementation of MS medium resulted in differentiation of roots as well as shoots and these developed into small plantlets.

NAA + GA_3: NAA (2 mg/l) + GA_3 (2 mg/l) supplementation of MS medium resulted in growth of callus only. Callus did not differentiate into roots or shoots even after 3 months.

NAA + IAA + KN: MS medium supplemented with NAA (2 mg/l) + IAA (2 mg/l) + KN (4 mg/l) resulted in formation of green coloured embryoid-like structure after about 35 days. However, it did not develop further.

Suspension cultures

A direct correlation was noticed among the number of cells in suspension cultures with increase in the number of days. The cells in suspension cultures were found to be rich in cytoplasm having potential for further division. The lowest number of cells observed was 12^2-10^3 after one week whereas the highest number was 10^4 after 2 weeks.

Colonies on control medium

The colonies appeared after 6-7 days on MS medium supplemented with 2,4-D (2 mg/l). The number of colonies increased following an increase in the number of inoculated cells per ml (Table 8).

TABLE 8: Number of colonies developed on control medium.

Control medium	Number of plates used	Number of cells inoculated per ml	Number of colonies developed per plate	Days after which the colonies appeared
MS + 2,4-D (2 mg/l)	10	5×10^3	700	6
	12	5×10^4	835	7
	11	5×10^4	915	7
	10	6×10^4	958	6

Effect of herbicide treatment

MS medium supplemented with 0.5% atrazine resulted in an increase in the number of colonies. However, with further increase in the concentration of atrazine (1%), a sudden reduction in the number of colonies was observed. Still higher concentrations of atrazine (1.5 and 2%) further reduced the number of colonies. The cell survival at 2.5% concentration of atrazine was nil, hence it proved lethal.

TABLE 9: Average number of colonies developed on herbicide-supplemented medium.

Medium	Percentage of atrazine	Number of plates	Average number of colonies developed per plate
MS + 2,4-D (2 mg/l)	0.5	10	124
	1.0	10	98
	1.5	10	18
	2.0	10	8
	2.5	-	-

The average number of colonies per plate with concentrations of agrazine (0.5, 1, 1.5 and 2%) was found to be 124, 98, 18 and 8 respectively (Table 9). In order to detect the resistance, the colonies growing in 2% concentration of atrazine were picked up and their suspensions having 10^2-10^3 cells per ml were spread on the modified MS medium [MS + 2,4-D (2 mg/l) + atrazine (2.5%)]. A few cells survived which further developed into colonies and the procedure was repeated several times.

Regeneration of whole plants

The resistant colonies were picked up and transferred to regeneration medium (Table 10) which resulted in normal growth of callus, but no differentiation occurred. Several trials were conducted employing different concentrations and combinations of growth regulators, but no success was achieved.

TABLE 10: Regeneration of resistant colonies.

Regeneration medium	Number of tubes inoculated	Callus growth	Regeneration
MS + KN (4 mg/l)	15	+	-
MS + 2,4-D (2 mg/l) + kn (4 mg/l)	15	+	-
MS + NAA (2 mg/l)	15	+	-

Regeneration medium	Number of tubes inoculated	Callus growth	Regeneration
MS + 2,4-D (2 mg/l) + NAA (2 mg/l)	15	+	-
MS + IAA (2 mg/l)	15	+	-
MS + 2,4-D (2 mg/l) + IAA (2 mg/l)	15	+	-

DISCUSSION

In recent years, plant cell cultures have become more important in research on the production of herbicide tolerant plants. These possibilities have been well demonstrated by bentazone resistant mutants of *Brassica napus* selected by Schenck and Robbellen in 1985. Herbicide tolerant cell lines *in vitro* have also been used to regenerate whole plants. Rey *et al.* (1990) produced atrazine and diuron resistant plants from photoautotrophic protoplasts-derived cultures of *Nicotiana plumbaginifolia*. OAC triumph, a canola quality summer rape (*Brassica napus* L.) cultivar, resistant to S-triazine and tolerant to S-triazinone has been produced by Beversdorf *et al.* (1990).

During the present investigation, atrazine tolerant cell lines of *Brassica campestris* were isolated and attempts were made to regenerate whole plants from these. The best callus induction and its subsequent proliferation occurred on MS medium supplemented with 2,4-D when roots attached to seeds were used as explant. This fact has been supported by the findings of Antonio *et al.* (1987). Murata and Orton (1987) and Jordan *et al.* (1990) also reported cell induction on MS medium supplemented with 2,4-D. Well proliferated calli were obtained from injured roots attached to seeds which is in contradiction to the findings of Antonio *et al.* (1990), who reported that callus growth was vigorous from cotyledonary segment. However, they reported that regeneration was most prominent in calli derived from injured roots attached to seeds.

The nature of explant, composition of the culture medium and addition of growth regulators were found to affect the frequency of differentiation. Direct root and shoot regeneration was noticed during the present investigation. KN promoted shooting whereas the addition of NAA promoting rooting. The best results were obtained when MS medium was supplemented with NAA (2 mg/l) + KN (4 mg/l). GA3 resulted in feeble differentiation either alone or in combination with NAA. Several workers viz. Pareek and Chandra (1981), Bajaj and Mohapatra (1987), Murata and Orton (1987), Narasimhulu and Chopra (1987), Vyvadilova and Zelenkova (1987), Fuchs *et al.* (1988), Jain *et al.* (1988a,b). Parkash *et al.* (1988) and Vilor *et al.* (1988) also reported a dominant role of genotype on shoot differentiation from callus cultures.

Suspension cultures showed a marked increase in the number of cells with increase in the number of days of inoculation. Lentini *et al.* (1986) prepared suspension cultures from callus of *Brassica napus*. A fine cell suspension was made by Soprasanna *et al.* (1987) from maize on MS medium supplemented with 2,4-D. Suspension cultures were also obtained directly from aseptically grown plant material (Mezentsev *et al.*, 1987).

Lower concentrations of atrazine (0.5 and 1%) were found to have a stimulatory effect on the formation of colonies whereas higher doses (2 and 2.5%) resulted in decreased number of colonies. Only a single resistant colony was found in 2.5% concentration of atrazine. Schenck and Robbellen (1985) and Rey *et al.* (1990) selected several mutants in the species of *Nicotiana* and *Brassica* after treatment with different herbicides employing *in vitro* cultures.

Attempts were made to regenerate whole plants from selected resistant cells. The resistant cells developed into callus, but further differentiation was not noticed. Rey *et al.* (1990) reported positive response of regeneration in *Nicotiana plumbaginifolia*. They obtained atrazine and diuron resistant clones and plants were regenerated from most tolerant calli. Similarly, Buschanan *et al.* (1990) obtained DPX-A7881 tolerant cell lines of cultivars of *Brassica juncea*, *B. napus* and *B. oleracea*, but no regeneration has been reported. However, the work is in progress and trials are being made to regenerate whole plants from resistant cells.

SUMMARY

The present investigation deals with the isolation of herbicide-resistant cell lines of *Brassica campestris*. Callus was mainly derived from injured roots attached to seeds using MS medium and from cotyledons using Gamborg B5 medium. Other explants like shoot tips, root tips, leaves, etc. did not respond well towards callus formation. MS medium was supplemented with different auxins viz. 2,4-D, NAA and IAA, etc. using different concentrations, but the best results were obtained when MS medium was supplemented with 2,4-D alone (2 mg/l). Callus was maintained by periodical subculturing on the same medium.

The modified MS medium was supplemented with different concentrations of atrazine (0.5, 1, 1.5, 2 and 2.5%). The concentration of 2.5% of atrazine was proved to be lethal dose for cell lines of *Brassica campestris*. These atrazine-resistant colonies were isolated. Callus was raised from atrazine-resistant colonies and attempts were made to regenerate whole plant from resistant colonies. Trials are being carried out to regenerate the whole plant from resistant colonies.

REFERENCES

Antonio, B.A., Namai, H. and Kikuchi, F. 1987. Tissue culture ability of vegetative organs from different cultivars of *Brassica*. Sabrao J. 19: 73-79.

Bajaj, Y.P.S. and Mohapatra, D. 1987. *In vitro* plant regeneration in *Brassica carinata*- An oilseed crop. Ind. J. Exp. Biol. 25: 161-163.

Beversdorf, W.D., Hume, D.J., Gostouic, P., Chu, G., Montminy, W. and Muscat, J. 1990. OAC triumph summer rape. Can. J. Plant Sci. 70: 857-860.

Buschanan, F.S., Swanton, C.J. and Gillespie, T.J. 1990. Differential response of selected species of Brassicaceae to DPX-A 7881. Can. J. Plant Sci. 70: 873-878.

Fuchs, A., Pannem, S.R. and Al-Chabak, M. 1988. Callus induction and shoot regeneration in several tropical plant species. Bert. Z. Trop. Land. W. Vet. Med. 26: 11-18.

Jain, R.K., Chowdhury, J.B. and Friedt, W. 1988a. Organogenesis and plant formation from cotyledon explant cultures of wild turnip rape (*Brassica tournefortii* L.). Plant Cell Tissue Org. Cult. 15: 107-111.

Jain, R.K., Chowdhury, J.B. and Friedt, W. 1988b. Genotypic and media effects on plant regeneration from cotyledon explant cultures of some Brassica species. Plant Cell Tissue Org. Cult. 14: 197-206.

Jordan, A.M., Maria, C.C. and Segura, J. 1990. Morphogenesis in callus and single cell cultures of *Lavandula latifolia* Medicus. J. Hort. Sci. 65: 49-54.

Lentini, Z., christey, M., Earle, E.D. and Mutschler, M.A. 1986. Cell suspension culture of rapid cycling *Brassica campestris* L. Cruciferae Newslet. 11: 86-87.

Mezentsev, A.V. and Kostenko, S.I. 1987. A universal method of producing plant cell suspension cultures. Doklady Vsesoyuznoi Ordena Leneina i. Skokho Zyaistvehmykh Nauki Imehi V.I. Lenina, 9: 3-5.

Murata, M. and Orton, T.J. 1987. Callus initiation and regeneration capacities in *Brassica* species. Plant Cell Tissue Org. Cult. 11: 111-123.

Narasimhulu, S.B. and Chopra, V.L. 1987. Plant regeneration from callus cultures of *Brassica carinata* A. Br. and its implications to improvement of oilseed Brassicas. Plant Breed. 99: 49-55.

Pareek, L.K. and Chandra, N. 1981. Induced regeneration in vegetative and flowering internodal segments and ovary explants of *Brassica juncea* L. Coss. Ind. J. Exp. Biol. 19: 874-875.

Parkash, S., Sharma, D.R. and Chowdhury, J.B. 1988. High frequency *in vitro* morphogenesis in *Brassica carinata* A. Br. Cruciferae Newslett. 13: 89.

Rey, P., Francoise, E. and Gilles, P. 1990. Atrazine and diuron resistant plants from photoautotrophic protoplast derived cultures of *Nicotiana plumbaginifolia*. Plant Cell Rep. 9: 241-244.

Schenck, H. and Robbellen, G. 1985. *In vitro* selection of herbicide tolerant rapeseed. Cruciferae Newslett. 10: 90-91.

Soprasanna, P., Rao, K.V. and Reddy, G.M. 1987. Initiation and maintenance of suspension cultures. Maize Genet. Coop. Newslett. 61: 58-59.

Vilor, T.A., Lepeshkov, A.G., Abramenko, O.A. and Boletski, Y.D. 1988. Tissue and protoplast culture in *Brassica juncea*. In: Biologiya Kul Tiviruemykh Kletok i Biotekhnologiya i (Ed. R.G. Butenko), Novosibirsk, USSR, Vol. 14, pp. 15-16.

Vyvadilova, M. and Zelenkova, S. 1987. The use of *in vitro* cultures in *Brassica napus* L. Scient. Agric. Bohemoslov. 19: 261-266.

Cotton Ovule Culture: A Tool for Understanding Development and Quality

C. P. Malik, Bir Singh and Pushp Sharma

INTRODUCTION

The cotton fibre is a single elongated epidermal cell of ovules of *Gossypium* sp. Fibre development is typically divided into two stages: primary wall formation and secondary wall synthesis. Both the stages comprise two phases each: initiation and elongation; secondary-wall thickening, and maturation (Basra and Malik, 1984). The rate and extent of primary and secondary wall synthesis are of great agricultural interest but are biochemical and developmental enigmas.

Developing cotton fibres provide a reliable model system for the study of cell extension and wall development. Interestingly fibre development in *Plantae* is extremely difficult because of the fragile nature of fibres which are concealed within the thick-walled ovary. However, ovules with fibres or fibres alone can be grown under culture conditions, thus permitting the examination of growth parameters and developmental sequences. Experiments have shown that *in vitro* fibres grown exhibit growth and developmental patterns that are comparable to fibres grown in *plantae*, though some differences in the timing of developmental events occur.

Because of the economic importance of cotton, concerted efforts are being made on cotton biotechnology. There are two approaches which are being pursued by cotton biotechnologists: the production of cotton which is increasingly environmentally benign, cheaper and easier (herbicide tolerant, disease and insect resistant, stress tolerant and high yielding). The second approach concerns improving the raw products in which biotechnology shall increase the value of cotton.

The most obvious target is the fibre. It is imperative to analyse the changes in gene expression, identify genes by experimental data related to wall characteristics and growth parameters of fibres, identify the factors which regulate the ultimate length of the cotton fibre, the amount of secondary wall deposition, and the transition between primary wall and secondary wall synthesis. Since a direct relationship exists between fibre quality and the economic value of cotton, lot remains to be investigated to determine the factors that regulate fibre development.

Cotton is native to tropical region and despite intensive breeding and selection, is sensitive to cool temperature. The effects of cool temperatures on the various steps of cotton wall synthesis are not known. Obtaining background data on normal gene expression during the primary and secondary wall deposition phases of fibre development is essential.

Using ovule culture techniques, various aspects of cotton fibre initiation, elongation etc. have been investigated. The review provides detailed information at the cellular level to determine hormonal effects on metabolism or cell morphology; provides information as to which processes are being influenced by the hormones, polyamines and the effects of cool temperature; cycling on rates of fibre growth, cellulose synthesis etc., role of Ca^{2+}, etc. in cotton ovule cultures.

IN VITRO CULTURING

Several attempts have been made by plant physiologists to grow the cotton ovules *in vitro*. Joshi (1960) described the *in vitro* culture of 6-day post anthesis (DPA) cotton ovules with vitamins, indole acetic acid, casein hydrolysate and gibberellic acid added to the culture medium. However, ovule growth displayed abnormalities and fibres failed to develop. Thereafter Mauney (1961) and Mauney *et al.* (1967) obtained limited success when the young proembryo and heart stage embryos were cultured.

Beasley (1971) reported extensive fibres development when fertilized ovules were floated on the high salt Murashige and Skoog's liquid medium. Beasley *et al.* (1971) reported that GA_3 markedly promoted the total amount of fibres produced from fertilized ovules. Beasley (1973) and Beasley and Ting (1974), Beasley and Ting (1974), Beasley and Egli (1976) further analysed the essentiality of indole acetic acid in addition to GA_3 to unfertilized ovules cultured on the basal medium to produced fibre.

Though the fibre development was extensive in the modified MS medium, callus formation and browning of ovules after two weak culture was reported. To overcome these artifacts Beasley and Ting (1973) made further modification in the MS medium, where sucrose was replaced by glucose and fructose and ammonium nitrate was added. By the use of variable concentration combinations of IAA and GA_3, Beasley and Ting (1973) quantified their optimum amount for maximum fibre production i.e. 0.5-5.0 mM GA_3 in fertilized ovules and 0.5 mM GA_3 in addition to 5.0 mM IAA in unfertilized cultured ovules. Haigler *et al.* (1991) cultured the ovules by floating them on the top of the medium according to the method of Beasley *et al.* (1974) except that 120 mM glucose was added to the medium as only carbon source in addition to 0.5 mM GA_3 AND 5.0 mM NAA to promote uniform fibre initiation and elongation.

Basal medium modified by Haigler (1991) was used by Malik and his colleagues (Malik *et al.* 1994) and also Sandhu (1994) with minor carbon source. Variation of hormonal constituents (22 g l^{-1} glucose, 1.0 mg l^{-1} IAA and 0.2 mg l^{-1} GA_3 at pH 5.0). Kaur (1988) modified the method of *in vitro* growth of cotton ovules (*hirsutum, arboreum*) and investigated developmental physiology of the effect of plant hormones on ovule and fibre growth (Nayyar *et al.* 1989; Kaur *et al.* 1990, 1991). Malik (1994) stressed that *in vitro* studies possibly could lead to an improvement of the number, length and maturity of cotton fibres by foliar application of some chemicals under field conditions.

Davidonis (1989) developed agar solidified MS culture method (2.3 mm kinetin + 0.45 mM 2,4-D with 0.23 mM kinetin + 10.7 mM NAA) to investigate the production of cotton (*Gossypium hirsutum* L. cv. Texas-Marker-I) fibres *in vitro*. Excised ovules from 3,5,7 and 9 DPA ovaries were laced on an agar solidified medium. After 8 weeks of culturing as long as 10 mm x 1-2 mm fibres developed. They reported that the ability of ovule derived cells to direct expansion in the longitudinal direction diminished, while lateral expansion increased with time in culture. Growing ovules on medium solidified with agar rather than on liquid medium possibly contribute to the formation of fibres on 9 days preanthesis ovules.

TOTAL FIBRE UNITS (TFU)

The amount of total fibres produced from the ovule surface was determined by Beasley and Ting (1983) calorimetrically from the intensity of a destaining solution to which the ovules were transferred following their 15-second immersion in a solution of toluidine blue O. Components other than cellulose in the primary cell wall also react with toluidine Blue O (Beasley *et al.*, 1974; Basra and Malik, 1984), but is a rapid method to compare quantitative changes occurring during fibre development. Total fibre units (TFU) value even though were not equivalent to fibre length yet was useful indication of fibre growth.

Beasley and Ting (1973), Nayyar *et al.* (1988) and Sandhu (1994) measured the TFU in cultured ovules in response to various plant growth regulators (GA_3, IAA and ABA) individually or their possible combinations. GA induced marked stimulation and kinetin and ABA a marked inhibition of fibre production from fertilized ovules. IAA afforded only a trend of stimulation of fibre production. GA overcame the inhibitory effects of both Kin and ABA. The most striking effect of IAA was its capacity to overcome the inhibitory effect of ABA. GA_3 consistently provided for a 2-3 fold increase in TFU compared to control. Nayyar *et al.* (1989) noticed the involvement of ABA and investigated the effect of fluoridone- a reported inhibitor of ABA biosynthesis, *in vitro*. Fluridone caused a marked promotion of fibre production as measured by TFU. Sandhu (1994) has reported that by 15 DAS, fluridone promoted fibre growth by 88% and 20% in short and long staple cultures, respectively. With ABA, TFU were reduced by 84% and 81% in the two cultivars. Compared with the control (TFU = 4.06) with DFMA and DFMO, inhibitors of polyamines biosynthesis, TFU were reduced to 0.135 and 0.125, respectively in *G. hirsutum*. In *G. arboreum*, similar trend was observed though the TFU were lesser than in *hirsutum*.

Growth and development of cotton fibres in a developmental mutant. Ligon-lintless and its near isogenic wild type, Texas Marker-I were compared in ovule and wall suspension culture (Triplett *et al.*, 1989). The pattern of growth of fibre cells in both organ and cell culture from the two genotypes mimicked the pattern of *in vivo* growth. The fibre amount as determined by Toluidine-blue-O (TBO) staining. In addition, fibre appearance was very similar in the two lines during the early stages of development. However, differential fibre amounts were observed for two lines within the next four days. Linear increase in TFU until 9 and 13 DPA was reported in Ligonlintless and Texas Marker-I respectively. They suggested that the cause of shorter length than TM-I *in vitro* cultures is due to affected uptake and the mutation not due to hormonal or nutritional deficiency, since all components were provided in the medium. Secondly several factors possibly control the length of the fibre finally attained. Only some of these factors appear to be expressed in culture system since these cells are 100-times shorter than in plants fibres.

The effect of 5-Bromo-2-deoxyuridine (BuDR) thiamidine analogue), AMO-1618 (2-isopropyl-4-dimethylamino-5-methyl phenol-1-piperidine carboxylate methyl chloride) a growth retardant and p-chlorophenoxy isobutyric acid (PCIB, antiauxin) on fibre development in unfertilized cotton (*Gossypium hirsutum* L.) ovules grown in *in vitro* was studied (Dhindsa, 1978). Supplementation of medium with BuDR (5 mM) reduced the fibre production by 70% after 6-day culture. The decrease in the fibre production by 5 mM BuDR is probably based upon the interference by BuDR in the differentiation of fibre initials. The fibres produced in the continued presence of BuDR presumably grow from fibre initials present at the time of BuDR treatment.

AMO-1618 when used with GA3 alone and in combination with IAA cause only a 25-40% reduction in fibre production. This is due to reduced availability of IAA to the ovules. PCIB completely inhibits fibre growth, but has little effect on dry weight in the presence of GA3. PCIB, an antiauxin, indicates an absolute requirement of IAA during fibre growth. The need for the continuous presence of IAA during fibre growth on unfertilized ovules is evident. Fluridone and other pyridinone compounds are shown to inhibit the carotenoid biosynthesis and thus inhibit ABA biosynthesis. Fluridone increased fibre growth more in SS cultivar compared with LS cultivar and this difference could be attributed to variable level of endogenous ABA in the two cultivars. Fluridone, thus overcomes the inhibition caused by ABA.

CARBOHYDRATE METABOLISM

The developing cotton fibres are used as experimental system for studying biochemical changes associated with cell elongation and/or maturation. Mature fibre contains, on an average (percentage of absolute dry matter basis), cellulose (94.0), was like substances (0.6), pectin and pectic acid (0.9), organic acid (0.8), nitrogenous substances (as proteins) (1.133), ash (1.2), non-cellulosic polysaccharides (0.3) and unidentified substances (0.9%) (Basra and Malik, 1984).

Cotton fibres accumulate good amount of carbohydrates, mainly glucose, fructose and sucrose during its development (Cooner *et al.*, 1972; Basra *et al.*, 1990). Early work showed that the level of reducing sugars in fibres is appreciably high during elongation, while decreased during secondary wall formation. The absolute amount of fructose, galactose, mannose, rhamnose, arabinose, uronic acid, and however, non-cellulosic glucose and of arabinose residues decreased during secondary wall formation possibly indicating a turnover of atleast some of the hemicellulose control.

The steady state levels of nucleotide sugars, were analysed from developing (*Gossypium hirsutum* L.) fibres harvested between 13 and 36 days post anthesis (DPA) (Carpita and Delmer, 1981). The UDP sugars were the predominant nucleotide sugars, GDP sugars were not detectable in significant quantities. The UDP glucose level increased by 10-fold from 13 DPA to 24 DPA. Just prior to the maximum rate of secondary wall cellulose synthesis ceased in ovules with associated fibres, cultured *in vitro* possessed nucleotide sugar composition similar to that of plant growth fibres (Carpita and Delmer, 1981). Pulse chase experiments with (^{14}C)-glucose showed that UDP-glucose was the predominant nucleotide sugar labelled suggesting that UDP-glucose was a precursor of secondary wall cellulose in the cotton fibre. Previous studies by Delmer *et al.* (1977) and Heimiger and Delmer (1977) *in vitro* showed that sucrose, sterile glucosides and ß-1,3-glucan were the predominant sugars accumulated from UDP- glucose and also concluded that glucose and sucrose are the only major neutral sugars found in the fibres.

Carbohydrate metabolising enzymes have been investigated in cotton at different developmental stages (Joshi *et al.*, 1985, Richa *et al.*, 1986, Nayyar, 1988, Basra *et al.*, 1990 and Hendrix, 1990). The effect of different hormones on the incorporation of ^{14}C sucrose into sugars was studied by Basra *et al.* (1992). They reported that ABA caused enhanced ^{14}C sucrose incorporation into sugars at 15 DAA but markedly inhibited it as 35 DAA in *G. hirsutum*. GA_3 promoted the incorporation at both the stages i.e. primary and secondary wall formation but IAA did so only at the primary wall stage (15 DAA). Studies on sucrose hydrolysis in developing cotton fibres revealed that an increase in glucose and fructose contents corresponded to the transition phase from primary to secondary wall formation or to the period of extensive secondary wall formation (Basra *et al.*, 1990). Most of the work in this regard is restricted to the *in vitro* cotton ovules/fibres. However some studies have been extended to *in vitro* cultures.

Recently Sharma *et al.* (communicated) and Singh *et al.* (communicated) analysed the activities of the different carbohydrate hydrolyzing enzymes in controls along with growth regulators incubated cotton ovules/fibres at various stages of development. The effect of a novel plant growth retardant (E-1(4-chlorophenyl)-4, 4-dimethyl-2-(1,2,4-triazole-ly-I)-1 penetene-3-ol) (S-3307) on hydrolysing enzymes namely acid invertase, ß-glucosidase and ß-galactosidase was studied (at 1,15 and 35 DAA). The relative effect of S-3307 on the activity of acid invertase on the soluble fraction was much greater than on the cytoplasmic activity especially at 15 DAA. Considerable decline in the activity of acid invertase by uniconazole, an antigibberellin compound adversely affected the initiation of fibres and suggested the role of GA in regulating the activity of acid invertase and hence high apoplastic hydrolysis by preventing sucrose accumulation. The role of acid invertase during the fibre elongation has already been speculated by Basra *et al.* (1990). From the information available it is inferred that GA via invertase activity has a profound influence on the immediate metabolism of sucrose entering the fibres during primary and secondary wall formation.

The activities of both wall-bound and soluble fractions of ß-glucosidase and ß-galactosidase were also negatively affected by uniconazole during all the three studied stages of fibre development. Since these enzymes hydrolyze non-cellulosic polysaccharides, hence they appear to be involved in cell wall loosening by cleaving wall linkage and thus facilitating cell growth though GA activation during the later stages enhanced the activity of acid invertase at all the three stages of development i.e. -1, 15 and 35 DPA is suggestive of its role in high apoplastic hydrolysis preventing sucrose accumulation (Singh *et al.*, communicated) increased activity of ß-glucosidase at 35 DPA was also an additive factor to wall biogenesis. The effect of this compound in counteracting adverse effect of low temperature will be discussed.

ENDOGENOUS LEVEL OF HORMONES AND THEIR POSSIBLE RELATIONSHIP

The decisive role of phytohormones in the development of cotton fibres raised *in vitro* has been demonstrated. Both fertilized and unfertilized ovules have revealed that hormones regulate the extent of fibre elongations. Kaur (1988) has concluded that the fertilized isolated cotton ovules were (i) deficient in their capacity to synthesize optimum levels of gibberellins (ii) sufficient in the production of cytokinins (iii) optimal or near optimal in the production of auxins (IAA) and (iv) ABA was most essential for fibre elongation.

Only a few studies have reported on the endogenous levels of growth substances during fibre elongation (Bhardwaj and Lad, 1977); Naithani *et al.*, 1982; Bhardwaj and Verma, 1985; Nayyar *et al.*, 1989a and 1989b). Varieties having relatively longer fibres contained higher levels of auxins and gibberellins than their counterpart with shorter fibres (Bhardwaj and Verma, 1985, Nayyar *et al.*, 1989b). The short staple cultivar (*G. arboreum*) as compared with the long staple cultivar (*G. hirsutum*) possessed high IAA and GA_3 levels before fertilization and just after fertilization (Nayyar *et al.*, 1989b) while converse was true during active phase of fibre elongation. The amount of ABA was considerably high at all phases of fibre developmental in short staple cultivar than long staple cultivar. Correlation coefficient and regression values indicated the involvement of ABA in the regulation of fibre elongation in genetically different cotton cultivars (Nayyar *et al.*, 1989b, Malik, 1994).

The endogenous level of hormones *in vitro* and *in vivo* was compared by Nayyar *et al.* (1989a) at 20 days after culturing field grown cotton fibres at the same stage. Quantitatively no variation in the ABA content in cultured and field grown fibres were observed, although IAA contents decreased by 40% in cultured fibres though enormous increase in the GA content was noticed. Supplementation of the culture medium with fluridone (inhibitor of ABA biosynthesis) increased the endogenous levels of IAA and GA_3 than the cultured control ones, however, markedly decreased the ABA content. Since GA and ABA are known to be antagonist (Dhindsa *et al.*, 1976; Dhindsa, 1978b), a reduced amount of ABA per se or coupled with altered IAA and GA_3 levels could allow far greater fibre elongation. The reduction of ABA content by fluridone has been shown previously (Moore and Smith, 1984).

In vivo studies of auxin metabolising enzymes i.e. IAA oxidase and peroxidase showed that IAA catabolism was low during the elongation phase, while during the phase of secondary thickening a four-fold increase in their activity was reported (Jasdanwala *et al.*, 1977; Rama Rao *et al.*, 1982; Basra *et al.*, 1992). It is suggested that IAA-oxidase *in vivo* possibly regulates the levels of IAA. The close correlation between the cessation of elongation growth and increase in wall peroxidase point to an important role of this enzyme in the termination of the elongation phase.

Basra *et al.* (1992) investigated the influence of ABA and calcium on soluble and wall-bound activities of peroxidase and O-diphenol oxidase of cotton (G. hirsutum L. Cv. F414 and G. arboreum L. cv. LD-230) fibres at the stage of primary and secondary wall development. ABA markedly inhibited the activities of both the fractions, whereas calcium was promotory. The incorporation of EGTA (a calcium chelator) and chlorpromazine (a calmodulin antagonist) to incubation medium caused decreased enzyme activity suggesting regulation of enzyme synthesis and/or secretion to the cell wall by calcium calmodulin.

Generally ABA inhibited both the enzyme activities though the maximum inhibition was observed in soluble peroxidase. In contrast the Ca^{2+} treatment markedly stimulated both the soluble and wall bound activities of peroxidase as well as o-diphenyl oxidase, suggesting the involvement of Ca^{2+} as a second messenger in mediating the inhibitory effects of ABA as reported for other systems (Mansfield *et al.*, 1990). The inhibition of enzyme activities caused by EGTA and chloropromazine provides supportive evidence for the involvement of calcium-calmodulin in their regulation.

The enzymic hydrolysis of the ß-D-glucopyranosyl ester of ABA by ß-glucosidases and esterases was studied in cotton (*G. arboreum* L. Cv. LD-327) fibres at 15 and 35 days after anthesis (Sharma *et al.*, communicated). The ß-D-glucopyranosyl ester of abscisic acid (ABA-Glc) is the chief conjugate of the plant hormone assayed in many plants as an endogenous substance or as a metabolite of exogenously applied ABA.

Since the activities of two hydrolases at primary wall formation (15 DAA) did not show any correlation with ABA content, it is concluded that the high content of free ABA at this stage was not a consequence of the enhancing activities of ABA-Glc splitting enzymes (Sharma *et al.*, communicated). However, the fibres harvested at 35 DAA showed increased activity of these enzymes and reduced ABA content. Contrarily, the addition of ABA decreased the activities of two enzymes and it is suggested that ABA conjugate is the final product of the ABA metabolism under different stages of fibre development.

The influence of phytohormones partitioning of ^{14}C-sucrose into other metabolic fractions of cotton fibres, i.e. amino acids and organic acids was also investigated (Basra *et al.*, 1992). ABA application decreased incorporation at primary and secondary wall formation stages (15 and 35 DAA) whereas GA_3 and IAA were stimulatory only at 35 DPA. Amino acid metabolism and quantitative and qualitative changes in proteins (Graves and Steward, 1988) are an integral part of fibre development especially during the elongation phase. Hence ABA also seems to control carbon partitioning into amino acids checking thereby the growth rate of fibres. The label incorporation into organic acids showed a similar influence of growth regulator with a bit variation in pattern of influence. GA_3 was stimulatory at both the stages. IAA only at 15 DAA, however, ABA inhibited incorporation at both the stages with marked response at 15 DAA. The increased incorporation into amino acids suggestively be due to the activation of dark metabolism of CO_2 by GA_3 (Dhindsa, 1978) for supporting fibre growth.

MINERAL NUTRITION

Cotton fibres require constant supply of B^+ to prevent callusing of epidermal cells (Birnbaum *et al.*, 1974). Boron deficiency symptoms were related to reduced activity in the pyrimidine biosynthesis pathway, hence may cause reduced synthesis of UDP-glucose and other UDP-sugars involved in cell wall composition of fibre. Boron has also been shown to maintain membrane integrity in cotton fibres (Wainright *et al.*, 1980). Second major effect of B deficiency in the *in vitro* cultured cotton fibre system is that the pyrimidine synthesis intermediates are shunted away from UDP-glucose synthesis and channeled preferentially into RNA synthesis.

Another important nutrient required for the growth and development of cultured ovules is NH_4^+. At low temperature (i.e. 20°C) either GA_3 or NH_4^+ is required along with IAA in the medium for fibre development whereas ovules cultured at 34°C only require IAA. The mechanism of NH_4^+ and high temperature in regulation of fibre development remains obscure.

DARK METABOLISM OF CO_2

It is well known that CO_2 is a physiologically active gas and plays an important role in the regulation of various plant processors (Basra and Malik, 1985; Malik *et al.*, 1991). Cotton fibres grow in the dark interior of the boll protected against the environmental perturbations. Presumably cotton boll growth results in the production of elevated levels of CO_2 inside the boll. This represents the external CO_2 inside the boll. This represents the external CO_2 pool. As fibres are growing, internal CO_2 pool will also exist in them (Basra and Malik, 1984), these two pools are exchangeable as CO_2 can diffuse freely across the membranes.

In vitro studies have been carried out on non-photosynthetic carbon metabolism (Dhindsa *et al.*, 1978; Basra and Malik, 1984). The inhibition of fibre growth is evident in the absence of K^+ and CO_2 in the culture medium. They also reported malate production via non-photosynthetically CO_2 assimilation, which serves as an indispensable metabolite/anion for growth of fibres. It was shown that fibre growth is dependent on the turgor pressure in the fibre and K^+ and malate are the osmotically active solutes, which are largely responsible for the production of turgor in the fibre. Phytohormones (i.e. IAA, GA_3 and ABA) regulate fibre development *in vitro* by commanding malate synthesizing enzymes. In elongating fibres K^+ and malate levels fluctuate in correlation with the growth rate, reach peak levels when growth rate is maximum (Basra *et al.*, 1993). ABA caused inhibition of fibre in *in vitro* by inhibiting malate synthesis and increasing thereby the K^+ malate ratio. Thereafter the levels of K^+ and malate relative to each other may influence the rate of fibre growth in the two cultivars (Basra and Malik, 1984).

During the period of rapid fibre growth high activities of PEPc, MDH and GOT were observed which declined during slow down period (Basra and Malik, 1984) indicating the channeling of PEP to malate through OAA. The enhanced activity of MDH over GOT and unchanged activity of ME, further supported the accumulation of malate by reduction of oxaloacetate is the main route of the metabolism of fixed carbon. Higher activity of GOT during rapid fibre elongation also suggested that some of the OAA formed may serve to provide carbon skeletons for the synthesis of aspartate. ME coupled with decreased activity of PEPc levels of K^+ and malate was reported as extension growth levels off (Basra and Malik, 1984). Increased malate catabolism during later stages of fibre growth due to enhanced ME activity is of physiological relevance to meet special requirements of fibre growth. At these stages, a large proportion of the carbon supply to the fibre is used for synthesis of thick secondary wall comprised predominantly of cellulose and to a lesser extent of 1,3ß-glucan. During the transhydrogenase steps of malate metabolism NADH is converted to NADPH. Pyruvate is produced concomitantly with NADPH by ME reaction and both are important metabolites for biosynthesis. In this respect, it is readable to say that malate serves as an osmoticum as well as carbon-energy source during fibre growth.

The development of cotton fibres is not necessarily by one mechanism i.e. the action of CO_2 but it can also be effected by several other factors/pathway. Some of the CO_2 effects may include: regulation of activities of several enzymes, regulation of cellular metabolism by changes in pH associated organic acid content, triggering of phytohormone changes and possibly other mechanisms. For details a reference may be made to the review of Basra and Malik (1984) and Malik *et al.* (1991).

Incubation of elongating fibres of *G. hirsutum* L. cv. F414 and *G. arboreum* L. cv. LD-230 with ^{14}C-labelled bicarbonate showed an appreciably higher incorporation of label into lipids, particularly the polar lipids at 15 DPA and 20 DPA (Nayyar *et al.*, 1988). The low incorporation of label at 25 DPA was possibly due to cessation of fibre elongation functions mainly in the synthesis of membranes and maintenance of their functional integrity by keeping turgor pressure up required for fibre expansion. Similar results on label incorporation with labelled acetate were observed in cotton fibres by Basra and Malik (1986).

Kaur (1988) and Kaur *et al.* (1990) studied dark metabolism of CO_2 in fibres raised *in vitro* in two cultivars of cotton when ABA or fluridone was added for 20 DAC. It is inferred that activities of malate synthesizing enzymes in ABA-treated ovules decreased. Thus, ABA lowered the ability of seeds to fix CO_2 in the dark (Nayyar *et al.*, 1989 and Kaur *et al.*, 1990). The presence of fluridone in the medium markedly enhanced the malate synthesizing system by increasing levels of PEPc, NAD, malate DH, GOT and NADP-malic enzyme. A concomitant increase in fibre growth capacity of the ovules was also observed. On the contrary ABA lowered the ability of ovules for non-photosynthetic C_4 metabolism by causing enzyme inhibition and consequently the fibre growth was retarded. In a more recent study Basra *et al.* (1993) investigated the influence of ABA and Ca^{2+} on dark fixation of CO_2 by fibres of *G. hirsutum* at 15 DAA. Both ABA and Ca^{2+} inhibited activity of PEPc and fixation of $H^{14}CO_3$ into organic acids. Ca^{2+} promotes carbon incorporation in amino acids though ABA was inhibitory. Apparently both the compounds could mediate the fibres capacity for dark fixation of CO_2 and hence rate of fibre elongation. However, we need to study the regulatory aspects in this biological system.

COTTON FIBRE DEVELOPMENT IN RESPONSE TO COOL TEMPERATURE

Due to the tropical habitat of cotton it is highly sensitive to cool temperature (Thaker *et al.*, Haigler *et al.*, 1991, Roberts *et al.*, 1992; Malik *et al.*, in press). Temperatures less than 16 °C essentially inhibit fibre growth (Haigler *et al.*, 1991; Malik *et al.*, in press) and 32 °C is ideal for fibre development. It has been known for many years from field studies that cool nights (< 20 °C) results in suboptimal cotton yield and reduce fibre maturity (Gipson, 1986).

FIBRE INITIATION AND ELONGATION

Gipson and Johan (1969) tested four different night temperatures (10, 15, 20, 27 °C) regimes on two varieties of field-grown *G. hirsutum*. Elongation of fibre was closely associated with both temperature and variety. As night temperature was lowered, fibre elongation rates declined and fibre elongation periods enhanced for both varieties. Fibre age and night temperature determined rate of fibre elongation, temperature coefficient of elongation decreased with progressive fibre age and night temperature. Subsequent studies by Xie *et al.* (1993) and Malik *et al.* (in press) have revealed that the initial stages of fibre elongation were highly temperature sensitive, while the later phases (elongation and secondary wall forming) tended to become temperature independent. Ovules were cultured *in vitro* at 34 °C constant to 34/15 °C cycling (12/12 h) and 32 °C constant to 32 °C/15 °C cycling (12/12 h) to mimic a typical diurnal temperature cycle at Texas (USA) and Ludhiana (India), respectively. Fibre initiation and very early elongation were analyzed through SEM and the subsequent fibre elongation was estimated using divider or ruler. Both the works demonstrated that the fibre initiation and early elongation (up to nearly 0.5 mm fibre length) were separately delayed by cycling cool temperature though later phases were temperature independent. Malik *et al.* (in press) showed that early delay in fibre initiation by cool temperature, extended the elongation period. On the basis of the response of fibres at cool temperature, three stages of fibre elongation could be recognized: fibre initiation, very early elongation and subsequent elongation. Of these, the final two stages were highly sensitive to cool temperature. Xie *et al.* (1993) and more recent study by Malik *et al.* (in press) inferred that the field temperature during the early fibre initiation and elongation phases had a marked effect on the final fibre length.

Haigler *et al.* (1991) have demonstrated that fibres of plants exposed to adverse diurnal temperature fluctuation elongate at reduced rates for an extended period, though attain the comparable final length as those grown under favourable conditions. Briefly, ovules exposed to initial low temperature require prolonged periods to attain fibre length equaling controls.

The adverse response of cotton fibre development to cool temperatures was not dependent on whole plant physiology, contrarily it is ovule/fibre specific event. Studies of Roberts *et al.* (1992) have demonstrated that low temperature alters fibre development even though sufficient carbon source is available. These authors have suggested that secondary wall synthesis was non-sensitive to cool condition than primary wall synthesis.

Incorporation of [$U^{14}C$] sucrose (54 mM) into callose and cellulose after 1 h was 79% and 21%, respectively. However, its presence in the secondary wall is doubtful, suggests its preferential incorporation into the wall during periods of two temperature, then degraded by the ß-1,3 glucnases in cotton fibres (Meier, 1983) or the enhanced callose synthesis at the plasma membrane/wall interface during cool temperature changes the physical properties of cellulose causing dense ring; control fibre length and weight were achieved under 34°C/28°C and 34°C/22°C cycling but both parameters were reduced under 34°C/15°C cycling temperature.

Recently Roberts *et al.* (1992) have shown that cellulose synthesis and respiration have different responses to cool temperatures. A decrease in temperature from 28°C to 15°C concomitantly caused a decline in respiration and cellulose synthesis and increased in a roughly linear manner between 18°C and at 28°C. Above 28°C, however, the trend in the two processes diverged i.e. respiration continued to increase and remained high at 40°C, whereas rate of cellulose synthesis plateued up to about 37°C and dropped off at 40°C, suggesting that more energy was released through respiration above 28°C but was not necessarily used for increased cellulose synthesis. Furthermore, the restriction in available energy due to cool temperature reduction in the rate of respiration is not likely the only explanation for reduced cellulose synthesis under cool temperatures. Cool temperature data suggested at least three possible methods for improvement in the cellulose biosynthetic pathway (Haigler, 1992) (a) to optimize use of increased metabolic energy at high temperature for increased rates of cellulose synthesis (b) to optimize fast recovery of cellulose synthesis after exposure to cool temperatures and (c) transfer genes and confer cool temperature tolerance for cellulose synthesis in existing cotton varieties to commercial cotton cultivars.

COOL TEMPERATURE EFFECT ON INITIATION AND EARLY ELONGATION OF COTTON FIBRE DEVELOPMENT AS ASSESSED FROM SEM

Xie *et al.* (1993) and Malik *et al.* (in press) observed pattern of fibre initiation as affected by cool temperature. *In vitro* cultured cotton ovules at -1 DPA and O DAA were obtained by Malik *et al.* (in press) who have suggested a developmental link between fibre initiation and time of anthesis *in situ* that could not be overcome by culture manipulations in *in vitro* fibre development suggested that by -2 DPA the ovules possibly acquire an internal signal that permitted initiation only on the normal day of anthesis.

Xie *et al.* (1993) and Malik *et al.* (in press) have shown that initiation and early fibre elongation were two different physiological processes and were independently affected by cool temperatures. They showed that cycling to cool temperatures of 32°C/15°C delayed fibre initiations and early elongation by 1 and 2 day respectively which was consistent with the 3-day extension in the elongation period, however, the latter elongation (beyond 0.5 mm length) proceeded in a temperature independent manner. The early delay in fibre development caused by cycling temperature was associated with elongation period required to reach the control length.

Recent studies conducted by Malik *et al.* (Communicated) have shown similar developmental response of *in vitro* cultured ovules to cool temperature ovules subjected to 32°C/15°C cycling temperature delayed initiation of ballooning by 1 to 2 d followed by slow initiation and fibre elongation. Once the fibres attained a length of 0.4 mm their further elongation was independent of temperature. Furthermore, *in vitro* culturing of ovules with fibres 0.4 mm length at 32°C/15°C cycling and 35°C constant had comparable rates of fibre elongation suggested that -1 DPA to -2 DAA stages of fibre development were highly sensitive to cool temperature and hence the developmental delay in fibre length under cool cycling compared with 35°C continuous temperature, was explained.

Our recent studies have shown that delay in fibre development induced by cool temperature could be overcome by the supplementation of guaianolide derivative-I into the culture medium. This compound caused preponderance of bulging of epidermal cells and was highly promotive. Furthermore, the effect of S-3307 (antigibberellin) has also been studied. SEM studies showed that gibberellin inhibitor completely suppressed the initiation phase of fibre development and no fibre formation was observed even after many days in cultured ovules. The basic possible mechanism behind the guaianolide enhanced bulging and S-3307 inhibited fibre formation will be discussed in this review.

CONCLUSION

Physiology Behind Cool Temperature

Cool temperature effects are not only limited to the observable morphological parameters of fibre but alteration in biochemical/physiological processes also occurred. Malik *et al.* (communicated) observed reduced levels of sucrose-glucose in cool temperature in cultured ovules/seeds even though glucose was not a limiting factor. The cycling effect of temperature (35°C/22°C) on sugar accumulation was reported up to the phase of fibre elongation, though the content was more at the 35°C/22°C (Malik *et al.*, communicated). However, reverse trend was observed during secondary wall thickening with higher amount of total sugar accumulated in seeds/ovules subjected to 32°C/15°C compared with 35°C/22°C suggested reduction in deposition of cellulose in the secondary wall during fibre maturity due to the limited rate of synthesis of precursor substrate. Even at higher concentration glucose was not used efficiently to synthesize and deposit secondary wall around the fibres clearly depicts that the causal mechanism of low temperature reduction of cellulose synthesis would be factors other than availability of glucose during secondary wall formation while during the pre-anthesis, however, fibre primordia initiation was delayed by the available resources which act as limiting factors.

Enzymic studies have shown differences in extent of response at -1 DPA, 15 DAA and 35 DAA stages supporting the contention that low temperature has independent effects on the initiation, elongation and secondary wall deposition process. Cool temperature diversely affected the activities of ß-glucosidase, acid invertase and o-diphenol oxidase at three mentioned stages (Malik *et al.*, communicated) with marked decrease in the activities of all three enzymes at -1 DPA suggesting their role during initiation and early elongation of fibres. Cool temperature caused marked decrease in the activity of o-diphenol oxidase was correlated with irregular and delayed initiation of fibres. The role of o-diphenol oxidase pertinent to auxin turnover and redox balance of developing cotton fibres has already been suggested (Naithani *et al.*, 1982; Basra *et al.*, 1992). The decreased activity of acid invertase at -1 DPA showed that enzyme has a considerable influence on the immediate influence on the immediate metabolism of sucrose entering the fibre primordia and cool temperature adversely affected it.

In cotton ovules, low temperature caused delayed and erratic initiation of fibre primordia was overcome by the supplementation of culture medium with guaianolide derivative I possibly through the two oxidases i.e. o-diphenol oxidase (DPO) and IAA oxidase (IAAO). At cool temperature reduced activities of oxidases at -1 DPA ovules and their subsequent increase till 5 DAA is correlated with essentially of auxin and fibre initiation (Singh *et al.*, communicated). Decrease in IAA oxidase activity and increase in o-diphenol oxidase activity at -1 DPA was reported with guaianolide derivative, however, both the oxidizing systems showed marked increase over the cool temperature affected during elongation and secondary thickening period. Possibly guainaolide derivative I regulated auxin degrading system during the entire period of fibre-growth. Low activity of IAA oxidase at -1 DPA is further maintained by o-diphenol oxidase via o-diphenols which inhibit its activity hence IAA level is enhanced causing fibre initiation.

Enhancement in auxin degradation possibly reduced the availability of auxin during elongation and secondary wall thickening phases. O-diphenol oxidase indicated a shift in redox balance from a reduced state to an oxidative in developing ovules (Basra *et al.*, 1992; Naithani *et al.*, 1981; Sharma *et al.*, communicated). In fact Stonier and Yong (1972) have implicated o-dihydroxyphenols in redox potential pertinent to cell division (reduction) and cell differentiation and senescence (oxidation). The low o-diphenol oxidase activity with guaianolide derivative compared with during the initiation suggested that dihydroxy phenols maintained -SH groups in a reduced state hence aid initiation and early elongation phases of fibre development. The enhanced activity of o-diphenol oxidase during the secondary wall deposition possibly shifts the redox balance towards oxidation resulting in an increase in -SS bonds which check the process of wall loosening. Marked increase in the level of acid invertase at -1 DPA with guaianolide derivative I points towards its influence in immediate metabolism of sucrose entering the fibre primordia, thus checking sucrose accumulation through its enhanced apoplastic hydrolysis.

A novel plant growth retardant (E)-1 (4-chlorophenyl)- 4,4-dimethyl-2 (1,2,4-triazo (-ly-l)-1 pentene-3-ol (S-3307) an antigibberellin completely suppressed fibres on ovules suggesting role of GA in differentiation of initials (Sharma *et al.*, communicated). Considerable decline in the activity of acid invertase by S-3307 is indicative of the role of GA in regulating the activity of acid invertase and hence high apoplastic hydrolysis by preventing sucrose accumulation. The higher influence of compound on wall-bound acid invertase suggested that wall bound invertase possibly concerns with sucrose uptake from apoplast of the cotton fibre by preventing phloem-reloading. Relationship between acid invertase and the rate of fibre elongation has already been mentioned (Basra *et al.*, 1990).

Enzymes hydrolysing non-cellulosic polysaccharides like ß-glucosidase and ß-galactosidase also depicted reduced levels with S-3307 during all stages of cotton fibre development. Their role to provide precursor for cellulose biosynthesis has been postulated (Sharma *et al.*, communicated). S-3307 reduced the activity of o-diphenol oxidase at -1 DPA but enhanced during the secondary wall thickening stages and its role in auxin catabolism and redox balance has already been documented.

In vitro studies have revealed that the developing fibres have growth characteristics comparable to systems exhibiting intercalary or tip growth (Fig. 1). In fibres both the processes are operative though to what extent these two forms of growth contribute to the final length of the fibre is not clear.

It may be stated that the precise mechanism of tip synthesis in cotton fibre is variable from that reported in pollen tubes, fungal hyphae and root hair. Cotton fibres exhibit tapered and obtuse tips, and their significance remains to be deciphered. Unlike the systems mentioned previously cotton fibres synthesized secondary cell wall thickening which limits intercalary growth and cause-effect relationship between the secondary thickening and cessation of fibre elongation remains to be worked out. Detailed studies at the cellular level to determine hormonal effects on metabolism and cell morphology have just begun (Malik, 1994) and information obtained will focus on the processes which are influenced by the hormone (s) and also the sequence of events that ultimately lead to the changes in fibre development. It must be realized that several factors regulate the ultimate length of the cotton fibre, the extent of secondary wall thickening, and the transition of primary wall and secondary wall synthesis ultimately determine the quality of the fibre. In brief, developing cotton fibres raised *in vitro* provide a good tool for the study of several fundamental problems in plant cell biology.

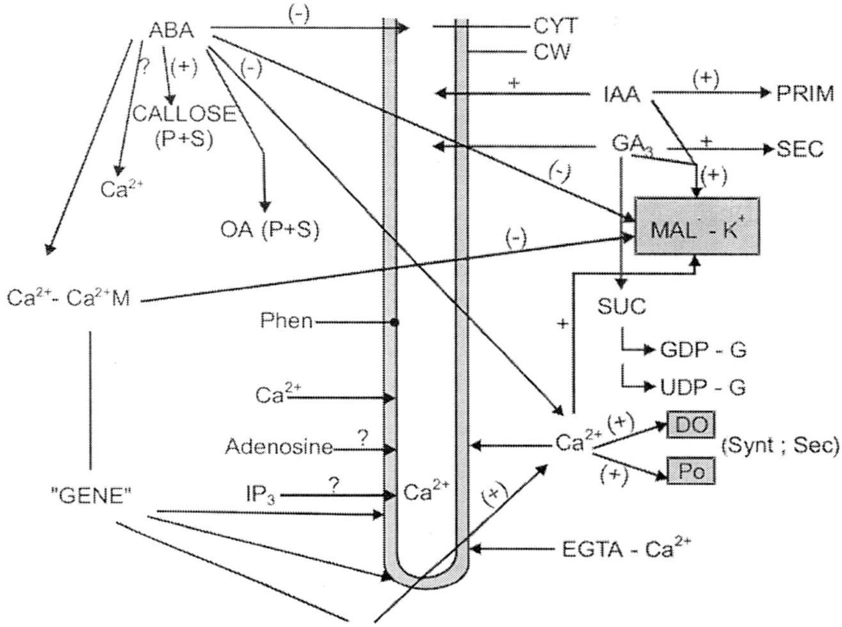

FIG. 1. Diagrammatic representation of cotton fibre. Note Ca^{2+} dependent as well as independent modes of ABA action. ABA having multiple receptors is surmised. The involvement of Ca^{2+} as a regulatory switch from fibre elongation to the onset of secondary wall deposition is suggested. + indicate promotory effect; - indicates inhibition of the process, shows conjunctural and lacking specific evidence.

REFERENCES

Basra, A.S. and Malik, C.P. 1984. Development of the cotton fibre, International Review of Cytology 89: 65-109.

Basra, A.S. 1990. Sucrose hydrolysis in relation to development of cotton (*Gossypium* spp.) fibres. Indian Jour of Experimental Biol. 28: 985-988.

Basra, A.S., Sarlaach, R., Dhillon, S., Grewal, R. and Malik, C.P. 1992. Calcium mediated changes in peroxidase and o-diphenol oxidase activities of cotton fibres (*Gossypium* spp.) and its possible relationship to ABA, Plant Growth Regulation 11: 159-164.

Beasley, C.A. and Egli, M.A. 1976. Development of cotton fibre. International Review of Cytology, Plant Physiol. 63: 57-60.

Beasley, C.A. and Ting, I.P. 1973. The effects of plant growth substances on *in vitro* fibre development from fertilized cotton ovules, Amer Jour Bot. 66: 771-777.

Beasley, C.A. and Ting, I.P. 1973. The effects of plant growth substances on *in vitro* fibre development from fertilized cotton ovules, Amer Jour Bot. 60: 130-139.

Beasley, C.A., Birnbaum, E.H., Dugger, W.M. and Ting, I.P. 1974. A quantitative procedure for estimating cotton fibre growth, Stain Tech 49: 85-92.

Bhardwaj, S.N. and Lad, B.C. 1977. Hormonal regulation of fibre elongation in lintless and linted genotypes of desi cotton (*Gossypium arboreum* L.). Indian J. Plant Physiol 20: 140-150.

Bhardwaj, S.N. and Verma, G.P. 1985. Correlation between endogenous hormone level and fibre elongation in cotton. Indian J. Plant Physiol. 28: 249-258.

Birnbaum, E.H., Beasley, C.A. and Dugger, W.M. 1974. Boron deficiency in unfertilized cotton (*Gossypium hirsutum*) ovules grown *in vitro*. Plant Physiol. 54: 931-935.

Carpita, N.C. and Delmer, D.P. 1981. Concentration and metabolic turnover of UDP-glucose in developing cotton fibres. J. Biol. Chem. 256: 308-315.

Conner, T.W., Kreig, D.R. and Gipson, G.R. 1972. Accumulation of simple sugars in developing cotton bolls as influenced by night temperature. Crop Sci. 12: 752-754.

Davidonis, G. 1989. Early events in fibre cell development. In Proc. Beltwide Cotton Prod. Res. Conf. New Orleans, Louisiana U.S.A.

Dhindsa, R.S., Beasley, C.A. and Ting, I.P. 1976. Effects of abscisic acid on *in vitro* growth of cotton fibre. Planta 130: 197-201.

Dhindsa, R.S. 1978a. Hormonal regulation of cotton ovule and fibre growth. Effects of bromodeoxyuridine, AMO-1618 and p-chlorophenoxyisobutyric acid. Planta 141: 269-272.

Gipson, G.R. 1986. Temperature effects on growth, development and fibre properties. p. 47-56. In J.R. Mauney and J.M. Stewart ed., Cotton Physiology. The Cotton Foundation, Memphis. TN.

Graves, D.A. and Steward, J.M. 1988. Chronology of the differentiation of cotton (*Gossypium hirsutum* L.) fibre cells. Planta 175: 254-258.

Haigler, C.H., Ramma Rao, N., Robert, M.E., Huang, Y., Upchurch, R. and Trolinder, N. 1991. cultured ovules as models for cotton fibre development under low temperatures, Plant Physiol 95: 88-96.

Heiniger, U. and Delmer, D.P. 1977. UDP-glucose: glucan synthetase in developing cotton fibres. II. Structure of the reaction product, Plant Physiol. 59: 719-723.

Hendrix, D.L. 1990. Carbohydrates and carbohydrate enzymes in developing cotton ovules, Physiol Plant 78: 85-92.

Jasdanwala, R.T., Singh, Y.D. and Chinoy, J.J. 1977. Auxin metabolism in developing cotton hairs, J. Exp. Bot. 28: 111-116.

Joshi, P.C. 1960. *In vitro* growth of cotton ovules. In: Plant embryology a symposium. 199-204. Council of Scientific and Industrial Research, New Delhi.

Joshi, P.A., Steart, J.M. and Graham, E.T. 1985. Localization of ß-glycerophosphate activity in cotton fibre differentiation, Protoplasma 125: 75-85.

Kaur, K. 1988. Physiological and biochemical mechanism underlying initiation of cotton fibres in *Gossypium hirsutum* L. and *G. arboreum* L. Ph.D. thesis, PAU, Ludhiana.

Kaur, K., Nayyar, H., Basra, A.S. and Malik, C.P. 1990. Stimulation of enzymes of non-photosynthetic C_4 metabolism in cultured cotton ovules by fluridone, Acta Physiol Plantarum 12(1): 3-6.

Malik, C.P., Singh, Bir, Sharma, P. and Basra, A.S. 1994. Fibre initiation and early elongation *in vitro* of cotton ovules as influenced by low temperature: SEM comparison and biochemical changes, Physiol Mol Biol of Plants 2 (in press).

Malik, C.P. 1994. Impact of plant physiology on Indian agriculture. National Seminar organised by the Indian Society for Plant Physiology, New Delhi.

Mauney, J.R. 1961. The culture of immature cotton embryos. Bot Gaz 3: 301-307.

Mauney, J.R., Chapopel, J. and Ward, B.J. 1967. Effect of malic acid salts on growth of young cotton embryos *in vitro*. Bot Gaz 128: 198-200.

Moore, R. and Smith, J.D. 1984. Growth graviresponsiveness and abscisic acid content by *Zea mays* seedling treated with fluridone, Planta 162: 342-344.

Naithani, S.C., Rama Rao, N., Krishnan, P.N. and Singh, Y.D. 1981. Changes in o-diphenol oxidase during fibre development in cotton. Ann Bot 48: 479-485.

Nayyar, H. 1988. Metabolic changes associated with elongation of cotton fibres. Ph.D. Dissertation, Punjab Agricultural University, Ludhiana.

Nayyar, H., Kaur, K., Malik, C.P. and Basra, A.S. 1989. Regulation of differential fibres development in cotton by endogenous plant growth regulators. Proc. Indian Natn. Sci. Acad. B55: 463-468.

Pillonel, C.H. and Meir, H. 1985. Influence of external factors on callose and cellulose synthesis during incubation *in vitro* of intact cotton fibres with [^{14}C] sucrose, Planta 165: 76-84.

Rama Rao, N., Naithani, S.C. and Singh, Y.D. 1982. Physiological and biochemical changes associated with cotton fibre development II. Auxin oxidising system, Physiol Plant 55: 204-208.

Richa, Basra, A.S., Ahluwalia, K.K. and Malik, C.P. 1986. Hydrolytic enzymes in cotton ovules during early development, Proc. Ind. Acad. Sci. (Plant Sci) 96(1): 49-53.

Roberts, M.E., Rama Rao, N., Huang, Y., Trolinder, L. and Haigler, H.C. 1992. Effects of cycling temperatures on fibre metabolism in cultured cotton ovules. Plant Physiol 100: 979-986.

Sandhu, R. 1994. Hormonal mechanisms of growth regulation in cotton fibre. Ph.D. Dissertation, Punjab Agricultural University, Ludhiana.

Sharma, P., Singh, B., Malik, C.P. and Basra, A.S. 1994. Hormonal regulation of fibre elongation and the enzymic hydrolysis of abscisic acid conjugate in developing cotton (*G. arboreum* L.) fibres. Proceed Indian Nat Sci Acad (Communicated).

Sharma, P., Singh, B., Malik, C.P. and Bajaj, K.L. 1994. Effect of a new plant growth retardant (E)-1-(4-chlorophenyl)-4,4-dimethyl-2-(1,2,4-triazol-1-yl)-1-penten-3-ol (S-3307) on the fibre initiation and enzyme activities of cotton ovules (*Gossypium arboreum*). Jour Plant Physiol. (Communicated).

Singh, B., Sharma, P., Malik, C.P., Kalsi, P.S. and Sanju, 1994. Guaianolide derivative I mitigates the adverse effects of cool temperature on fibre initiation and elongation in *Gossypium arboreum* L. Plant Growth Regulation (In press).

Thaker, V.S., Saroop, S., Vaishnav, P.P. and Singh, Y.D. 1989. Genotypic variations and influence of diurnal temperature on cotton fibre development, Field Crops Research 22: 129-141.

Triplett, B.A., Williams, B. and Wilton, J.R. 1989. Ovule and suspension culture of a cotton fibre development mutant *in vitro*, Cellular and Developmental Biology.

Wainright, I.M., Palmer, P.L. and Dugger, W.M. 1980. Pyrimidine pathway in boron deficient cotton fibre, Plant Physiol 65: 893-896.

Xie, W., Trolinder, L.N. and Haigler, H.C. 1993. Cool temperature effects on cotton fibre initiation and elongation clarified using *in vitro* cultures, Crop Sci. 33: 1258-1264.

Microspore Embryogenesis in *Brassica juncea*

Abha Agnihotri, G. Seguin-Swartz and R.K. Downey

INTRODUCTION

Brassicas include important crop species grown for vegetables, condiments and edible oils. Rapeseed-mustard is the second most important oilseed crop in India after groundnut. More than 70% of the edible oils in India are produced from these two crops.

The fatty acid composition of the oil determines its quality and nutritional value. Rapeseed-mustard contains neutral oleic acid, which is desirable for a longer shelf life, and considerable amounts of linoleic and linolenic acids, two essential fatty acids not synthesized by the human body. However, rapeseed-mustard oil also contains erucic acid which may cause fibrotic myocardium in humans. Increased levels of oleic and linoleic acids are desirable for improving the nutritional quality of the oil while a reduced level of linolenic acid is desirable for increasing oil stability (Ahuja and Banga, 1992).

Rapeseed-mustard meal is a rich source of protein with a well balanced amino acid composition. It is used as animal feed and fertilizer. However, the presence of glucosinolates, sulphur containing compounds which are the precursors of goitrogenic active and toxic chemicals, gives the meal a pungent smell and limits its uses in animal nutrition. Therefore, the development of rapeseed-mustard varieties with low erucic acid and low glucosinolate content is a desirable goal.

The term canola refers to Brassica species in which the seed oil contains less than 2% erucic acid and the seed meal contains less than 30 umoles of aliphatic glucosinolates per gram of defatted meal (Downey, 1990). Genotypes of canola rapeseed-mustard developed in Europe and Canada show poor adaptability to Indian conditions. Therefore, the breeding mandate for rapeseed-mustard in India is to develop varieties which conform to the international canola standards.

Haploid plant production via anther or microspore culture has been shown to be most useful in the breeding of several crops, including tobacco and brassicas. First successful regeneration from microspore embryos in *B. napus* was described by Lichter (1982). Following this, microspore embryogenesis has been reported in several *Brassica* species (cited in Cao *et al.*, 1994).

In the present study, microspore embryogenesis was attempted with Indian *B. juncea* varieties to speed up the selection of canola quality rapeseed-mustard cultivars adapted to Indian growing conditions.

MATERIALS AND METHODS

B. juncea cultivars Pusa Bold, Kranti and Varuna were used as microspore donors. *B. napus* cultivar Topas (line no. 4079) was used in the control treatments. The plants were grown in Cornell soilless mix either in the greenhouse (20°/10°C, day/night cycle or in a growth cabinet (10°/5°C, day/night cycle) at 200 mEm-2s-1 light intensity and 16 h photoperiod. Plants were watered and fertilized daily with 20-20-20 (N-P-K) fertilizer.

Pollen development was examined to determine the size of flower buds containing late uninucleate or early binucleate microspores. The microspore isolation and culture protocol was based on the procedure described by Keller *et al.* (1987) and modified by Seguin-Swartz (Personal communication).

About 50 flower buds were surface-sterilized with 70% ethanol for 2 min. and 5.6% sodium hypochlorite for 10 min. and rinsed with sterile distilled water three times. The buds were crushed with a glass rod in medium B5 (Gamborg *et al.*, 1968) containing 13% sucrose. The microspore suspension was filtered through a sterile 44 um nylon membrane into 50 ml graduated tubes; the microspores were collected by centrifugation at 4000 rpm for 5 min. the supernatant was decanted and the microspore pellet was washed twice with medium B5.

The microspores were suspended in medium NLN (Lichter, 1981) containing 13% sucrose with or without growth regulators. The microspore density was adjusted to 8 x 10^4 microspores/ml. Ten ml of microspore suspension was plated per 60 x 15 mm disposable petri plates and the plates were sealed with Paraffin.

The cultures were incubated in the dark, first at 32°C for 24-96 h, and then at room temperature (about 25°C) for 2-3 weeks at a slow rotating shaker (200 rpm). Plates with embryos were transferred to light (18 h photoperiod, 50 mEm-2s-1), fresh medium was added, and the cultures were incubated at room temperature for about one week prior to being transferred to solid medium. The embryos were subcultured either to basal medium B5 supplemented with 3% sucrose or to medium E4 (MS medium supplemented with growth regulators, Agnihotri, 1993). In some cases plantlets were subcultured for 7-10 days on medium E4 supplemented with 0.5 ppm IBA to initiate rooting.

Plantlets were transplanted into soil and grown to maturity in the greenhouse. Nodal explants from haploid plants were treated with rooting mixture and subsequently treated with colchicine using a method developed at the Agriculture and Agri-Food Canada Research Centre to double the chromosome number and restore fertility.

RESULTS AND DISCUSSION

In *B. napus*, the effect of the donor plant growth conditions, particularly temperature, on microspore development has been well documented (Lo and Pauls, 1992). Microspores at the late uninucleate or early binucleate stage have been shown to be embryogenic (Pechan and Keller, 1987, Fan *et al.*, 1988).

Microspore-derived embryos were obtained only in cultivar Kranti and from microspores obtained from buds of the plants grown at cool temperatures (10°/5°C). Baillie *et al.*, 1992 have also reported genotypic differences in the responsiveness of the microspores to *in vitro* embryogenesis in *B. campestris*.

Lo and Pauls (1992) demonstrated that embryogenesis depends on the physiological stage of the donor plants and that lower growing temperatures (15°/12°C) were inducive to embryogenic response.

Hundreds of embryos were obtained with cv. Topas. Microspores from 3.0-3.5 mm and 3.6-4.0 mm long buds having microspores at late uninucleate or early binucleate stage produced embryos. The embryos derived from the larger buds showed a better response to develop into normal plantlets under the conditions used.

A total of 16 embryos were obtained in *B. juncea* cv. Kranti in experiments using 3.0-3.5 mm and 3.6-4.0 mm long buds. Embryos derived from the larger buds (3.6-4.0 mm) developed into normal plantlets when transferred to solid B5 basal medium containing 3% sucrose, while those from the smaller buds underwent further organogenesis producing shoots from the embryo axis. When transferred to medium E4 supplemented with 0.5 ppm IBA, the shoots rooted and whole plantlets were recovered.

It was necessary to add fresh medium (NLN) to the embryos when they were transferred to light as embryo growth was stunted and abnormal when the medium was not replenished. Transfer to medium containing a low concentration of sucrose (3 to 10%) was detrimental to embryo development until after about one week of growth under light.

A number of factors affecting microspore embryogenesis in *B. juncea* were studied including the temperature of the donor plant growth, and pH of the microspore isolation medium, the length of the incubation temperature, the sucrose concentration and hormone supplementation of the culture medium. The results of these studies will be discussed. Further standardization will be required to improve the efficiency of microspore embryogenesis in this species.

ACKNOWLEDGMENT

Grant of Biotechnology Overseas associateship by the Department of Biotechnology, Govt. of India, to Dr. Abha Agnihotri is highly appreciated.

REFERENCES

Agnihotri, A. 1993. Hybrid embryo culture. In: Plant Culture Manual. Edited by K. Lindsley, Kluwer Academic Publishers, Boston, MA, USA. E4: 1-8.

Ahuja, K.L. and Banga, S.S. 1992. Oil and Meal Quality. In: Labana, K.S., Banga, S.S., Banga, S.K. (eds.) Breeding Oilseed Brassicas, Narosa Pub. House, New Delhi, pp. 76-93.

Baillie, A.M.R., Epp, D.J., Hutcheson, D.S. and Keller, W.A. 1992. *In vitro* culture of isolated microspores and regeneration of plants in *Brassica campestris*. Plant Cell Reports 11: 234-237.

Cao, M.Q., Yan, Li, Fan Liu and Dore, C. 1994. Embryogenesis and plant regeneration of pakchoi (*Brassica rapa* L. ssp. *chinensis*) via *in vitro* isolated microspore culture. Plant Cell Reports 13: 447-450.

Downey, R.K. 1990. Brassica oilseed breeding-achievements and opportunities. Plant Breeding Abstracts 60(10): 1165-1170.

Fan, Z., Armstron, K.C. and Keller, W.A. 1988. Development of Microspores *in vivo* and *in vitro* in *Brassica napus* L. Protoplasma, 147: 191-199.

Gamborg, O.L., Miller, R.A. and Ojima, K. 1968. Exp. Cell Res. 50: 151-158.

Keller, W.A., Fran, Z., Pechan, P., Long, N., Grainger, J. 1987. Proc. Seventh International Rapeseed Congress, Poland 11-14 May, pp. 152-157.

Lichter, R. 1981. Anther culture of Brassica napus in a liquid culture medium. Z. Pflanzenphysiol 103: 229-237.

Lichter, R. 1982. Induction of haploid plants from isolated pollen of *Brassica napus* Z. Pflanzenphysiol. 105: 427-434.

Lo, Kwan-Hung Lo and Pauls, K.P. 1992. Plant growth environment effects on rapeseed microspore development and culture. Plant Physiology 99: 468-472.

Pechan, P.M., and Keller, W.A. 1988. Identification of potentially embryogenic microspores in *Brassica napus*. Physiol Plant 74: 377-384.

In Vitro Morphogenesis and Microcorm Production in Gladiolus[1]

H.S. Grewal, S.S. Gosal and J.S. Arora

INTRODUCTION

Gladioli are known over the world and commercially cultivated for cut flowers, corms and cormels production. Corms and cormels are the means of their vegetative propagation, however, the rate of multiplication through corms is one-to-one and through corm to cormel is variable depending upon the cultivar. Sometime rate of multiplication is very low in the cultivars of commercial interest and tissue culture technique has, therefore, been visualized as its alternative for fast multiplication of otherwise difficult to propagate varieties (Ziv *et al.*, 1970; Ziv, 1979; Bajaj *et al.*, 1982/83; Dantu and Bhojwani, 1987; Grewal *et al.*, 1990a, b; Gosal and Grewal, 1990; Ziv, 1990; Steinitz *et al.*, 1991; Grewal *et al.*, 1993a, b). Application of tissue culture technique holds a significant promise in the cut flower producing plants due to short commercial life of variety because of change in the consumer preference. The plantlets of new cultivar can be rapidly multiplied and released for commercial cultivation by the growers through tissue culture (Wilfret, 1971). Moreover, the tissue culture propagation aimed at disease-free plant production ensures greater longevity of a particular cultivar.

In Gladiolus, the technique of tissue culture has been exploited for the production of corms, microcorms/cormlets and cormels (Ziv *et al.*, 1970; Dantu and Bhojwani, 1987; Grewal, 1990a, b; Steinitz *et al.*, 1991). The microcorm developed *in vitro* resulted in the development of flowering grade corms within one year, thereby, saving two years as in case of cormel to corm propagation (Grewal *et al.*, 1993a). Further, the tissue culture technique (Grewal *et al.*, 1990a) may be useful in purification of cultivar from the systemic build up of pathogens as Gladioli are propagated vegetatively and pathogens keep on accumulating generation after generation.

Most of the research workers followed the pathways of either (i) *in vitro* plantlet production, or (ii) corm, microcorm/cormlet, cormel, production, for fast Gladiolus propagation (Table 1, see Ziv and Lilien-Kipnis, 1990 and Bajaj *et al.*, 1992). Therefore, accordingly in the present paper the work is being reviewed with more emphasis on Gladiolus propagation through microcorm/cormlet production *in vitro*.

---

[1] Abbreviations: BAP (6-benzylamino purine); GA (Gibberelic acid); IBA (ß-Indole butyric acid) Kin (6-Furfuryl amino purine); MS (Murashige and Skoog, 1962), NAA (α-naphthalene acetic acid)

TABLE 1: *In vitro* propagation of Gladiolus through plantlet, corm, microcorm/ cormlet and cormel production

Genotype	Explant or source	Response	Reference
Sans Souci, Jo Wagonaar	Corm, Inflorescence	Formation of callus root primordia, buds, cormlets	Ziv *et al.*, 1970
Firmament, Hit Parade	Cormel stem tips	Formation of callus, regeneration of shoots, rooted plantlets, cormels	Sinonsen and Hildebrandt 1971
Spic and Span	Vegetative shoots of corms	Formation of single plant, multiple plantlets	Wilfret 1971
White Friendship	Stolon tips	Formation of cormels	Ginzburg and Ziv 1973
Forest Fire, Elvira, unnamed hybrid "488/2"	Axillary buds excised from corms	Formation of plantlets and small corms	Hussey 1977
Eurovision	Axillary buds excised from corms	Formation of plantlets and cormlets upon transplantation.	Ziv 1979
Oscar, Snow Princess	(I) Inflorescence, flower stalks, denuded flower, bract, perianth, leaf	Formation of callus, roots, shoots	Bajaj *et al.*, 1982-83
	(II) Cormels, tips, axillary buds	Formation of plants	
Gladiolus X Homoglossum (GH hybrid)	Buds excised from corms	Formation of shoots and corms	Sutter 1986
Gladiolus flanaganii	Axillary buds excised from corms	Elongation of axillary buds and formation of roots	Dickens *et al.*, 1987
Friendship, Her Majesty, American Beauty	Axillary buds excised from corms	Shoot multiplication shoots elongation, *in vitro* rooting, induction and development of corm	Dantu and Bhojwani, 1987
Gladiolus hybridus	Flower buds, dormant buds excised from corms	Formation of shoots *in vitro* rooting	Zhuo and Sun 1986
6 miniature *Gladiolus* hybrid	Apical and lateral buds excised from corms and cormels	Bud proliferation, formation of shoots *in vitro* rooting, plantlets, development of corms and cormels upon transplantation	Lilien-Kipnis and Kochba, 1987
Firebrand, Topaz	Corm	Formation of callus, shoots, roots	Kim *et al.*, 1988
-	Plantlets	Formation of corm, cormels	Steintiz and Lilien-Kipnis, 1989
Eurovision	Apical buds	Formation of shoots, protocorms and corms.	Ziv 1990.
Mayur, Sylvia, Spic and Span	Nodal-internodal-rachis- and floret-segments excised from flower stalk	Formation of multiple buds, multiple shoots, plantlets, cormels upon transplantation	Grewal *et al.* 1989
Hunting Song, Mayur, Sylvia, Spic	Nodal segments excised from flower stalk	Formation of multiple buds, multiple shoots, corms	Grewal *et al.*, 1990a, b.

Genotype	Explant or source	Response	Reference
and Span			
Blue Isle Hunting Song	Inflorescence stalk	Formation of callus, plantlets	Kamo *et al.* 1990
Peter Pears, Rose Supreme, Jenny Lee	Cormel slices, Plantlets	Formation of callus, cell suspension, plantlets from cell suspension	Kamo *et al.*, 1990
Gladiolus hybridus	Buds excised from corms	Formation of plantlets	Rumynin *et al.*, 1990
Topaz, Firebrand	Callus	Formation of shoots, roots, plantlets	Kim *et al.*, 1991
Kinneret	Terminal and lateral buds excised from corms	Formation of shoots, corms	Steinitz *et al.*, 1991

In vitro Gladiolus propagation through plantlets

Attempts have been made for Gladiolus propagation through tissue culture with variable success by different workers (Table 1). Ziv *et al.* (1970) pioneered in tissue culture propagation of Gladiolus by using explants of inflorescence stalk for plantlet regeneration within 6-7 weeks through calli, however, she realized that the organ regeneration from the explants was dependent on growth substances supplemented in the basal medium with best combination of NAA (10.0 mg/l) and Kin (0.5 mg/l). The regeneration was step wise by formation of a thin callus layer and root primordia at the basal end and buds as well as cormlets at the distal end of the cultured explant. Simonsen and Hildebrandt (1971) used cormel stem tips for callus induction and obtained complete plantlets *in vitro*. Likewise, plantlets of Gladiolus have been developed using explants excised from cormels (Bajaj *et al.*, 1982, 1983; Kamo *et al.*, 1990), corms (Wilfret, 1971; Bajaj *et al.*, 1982, 1983; Zhuo and Sun, 1986; Dickens *et al.*, 1987; Kim *et al.*, 1988; Rumynin *et al.*, 1990) and inflorescence stalk (Bajaj *et al.*, 1982, 1983; Kamo *et al.*, 1990).

In vitro Gladiolus propagation through production of corms, microcorms/cormlets and cormels

The development of organs for vegetative propagation is difficult, however, Ziv *et al.* (1970) obtained a limited success in cormlet induction and development by culturing inflorescence disc on basal modified Murashige and Skoog (1962) medium supplemented with Kin (0.5 mg/l) after 8 weeks of initial culturing. In Gladiolus, organ development for vegetative propagation has been found successful. The corm, microcorm/cormlet and cormel formation has been achieved in Gladiolus either (a) by transplanting the *in vitro* raised plantlets, or (b) by prolonged culturing of plantlets *in vitro* in the modified medium.

Corm, microcorm/cormlet and cormel development through transplanting plantlets

Simonsen and Hildebrandt (1971) obtained cormels by transplanting *in vitro* raised plantlets of Gladiolus cv Firmament and Hit Parade after 4 months. The cormels cold stored for 2 months sprouted, developed plants and produced corms and cormels after three months. Ziv (1979) reported that the *in vitro* propagated Gladiolus cv. Eurovision plantlets continued to grow upon transplantation and developed larger cormlets under non-aseptic conditions than formed *in vitro*, thereby, shortening the propagation period for producing large-size cormlets. Lilien-Kipnis and Kochba (1987) first multiplied the shoots of six miniature Gladiolus hybrids through apical and lateral bud proliferation by culturing on MS medium supplemented with low levels of NAA and BA or Kin at varying concentrations. The shoots were rooted in medium lacking cytokinins and the plantlets were transplanted directly in greenhouse, however, it was observed that development of corms and cormels harvested from these plants was enhanced by the presence of NAA and activated charcoal in the medium when used separately. Grewal *et al.* (1989) obtained single shoot (Fig. 1A) from nodal segments of flower stalk by culturing on MS medium supplemented with BAP (1.0 mg/l) and multiple shoot buds (15-20 bud primordia) when explants were cultured on MS medium containing BAP (5.0 mg/l) (Fig. 1B). The multiple buds developed shoots on MS medium or MS containing GA_3 (2.0 mg/l) and the shoots exhibited rooting when individually cultured on liquid medium containing IBA (4.0 mg/l). The plantlets (Fig 1C) developed cormels (3-7 mm dia) upon transplantation after 7 weeks (Fig. 1D).

In vitro corm, microcorm/cormlet and cormel development

In Gladiolus, explants excised from inflorescence stalk (Ziv *et al.*, 1970; Grewal *et al.*, 1990a, b) and corm or cormel (Ginzburg and Ziv, 1973; Hussey, 1977; Sutter, 1986; Dantu and Bhojwani, 1987; Steinitz and Lilien-Kipnis, 1989; Ziv, 1990; Steinitz *et al.*, 1991) have been used for *in vitro* corm, microcorm/cormlet and cormel production, however, the development of such organs has been found to be dependent on growth regulators (promoters or retardants) and sucrose.

Role of growth regulators (promoters or retardants)

Ziv *et al.* (1970) induced callus on inflorescence stalk discs by culturing on modified MS medium supplemented with NAA (10.0 mg/l) and Kin (0.5 mg/l). The subcultured explants underwent primordial development leading to formation of roots some of which resembled stolons. The swollen tips of such stolons was similar to the small cormlets. Cormlet and a sprout developed from the original inflorescence stem disc subcultured on modified MS (1962) medium supplemented with Kin (0.5 mg/l) 8 weeks after initial culturing. Ginzburg and Ziv (1973) studied the influence of NAA, Kin and GA on cormel development in Gladiolus stolon tips grown *in vitro* in cv. White Friendship. It was observed that Kin (0.5 mg/l) supplemented in the MS (1962) medium induced the stolon tips to differentiate into cormel with improvement in size at tips to differentiate into cormel with improvement in size at Kin (5.0 mg/l). The presence of NAA (0.5 mg/l) along with Kin further improved the growth of explant and cormel formation. However, GA_3 (0.01-5.0 mg/l) inhibited cormel induction and formation at low Kin levels.

Hussey (1977) observed elongation of axillary buds excised from corms of Gladiolus hybrid Forest Fire, Elvira and unnamed hybrid "488/2" when cultured on MS medium supplemented with low BAP (0.12-0.5 mg/l) and proliferation of buds at BAP (2.0 mg/l). However, little or no growth was observed at BAP (0.03 mg/l) or lower level supplemented in the medium. Grewal *et al.* (1989) also observed quick shoot elongation from the nodal segments of Mayur, Sylvia, Spic and Span excised from the flower stalk by culturing on low BAP (1.0 mg/l) supplemented in MS medium (Fig 1A). However, Dantu and Bhojwani (1987) obtained shoots of cvs. Friendship, Her Majesty and American Beauty by culturing axillary buds excised from the corms on MS medium supplemented with BA (0.5 mg/l) which multiplied upto 5 folds by subculturing at 4 weeks interval.

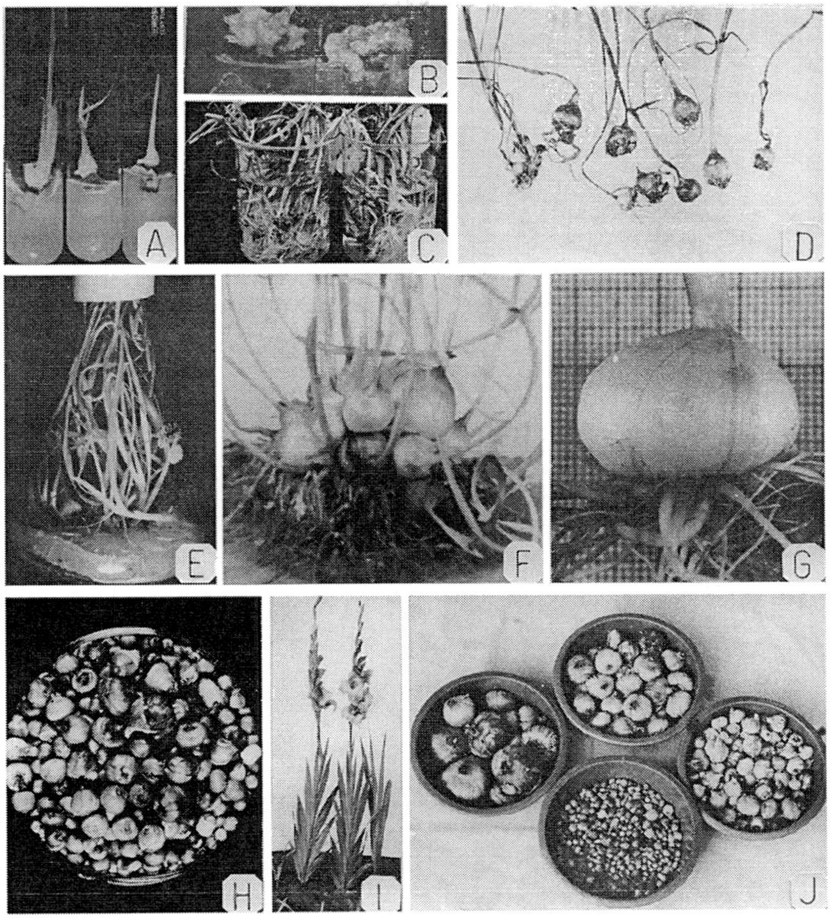

FIGURE 1(A-J): *In vitro* morphogenesis and microcorm production in Gladiolus. (A) Single shoot elongation of nodal buds cultured on MS+BAP (1.0 mg/l) in cultivar Mayur, 30 days after culturing. (B) Multiple shoot buds formation from nodal buds cultured on MS+BAP (5.0 mg/l) in cultivar Hunting Song, 30 days after culturing. (C) Rooted plantlets of cultivar Sylvia and, Spic and Span. (D) Cormel formation on the transplanted rooted plantlets after 7 weeks. (E) Multiple shoot elongation of proliferated shoot buds upon culturing on MS (basal) medium in cultivar Mayur, 28 days after culturing. (F) Development of microcorms at the basal region of shoots, in bunches, by the addition of liquid 1/2 MS+IBA (4.0 mg/l) + sucrose (6%) medium in the flasks containing multiple shoots in cultivar Hunting Song, 8 weeks after culturing. (G) Largest (27mm dia.) microcorm ever reported *in vitro* cultivar Hunting Song. (H) *In vitro* produced microcorms of cultivar Hunting Song. (I) Flowering of *in vitro* produced microcorms of cultivar Hunting song in the first year of sowing. (J) Flowering grade corms and small cormels of cultivar Hunting Song obtained from the microcorms in the first year of sowing.

Shoot elongation occurred only on medium either lacking BA or containing very low BA (0.1-0.2 mg/l). The medium modified by reducing NH_4 NO_3 concentration to half strength and replacing KH_2PO_4 by NaH_2PO_4 (300 mg/l) and containing BAP (0.5 mg/l) exhibited 8-fold multiplication of shoots obtained by culturing axillary buds excised from the corms (Bhojwani & Dantu, 1993). Steinitz *et al.* (1991) obtained shoots in Gladiolus cv. Kinneret from terminal and lateral buds which exhibited proliferation upon culturing on MS medium supplemented with BA (0.3 μM) and NAA (0.1 μM). Lilien-Kipnis and Kochba (1987) also obtained repeated proliferation of apical and lateral buds excised from corms and cormels of six miniature Gladiolus hybrids by culturing on MS medium supplemented with low levels of NAA and BA or Kin. Grewal *et al.* (1990a) obtained rapid bud proliferation from nodal buds excised from flower stalk cultured on MS (1962) medium containing BAP (5.0 mg/l) (Fig. 1B).

Besides growth promoters, the role of growth retardants i.e. ABA (Ginzburg and Ziv, 1973), paclobutrazol (Steinitz and Lilien-Kipnis, 1989; Ziv, 1990; Steinitz *et al.*, 1991), daminozide, ancymidol and uniconazole (Ziv, 1990) on shoot proliferation and morphogenesis in Gladiolus has been studied.

Ginzburg and Ziv (1973) observed in the Gladiolus cultures of White Friendship that Kin induced Cormel differentiation was unaffected by the presence of ABA (0.05, 1.0 or 5.0 mg/l) in the medium, however, ABA inhibited the cormel growth and leaf elongation. Steinitz and Lilien-Kipnis (1989) reported that paclobutrazol induced cormels and mediates its influence by reducing endogenous GA level and paclobutrazol dependent corm and cormel development could be substituted by adding GA_3 to a poaclobutrazol containing medium. Likewise, Ziv (1990) observed that growth retardants (paclobutrazol, daminozide, ancymidol, uniconazole) caused extensive bud proliferation, inhibited leaf growth and induced protocorm development, however, paclobutrazol (1.0 mg/l) induced the highest number of buds per explant which developed into protocorms, whereas, largest size of protocorm was observed in ancymidol (2.5 mg/l) and uniconazole (1.0 mg/l) treated explants. Steinitz *et al.* (1991) reported that paclobutrazol (10.0 mg/l) supplemented in the liquid media promoted corm formation *in vitro* by shifting assimilate allocation towards the growing corms. Further, they observed that the effect of high sucrose level (60.0 g/l) on corm development was accentuated due to the presence of paclobutrazol (10.0 mg/l) in the medium, thereby, improving the per cent success and corm fresh weight.

Role of Sucrose

Gladiolus corms and cormels are starchy and enriched with carbohydrates. That might be the reason for the beneficial influence of increased sucrose level (6-10%) supplemented in the culture medium on *in vitro* microcorm or cormlet induction and development. Dantu and Bhojwani (1987) reported that high sucrose level (6 or 10%) favoured formation of large corms (10-23 mm dia) *in vitro*. Grewal *et al.* (1990a) established a protocol for efficient *in vitro* corm production of Gladiolus. The shoots obtained through bud proliferation of nodal buds on MS (1962) medium containing BAP (5.0 mg/l) in Step I (Fig 1B) and elongated on MS (1962) medium (Fig 1E) were individually subcultured on liquid half-strength MS medium supplemented with IBA (4.0 mg/l) and varying sucrose level (3%, 6% and 9%) in Step III and the results are described (Table 2). Among the three media, the highest per cent cormlet formation was observed in MS_2 medium being maximum in cultivar Mayur (61.11%), followed by Sylvia (60.00%) and Spic and Span (55.55%) within 4-5 weeks. The shoots failed to induce cormlets in MS1 medium, whereas, the higher sucrose level (9%) supplemented in MS3 medium failed to improve percentage and size of cormlets.

TABLE 2: *In vitro* cormlet formation in three cultivars of Gladiolus on tissue culture derived shoots upon culturing in three liquid media composition varying in sucrose level

Genotype	Medium	Cormlet formation	Days required for cormlet formation	Size of cormlet
Mayur	MS_1	0.00	---	--
	MS_2	61.11	28-32	3-13
	MS_3	60.87	30-32	4-12
Sylvia	MS_1	0.00	--	--
	MS_2	60.00	28-32	3-10
	MS_3	59.99	28-32	3-11
Spic and Span	MS_1	0.00	--	--
	MS_2	55.55	28-32	3-12
	MS_3	52.63	30-32	3-12

MS_1 : 1/2 MS + IBA (4.0 mg/l) + sucrose (3%), MS_2 : 1/2 MS + IBA (4.0 mg/l) + sucrose (6%) MS_3 :1/2 S + IBA (4.0 mg/l) + sucrose (9%).

The cormlets obtained in MS2 medium varied in size in different cultivars i.e. Mayar (3-13 mm), Sylvia (3-10 mm) and Spic and Span (3-12 mm) due to differences in health of the shoots. The developed protocol was modified for efficiency, simplicity and economics IBA (4.0 mg/l) and sucrose (6%) into the flasks/jars containing multiple shoots rather than culturing the shoots individually on the filter paper bridges in the test tubes (Grewal *et al.*, 1990b). The microcorms thus, developed at the basal regions of the shoots in bunches and matured within 8-10 weeks indicated by brown tunica formation (Fig 1F). Gosal and Grewal (1990) reported that the size of microcorms ranged from 4-22 mm diameter and the number of large (16 mm-22 mm) and medium (11 mm-15 mm) size microcorms was more than the small (4 mm-10mm) ones. During the further experiments, the largest (27 mm dia.) corm ever reported was developed *in vitro* in cv. Hunting song (Fig 1G). Steinitz *et al.* (1991) supported the use of high sucrose (60 g/l) in the culture media for corm production. Grewal *et al.* (1993a) using the established protocol developed the microcorms of cultivar Hunting Song *in vitro* (Fig 1H) and their germination percentage under field conditions (sown after 2-4 months storage at 4-10°C) was 73.47% (16-23 mm dia), 70.00% (11-15 mm) and 64.28% (4-10 mm dia) (Table 3) and flowering was 12.6% in the first generation (Fig 1I).

The microcorms increased to the flowering grade size in the very first generation (Fig. 1J) and germination was 62.50% (<1cm diameter corms), 76.11% (1-2 cm), 88.23% (2-3 cm), 95.00% (3-4 cm) and 100.00% (4-5 cm and 5-6 cm) during the second generation, 30 days after sowing (Table 4).

TABLE 3: Field performance of *in vitro* obtained microcorms of Gladiolus cv. Hunting song in the first year.

Size in dia. (mm)	Microcorms harvested *in vitro*	Microcorms sown in the Ist Year	Microcorms germinated in Ist Year	Germination (%)	Corms harvested from field in Ist year
4-10	61	56	36	64.28	36
11-15	149	140	98	70.00	98
16-23	59	49	36	73.47	36

TABLE 4: Field performance of corms obtained from microcorms of Gladiolus cv. Hunting song in the second year.

Size in (cm)	Number of corms	Corms sown	Corms germinated after 30 days of sowing[a]	Number of plants exhibiting flowering[b]	Average number of florets per spike
<1	8	8	5 (62.50)	0.00 (0.00)	--
1-2	68	67	51 (76.11)	0.00 (0.00)	--
2-3	60	51	45 (88.23)	14 (31.11)	9.42
3-4	28	20	19 (95.00)	13 (68.42)	12.0
4-5	3	3	3 (100.00)	3 (100.00)	16.66
5-6	3	3	3 (100.00)	3 (100.00)	18.00

[a] Figures given in brackets indicate per cent corm germination, [b] Figures given in brackets indicate per cent flowering.

However, the percentage of flowering ranged from 31.11% (2-3 cm corm size) to 100.00% (5-6 cm), 105 days after sowing. The data indicates that average number of florets ranged from 9.42 (2-3 cm dia. corms) to 18.0 (5-6 cm) per spike. The described protocol helped in obtaining flowering grade corms developed from microcorms within 1 year, thereby, saving 2 years as in the case of cormel to corm production which require 3 years period. The realities unsolved are still under investigations.

REFERENCES

Bajaj, Y.P.S., Sidhu, M.M.S. and Gill, A.P.S. 1982-83. Some factors affecting the *in vitro* propagation of Gladiolus. Scientia Horticulturae 18: 269-275.

Bajaj, Y.P.S., Sidhu, M.M.S. and Gill, A.P.S. 1992. Micropropagation of Gladiolus. In Y.P.S. Bajaj (Ed.) Biotechnology in Agriculture and Forestry, Vol. 20. High-Tech and Micropropagation IV, Springer-Verlag, Berlin Heidelberg, pp. 135-143.

Bhojwani, S.S. and Dantu, P.K. 1993. *In vitro* clonal multiplication and field performance of micropropagated plants of three cultivars of Gladiolus. In: Abstracts, International Plant Tissue Culture Conference, Dec. 19-21, 1993, Dhaka, Bangladesh, pp. 13.

Dantu, P.K. and Bhojwani, S.S. 1987. *In vitro* propagation and corm formation in Gladiolus. Gartenbauwissenschaft 52: 90-93.

Dickens, C.W.S., Kelly, K.M., Manning, J.C. and Van Staden, J. 1986. *In vitro* propagation of *Gladiolus flanaganii*. South African J. Bot. 52: 485-487.

Ginzburg, C. and Ziv, M. 1973. Hormonal regulation of cormel formation in Gladiolus stolons grown *in vitro*. Ann. Bot. 37: 219-224.

Gosal, S.S. and Grewal, H.S. 1990. Tissue culture propagation: Problems and Potentials. In: Horticulture- New Technologies and Applications. J. Parkash and R.L.M. Pierik (Ed.), Kluwer ACademic Publ., The Netherlands, pp. 197-200.

Grewal, H.S., Arora, J.S., Gosal, S.S., Singh, Kushal and Malhotra, A. 1989. Mass propagation of gladiolus through tissue culture. In Tandon, P. (Ed.) Proc. Advances in Plant Tissue Culture in India. Pragati Prakashan, Meerut, India.

Grewal, H.S., Gosal, S.S. and Arora, J.S. 1990a. Efficient *in vitro* corm production of Gladiolus In: Abstracts, VIIth International Congress on Plant Cell and Tissue Culture, June, 25-29, 1990, Amsterdam, The Netherlands, pp. 102.

Grewal, H.S., Arora, J.S. and Gosal, S.S. 1990b. Propagation of Gladiolus and Chrysanthemum through tissue culture. In: Abstracts, International Seminar on New Frontiers in Horticulture, Nov. 25-28, 1990, Bangalore, India, pp. 139-140.

Grewal, H.S., Gosal, S.S., Arora, J.S. and Dhaliwal, H.S. 1993a. *In vitro* corm production of Gladiolus and their field performance. In: Abstract, Ist regional conference on prospects of Biotechnology in Nepal, April 11-13, 1993, Birgunj, Nepal, pp. 19.

Grewal, H.S., Gosal, S.S., Arora, J.S. and Singh, Kushal 1993b. Propagation of ornamental plants through tissue culture. In: Abstracts, International Plant Tissue Culture Conference, Dec. 19-21, 1993, Dhaka, Bangladesh, pp. 40.

Hussey, G. 1977. *In vitro* propagation of Gladiolus by precocious axillary shoot formation. Scientia Horticulturae 6: 287-296.

Kamo, K., Chen, J. and Lawson, R. 1990. The establishment of cell suspension cultures of Gladiolus that regenerate plants. *In Vitro* Cellular and Developmental Biol. 26: 425-430.

Kim, K.W., Choi, J.B. and Kwon, K.Y. 1988. The rapid multiplication of Gladiolus plants by callus cultures. J. Korean Soc. Hort. Sci. 29: 312-318.

Lilien-Kipnis, H. and Kochba, M. 1987. Mass propagation of new Gladiolus hybrids. Acta Hort. 212: 631-638.

Murashige, T. and Skoog, F. 1962. A revised medium for rapid growth and bioassays with tobacco tissue cultures. Physiol. Plant 15: 473-497.

Rumynin, V.A., Agadzhanyan, I.V. and Slyusarenko, A.G. 1990. Mass Clonal Propagation of Gladiolus Plants. Byulleten Glavnogo Botanicheskogo Sada 156: 68-72.

Simonsen, J. and Hildebrandt, A.C. 1971. *In vitro* growth and differentiation of Gladiolus plants from callus cultures. Canadian J. Bot. 49: 1817-1819.

Steinitz, B., Cohen, A., Goldberg, Z. and Kochba, M. 1991. Precocious gladiolus corm formation in liquid shake cultures. Plant Cell, Tissue and Organ Culture 26: 63-70.

Steinitz, B. and Lilien-Kipnis, H. 1989. Control of precocious gladiolus corm and cormel formation through tissue culture. J. Plant Physiol. 135: 495-500.

Sutter, E.G. 1986. Micropropagation of Ixia virdifolia and a Gladiolus x Homoglossum hybrid. Scientia Horticulturae 29: 181-189.

Wilfret, G.J. 1971. Shoot-tip culture of Gladiolus: An evaluation of nutrient media for callus tissue development. Florida State Hort. Soc. 84: 389-393.

Zhuo, W. and Sun, Z. 1986. Flower bud and dormant bud culture and plant redifferentiation of *Gladiolus hybridus in vitro*. Acta Agri. Univ. Pekiensis 12: 191-194.

Ziv, M. 1979. Transplanting Gladiolus plants propagated *in vitro* Scientia Horticulturae 11: 257-260.

Ziv, M. 1990. The effect of growth retardants on shoot proliferation and morphogenesis in liquid cultured Gladiolus plants. Acta Horticulturae 280: 207-214.

Ziv, M., Halevy, A.H. and Shilo, R. 1970. Organs and plantlets regeneration of Gladiolus through tissue culture. Ann. Bot. 34: 671-676.

Ziv, M. and Lilien-Kipnis, H. 1990. Gladiolus, In P.V. Ammirato, D.A. Evans, W.R. Sharp and Y.P.S. Bajaj (Eds.) Handbook of Plant Cell Culture, Vol. 5. McGraw-Hill, New York, pp. 461-478.

Callus Growth Responses of Sunflower Explants Cultured *In Vitro*

Puspita Samantaray, Gyanaranjan Rout and Prasanta K. Behera

INTRODUCTION

A comparative growth study was conducted on sunflower (*Helianthus annuus* L. cv. Morden) callus cultures induced from various explants on four culture media (White, MS, Street and B_5). Petiole, leaf and leaf mid-rib exhibited different callus growth depending on culture medium and hormonal treatments. Callus formation percentage, callus growth and 50% inhibition dose (I_{50}) for callus regeneration were recorded. MS and B_5 medium generated callus upto 50% and above with 2,4-D (3 mg l^{-1}) in all explant types. B_5 medium with IAA (2 mg l^{-1}) initiated 90% callusing in petiole and 95% callusing in leaf mid-rib within 3 weeks, and in first subculture, the leaf mid-rib within 3 weeks, and in first subculture, the leaf mid-rib callus showed 81% in growth followed by 70% in third subculture period. The passage time for subculturing leaf mid-rib callus is within 21 to 28 days. Callus pH value ranged between 5.2 to 5.8. It is suggested that sunflower leaf mid-rib tissues act as potential stimulator for vigorous callus growth at optimum level of IAA, may be due to endogenous IAA and/or other accumulated hormone levels involved as synergistic effect.

A callus-culture system offer many advantages as a model system for several biological investigations. The callus model of plant flowering was proposed by Chailakhyan *et al.* (1975). Callus cultures has been used widely in various physiological and related studies by Weinstein *et al.*, 1962 and Vesins *et al.*, 1972 in the genus *Rosa* and by Altman *et al.*, 1982; Tissert and Galleta, 1987 in the genus *Citrus*.

Production of callus from various plant tissue and organs of different species is now feasible but *in vitro* methodologies for callus initiation and maintenance are not well established. Behera and Tripathy (1980) reported on establishment of callus raised from two different explants of *Catharanthus roseus* L. *In vitro* studies dealing with callus cultures usually involve monitoring callus growth. Volume estimates can be used to determine growth and estimate callus fresh weight (Flower and Janick, 1947). The use of volume as a measurement of growth as well as the estimation of volume from one or more linear dimensions has been used for years in field studies of trees and shrubs (Wakimoto, 1977). Measuring linear dimensions of the callus through the test tube is non destructive and simple to perform.

The objectives of this study were to report growth responses of sunflower (*Helianthus annuus* L. cv. Morden) callus cultured under clearly specified *in vitro* culture conditions and long-term subculturing of these cell towards development of a system that promotes callus the ability to rapid regeneration, a prerequisite for the physiological studies *in vitro*.

MATERIAL AND METHODS

Explant preparation

The explant sources petiole, leaf, and leaf mid-rib of Helianthus annuus L. cv. Morden were treated with liquid detergent (2% teepol) for 5 min., washed in slow running tap water for 10 min. and surface sterilized with 0.1% mercuric chloride for 20 min. and were cut aseptically to 2 mm size in the Laminar air flow cabinet.

Callus media and incubation conditions

White's medium (1954); basal Murashige and Skoog (MS) medium (1962); Street's medium (1966) and Gamborg's (B_5) medium (1968) were used supplementing with different concentrations of the hormones alone or in combination of indole acetic acid (IAA), 2,4-dichlorophenoxyacetic acid (2,4-D) and kinetin (Kn). The final media contained 2% (w/v) sucrose; 0.8% (w/v) agar and pH 5.6 prior to adding agar. All media were autoclaved at 121°C (1055 g cm^{-2}) for 15 min. and kept for setting for 24 hours. All media and the plant hormones tested were from HiMedia Laboratories, Bombay, India.

The explants inoculated two per tube (25 x 150 mm) containing one of the treatment media (25 ml) and were grown in culture room maintaining temperature 25 ± 2°C, humidity 50-55 and 14 h photoperiod (2500 lux). Each treatment consisted 10-20 explant types. Experiments were repeated twice.

pH and growth measurements

pH changes *in vitro* were measured according to Skirvin *et al.*, 1986.

Non-destructive and continuous method of estimation of callus growth followed according of Fowler and Janick, 1974. For each callus, three dimensions were measured through the tube: a) diameter 1 (D1), linear dimension parallel to the medium at an arbitrary location; b) diameter 1 (D2), dimension parallel to the medium and perpendicular to D1; and c) height (H), dimension perpendicular to the medium. The dimension (callus diameter) of the hemispherical shape for the callus was measured calculating from the volume formula for a hemisphere:

$$\text{Hemisphere} = C \times [(D1 + D2 + 2H)/3]^3$$

Where

$$C = (1/6\pi)/2 = 0.262$$

D1 = diameter

D2 = diameter perpendicular to D1

H = height

Callus subculturing

Callus initially cultured on B_5 medium 2,4-D 3 mgl^{-1} were transferred to media with IAA at 2,3 and 4 mgl^{-1} and subcultured upto third passage.

Visual observations were recorded for characterizing the nature of the callus.

RESULTS

Influence of different media on explant types

Growth responses of explant types to callus production in different media showed that in B_5 medium the highest per cent of calli occurred with in 4 weeks in leaf mid-rib (Table 1). All three types of sunflower explants resulted in a relatively poor generation of callus in White and Street's medium in compare to MS and B_5 medium where the explants generated upto callus initiation of 50% and above. Generated calli were characterized by their friable and light green colour. Street's medium could not be tried for leaf mid-rib as several cultures were infected.

TABLE 1: Response of different explants of sunflower to callus initiation on four culture media fortified with Kn (0.5 mg l^{-1}) and 2,4-D (3 mg l^{-1}).

	Per cent of callus initiation		
Medium	Petiole	Leaf	Leaf mid-rib
White	41.5	30.4	42.7
MS	50.5	47.2	56.8
Street	43.6	41.5	Not done
B_5	57.0	51.4	68.5

Data based on 20 cultures. Culture period for each treatment is 4 weeks.

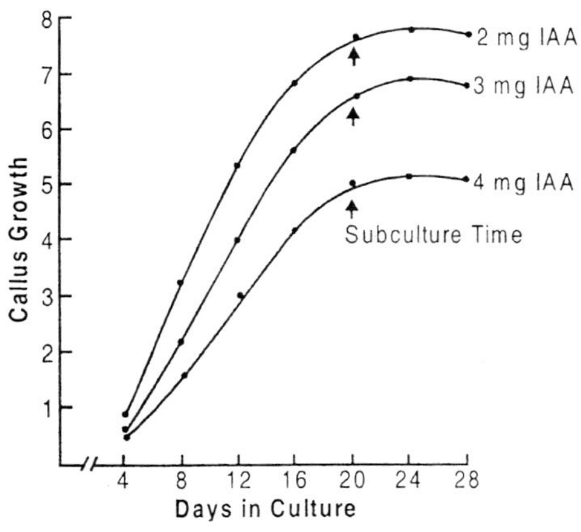

FIG. 1. Callus growth curves of sunflower leaf mid-rib callus calculated with volume formula. Values are means of 10 replicates.

Effect of IAA on callus induction

Of the evaluation of four media for callusing, 2,4-D responded better in Gamborg's B_5 salt mixture in all the three types of explants. Therefore, this medium was chosen for further assessing with IAA. Treatment of B_5 medium with IAA 2mg l^{-1} on petiole, leaf and leaf mid-rib increased induction of callus to the highest levels and IAA 4 gm l^{-1} responded decreased callus induction while verifying the effectiveness of different concentrations of IAA (Table 2). The results indicate that the 50% inhibition dose (I_{50}) for callus regeneration on B_5 medium at treatment with IAA in between 3 to 4 mg l^{-1} would probably be a critical limit of this hormone. Polar effects in callus response of petiole and leaf mid-rib explants were observed and callus appeared along the cut edge of the leaf. The calli were characterized by whitish brown to light green tissue.

TABLE 2: Response of IAA on B_5 medium with Kn (0.5 mg l^{-1}) for callus induction in different explants of sunflower.

	Per cent of callusing in		
IAA (mg l^{-1})	Petiole	Leaf	Leaf mid-rib
2	90.0	73.8	95.0
3	60.8	54.3	75.6
4	40.5	33.8	45.5

Data based on 20 cultures. Culture period for each treatment is 4 weeks.

Leaf mid-rib calli and subculture time

Leaf mid-rib responded maximum percentage of callusing to IAA on B_5 medium and this was further assessed for establishment of the subculture period. Percentage of callus formation in leaf mid-rib were higher (81%) in first subculture, but lower (70%) in third subculture 1 to 3 passages were much lower as 54 to 40% respectively (Table 3). The calli has a tendency towards browning. Growth curve of sunflower callus grown under subculture conditions showed an initial lag period followed by an exponential rise in callus growth volume and a linear increase in the callus population upto three weeks and thereafter the callus entered a stationary phase. The growth curves were analogous under different IAA concentrations. The passage time for sunflower leaf mid-rib callus is within about 21 to 28 days (Figure 1).

TABLE 3: Response of IAA on B_5 medium at each subculture (1-3) for sunflower leaf mid-rib after 3 weeks

	Per cent of calli growth in leaf mid-rib callus		
IAA (mg l^{-1})	Subculture 1	Subculture.2	Subculture.3
2	81	78	70
3	72	70	61
4	54	49	40

There are significant differences in pH before and after autoclaving, ranging in between 5.67 to 6.58 and with time the pH was 5.2 to 5.8 and without callus the pH value ranged from 6.2 to 6.7 in the four culture media (Table 4). The results indicate that maximum callus proliferation occurred when the pH of the media adjusted to 5.6 to 5.8 prior to inoculation.

TABLE 4: pH changes *in vitro* with and without sunflower callus in different media. Data represent the mean of 20 observations at 48 hours.

Medium	Pre-autoclave	Post-autoclave	+callus	-callus
White	5.81	6.38	5.8	6.2
MS	5.67	6.58	5.4	6.5
Street	5.71	6.38	5.7	6.3
B_5	5.76	6.43	5.2	6.7

DISCUSSION

Only a few or less successful efforts have been undertaken during the last five years to obtain large scale calli from sunflower explants. Prior to the work described here, there had been no reports of callus growth response of *in vitro* cultured sunflower explants. Natural auxin IAA was more effective than synthetic auxin 2,4-D for callus induction in sunflower explants (Table 1&2). This may be due to strong interaction between IAA and kinetin and also species.

The possible differences in degree of induction and callus growth may be due to result in enhanced variation in the state of explant tissue as proposed by Meins (1983). According to Flick *et al.* (1983) the physiological state of the parent plant is often important in determining the potential of callusing may also hold correct for sunflower callusing. The strong effect of IAA at optimum level suggests that endogenous IAA and/or other accumulated hormone levels may also be involved as synergistic effect in the high yield of callus. IAA in a higher concentration retarded callus growth remarkably (Table 2). This may be due to influence of IAA on the decreased uptake of macronutrients and water absorption from the culture media as proposed by Votrubova and Votruba (1986).

Leaf mid-rib callus was found as an indicator of the potential to generate callus profusely (Table 2). Therefore, one can speculate that, in the leaf mid-rib, tissues act as potential stimulators for vigorous callus growth.

The timing requirement of the callus response suggests that there is a specific developmental response occurring in the leaf mid-rib tissue between days 4 and 21 (Table 3 and Figure 1). Growing the callus in the subculture stage resulted in callus its regaining potential due to the absence of accumulated ethylene as suggested by Robinson and Adams (1987).

With time the pH of the medium drifts into the acid range (Table 4) supported the reports of Skirvin *et al.* (1986). This suggests that change in pH is associated with the nutrients of medium and culture conditions.

Attempts are underway to further optimize the callus production in subcultures for indefinite periods. We believe that callus cultured cells offer a valuable model system for a better understanding of the physiological functions controlling the cell life. Therefore, although rigorous experimental testing would be required before the technique become a feasible method for producing physiologically active callus tissues which could provided the basis for biotechnological approaches, it should be possible to introduce sunflower leaf mid-rib as potential cells to generate callus.

ACKNOWLEDGMENTS

Research supported in part by the University Grants Commission, New Delhi to one of us, P.K.B; the laboratory facilities provided by Prof. P.N. Das and the excellent technical assistance of Dr. B.K. Debata of Regional Plant Resource Centre, Bhubaneswar is gratefully acknowledged. The authors would like to thank Prof. P.A. Khan, Head of Plant Physiology Laboratory, Berhampur University for helpful discussions and Dr. A. Seetharam, Indian Council of Agricultural Research, Bangalore for supplying the sunflower pure line seeds.

REFERENCES

Atlman, A., Gulsen, Y. and Goren, R. 1982. Growth and metabolic activity of lemon juice vesicle explants *in vitro* Plant Physiol. 69: 1-10.

Behera, P.K. and Tripathy, P.C. 1980. Establishment of callus raised from two different explants of *Catharanthus roseus* L.G. Don. J. Orissa, Bot. Soc. 21: 44-45.

Chailakhyan, M.K.H., Aksenova, N.P., Konstantinova, T.N. and Bavrina, T.V. 1975. The callus model of plant flowering. Proc. Roy Soc. B. 190, 333-40.

Flick, C.E., Evans, D.A. and Sharp, W.R. 1983. Organogenesis, In handbook of plant cell cultures, Vol. 1. Techniques for Propagation and Breeding (Eds., D.A. Evans, W.R. Sharp, P.V. Ammirato, and Y. Yamada), pp. 13-81. Macmillan Publishing Co., New York, ISBN 0-02-949230-0.

Fowler, C.S. and Janick, J. 1974. Non destructive estimation of callus growth, Hort-Science 9: 552.

Gamborg, O.L., Miller, R.A. and Otimek, K. 1968. Nutrients requirement of suspension culture of soybean roots callus. Exp. Cell Res. 50: 151-158.

Meins, F.J.R. 1983. Heritable variation in plant cell culture, Ann Rev. Pl. Physiol. 34: 327-346.

Murashige, M. and Skoog, F. 1962. A revised medium for rapid growth and bioassay with tobacco tissue culture. Physiol Plant 15: 473-497.

Robinson, K.E.P. and Adams, D.O. 1987. The role of ethylene in the regeneration of *Helianthus annuus* (sunflower) plants from callus. Physiol Plant 71: 151-156.

Skirvin, R.M., Chu, M.C., Mann, M.L., Young, H., Sulliven, J. and Freenish, T. 1986. Stability of tissue culture medium pH as a function of autoclaving time and cultured plant material. Plant Cell Rep. 5: 292-294.

Street, H.E. 1966. Growth differentiation and organogenesis in plant tissue and organ cultures. In Wilmer, E.N. (ed.) cellus and tissues in culture, 3: 631-690.

Street, H.E. 1977. Cell (suspension) cultures-techniques In plant tissue and cell culture 2d ed., ed H.E. Street, Oxford, Blackwell Scientific Publications, pp. 61-102.

Tisserat, B. and Galleta, P.D. 1987. *In vitro* culture of lemon juice vesicles. Plant cell and organ culture 11: 81.

Vesins, K.K., Vehytil, L.E. and Fletcher, J.S. 1972. Minimal organic medium for suspension cultures of Paul's scarlet rose. Planta 106: 173-176.

Votrubova, O. and Votruba, M. 1986. The influence of IAA on the uptake of potassium, calcium, magnesium water absorption and growth in young maize seedlings. Biol. Plant 28(6): 460-464.

Wakimoto, R.H. 1977. Chaparral growth and fuel assessment in southern California, In: Proc. Symp. on the environmental consequences of fire and fuel management in Mediterranean ecosystems, US Dept. Agr For. Serv. Gen. Tech. Rep. WO-3 pp. 412-418.

Weinstein, D.H., Tulecke, W., Nickell, L.G. and Lahrencot, H.J. 1962. Biochemical and physiological studies of tissue cultures and plant parts from which they were derived III Paul's scarlet rose. Contrib. Boyce Thompson Inst. 21: 371-386.

White, P.R. 1954. The cultivation of animal and plant cells. The Ronald Press Co., New York, pp. 239.

Vitro Trees of Cashew (*Anacardium occidentale* L.) A Preliminary Evaluation

Smitha Hegde, L. D'Souza, Icy D'Silva, Alice C. Augustine

Cashew is an evergreen tree grown all over the tropics especially along the coastal regions. Since the tree grows well even on poor soils it is a popular plantation crop for development of the waste lands (Nayar, 1993). The cashewnut kernels fetch considerable amount of foreign exchange for the developing countries (Balasubramanian, 1987). Cashew supports a flourishing industry in our country which is highly labour oriented, providing a means of livelihood to thousands of unskilled labourers especially women.

The first cashew plants produced *in vitro* have been successfully transferred to soil and have flowered and borne fruit. The plants were obtained from cotyledonary nodes by a 4 step process: Induction of multiple buds on MS medium supplemented with 117 mM sucrose, 14.6mM maltose and 44.4 mM BAP; proliferation of buds through sequential culture on MS medium composing of 22.2 mM BA, coconut water 100 ml/l, L-arginine 474.6 mM and DL-valine 853.6 mM in succession (D'Silva and D'Souza, 1992a); elongation of shoots on reduced BAP (4.4 mM) and increase of sucrose (146 mM) and rooting of the shoots was obtained with IAA 2.9 mM and IBA 4.92 mM (D'Silva and D'Souza, 1992b). The rooted shoots were transferred to a mixture of sand, soil and coconut husk. After hardening in the glass house the plants were transferred to the field (D'Souza *et al.*, 1994).

The vitro trees were planted in 3 different places in Dakshina Kannada district together with seedlings obtained from seeds of the same season as those used for producing the vitro trees. A preliminary evaluation of the trees at the age of 6 years was done (Table 1). It is found that the average height of the vitro trees is not significantly different from the seed grown trees. The average girth of the trunk at 10 cms above the ground is around 42 cms in case of vitro trees whereas it is only 30 cms in case of seed grown trees. Cashew trees often exhibit plagiotropic branches when grown from rooted cuttings. The plagiotropic branches were not seen in the vitro trees. It is also found that the vitro trees flower earlier. One tree flowered at the age of 3 years others at the age of 4 years whereas seed grown controls flowered at the age of 5 years. The size and the colour of the fruits and nuts were similar to those of the parent plant.

TABLE 1: Height, girth and flowering age of the seed grown and *in vitro* produced cashew trees at the end of 6 years.

	Height (cms)	Girth (cms)	Time for first flowering (years)
Vitro trees	495	42	3-4
Seed trees	517	30	5

ACKNOWLEDGMENT

Sincere thanks are due to Mangalore Jesuit Education Society for the facilities provided and the DST for partially funding the work carried out on Cashew. Special thanks to Joseph Alexander for technical assistance.

REFERENCES

Balasubramanian, P.P. 1987. Cashew can be still promisive in India. The Cashes 1: 13-19.

D'Silva Icy and D'Souza, L. 1992a. *In vitro* bud proliferation of *Anacardium occidentale* L. Beitr. Biol. Pflanzen 67: 273-279.

D'Silva Icy and D'Souza, L. 1992b. *In vitro* propagation of *Anacardium occidentale* L. Pl. Cell Tiss. and Org. cult. 29: 1-6.

D'Souza, L., Smitha Hegde, Alice, C. Augustine, Rajendra, K. and Icy D'Silva. 1994. Micropropagation of an important tropical nut tree *Anacardium occidentale* L. (Cashew). Abstracts VIIIth International Congress of Plant Tissue and Cell Culture, Firenze.

Nayar, K.G. 1993. Cashew cultivation and wasteland development. Cashew Bulletin 30: 2.

Micropropagation of Plum (*Prunus domestica* L.)*

Atul Kumar, Vandana, A. Kumar and M. C. Nautiyal†

INTRODUCTION

Among all the stone fruits grown in hilly region, plum (*Prunus domestica* L.) is an important fruit crop of commercial importance occupying considerable area. Plum fruits are rich in carbohydrates and carotenes.

Most of the stone fruits (like plum) are propagated vegetatively by grafting as they are difficult to root from cuttings. This process is time consuming, requires skilled labour and success rate is also very low (Skirvin, 1984). As clonal root stocks are used for grafting which exhibit variable field performance since it is difficult to obtain highly uniform seed stocks. Various efforts made to improve root stock and scion rootability (Farmer and Besemann, 1974, Robitaille and Yu, 1980) did not prove useful enough to become a standard propagation method. To combat the problem, *in vitro* propagation can be effectively employed to substantially increase the rate of multiplication of clonal root stocks, commercial cultivars and hybrids generated through conventional breeding techniques.

Thus present investigation is attempted to standardize a protocol for *in vitro* propagation of plum.

MATERIALS AND METHODS

Preparation of plant material and sterilization

Nodal segments possessing 1-2 axillary buds were collected during active growth period (April/May) from 8 year old trees of a plum hybrid Kabugl Green Gage X Ramgarh Meynard. For sterilization and controlling exudation of phenolics, explants were treated under nine different treatment sequences (Fig. 1).

* Abbreviations: 2,4-D = 2,4-Dichlorophenoxy acetic acid, IAA = Indole-3-acetic acid, BAP = 6-Benzyl aminopurine.
Abbreviations: IBA = Indole butyric acid, IAA = Indole acetic acid, NAA = Naphthalene acetic acid, BA = Benzyl ademine, GA_3 = Gibberellic $acid_3$, MS = Murashige & Skoog

† *Research paper No. 8362 through the Experimental Station GBPU & AT, Pantnagar, Nanital (UP)

A. R.W.(3 hrs.) — Savlon (0.5%) 10 min. — (2) Sod. Hypo. (1%) 10 min. — (3) - EtoH (70% 90 sec. — Implantation
B. Ac. Charcoal + (100 mg 1^{-1}) in medium
C. Savlon + Ascorbic (100 mgl^{-1}) + Citric acids (150mgl^{-1}) 3 hrs — $HgCl_2$ (0.1%) 15 min.— (3) - EtoH - Implantation
D. $HgCl_2$ (0.1%) 15 min. — Ac. Charcoal + (1gl^{-1}) in medium.
E. Ascorbic + Citric acids - Savlon — (2) - Sod. Hypo. (1%) 15 min. — (3) Antibiotics [1]- 10 min. (3) - EtoH - Implantation
F. Sod. Hypo. (1%) 20 min.— Implantation
G. Sod. Hypo. (1%) 15 min. — Antibiotics[2] 10 min.— Implantation
H. Sod. Hypo. (0.6%) + Antibiotics[3] — (3) — EtoH -Implantation
I. R.W. — Savlon — (2) Sod. Hypo. (1%) 10 min. — (3) EtoH - Implantation

(Blank)*

Abbreviations: R.W. - Running water, Sod. hypo. - Sodium hypochlorite, $HgCl_2$ - Mercuric Chloride EtoH- Ethanol. - Washing with sterile water, on. of times indicated within bracket. Antibiotics1: Streptomycin + Penicillin (5 mg 1-1 each). Antibiotics[2]- Streptomycin + Penicillin + Nystatin (5 mg 1^{-1} each). Antibiotics[3] - Gentamycin + Amphotericin (5mgl+$^{-1}$ each) + - As addition of Ac. Charcoal had no significant effect on browning control, it was omitted from further treatments. *Blank- No treatment for browhing control was given to explants or in medium. All treatments, except blank contained Ascorbic + Citric Acids in medium. --- (broken lines) indicate continuation of that particular step, and .... dotted lines) indicate omission of that particular step.

FIG. 1. Treatments for sterilization and browning control.

Culture media

For establishment of explants (Stage 1), two culture media, viz. modified white's medium (1963) and Murashege and Skoog (1962) medium with half nitrates were used. Both the media were supplemented with 1 mg l^{-1} IBA + 2.27 mg l^{-1} BA.

Proliferation and shoot elongation (Stage 2) was attempted in MS medium supplemented with 3 different hormonal combinations i.e. (i) 2 mg l^{-1} each of BA and IBA + 0.1 mg l^{-1} GA_3. (ii) 0.1, 0.01 and 2.0 mg l^{-1} of GA_3 + NAA + Calcium pentothenate, respectively, and (iii) 1.0 mg l^{-1} IAA.

To initiate rooting in proliferated and elongated shoots (Stage 3), MS (half nitrates) was supplemented with (i) 1 mg l^{-1} each of IAA + IBA, and (ii) 1 mg l^{-1} each of IAA + IBA + NAA.

For further proliferation and elongation in axillary buds of stage 2, these were subcultured on MS (half nitrates) supplemented with four different hormonal combinations (stage 4), viz. (i) 1 mg l^{-1} each of IAA + IBA, (ii) 1 mg l^{-1} each of IAA + IBA + NAA, (iii) 1 mg l^{-1} each of NAA + BA and (iv) 1 mg l^{-1} each of IBA + Kinetin.

Before autoclaving at kg/cm^2 for 20 min. pH was adjusted to 5.7 and 10 ml medium was dispensed in 50 ml Borosil glass tubes.

Culture conditions

Cultures were incubated at 24 ± 2°C under 16 hr photoperiod (2000 lux) with fluorescent tubes.

In vitro rooted plantlets were acclimatized by transferring to soil (fumigated): sand: leaf mould mixture (4:1:4) in pots under diffused light (1000 lux) and high humidity. Occasional irrigation was provided with 1/5th strength of MS micronutrients.

RESULTS AND DISCUSSION

Explant establishment (Stage 1)

Data on the effect of different treatments on disinfection and control in browning are presented in Table 1. Though per cent survival of explants was comparatively more in modified White's medium than MS (half nitrates), no proliferation was observed in cultures even after 45 days of initiation in the former. Sequence of treatment C (Fig 1 and Table 1) was the most suitable for culture establishment.

Skirvin (1984) has reviewed different schemes for sterilization of explants from stone fruits showing varying degree of contamination control. However, treatment with 0.1% $HgCl_2$, for different durations depending upon the fruit crop, has been found to be very effective in controlling high contamination rates (Nekrosava, 1964). In our experiment, treatment with 0.1% $HgCl_2$ for 10 min was more effective than sodium hypochlorite alone or in combination with antibiotics. Though use of antibiotics has been found to reduce the bacterial contamination in cultures, certain antibiotics have also been reported to inhibit the growth of cultures (Skirvin, 1984). Addition of antibiotics in sterilization sequence did not produce desirable results in our experiment.

Pretreatment of explants with the mixture of ascorbic + citric acids and addition of this mixture in culture medium completely controlled exudation of phenolics. Addition of activated charcoal in the medium was less effective and explants showed yellowing in later stages. Inspite of encouraging reports, *in vitro* propagation of woody species still poses problems due to high contamination rates and exudation of phenolics (Dalal *et al.*, 1992). Many solutions have been proposed to solve these problems but untill now no generally applicable solution is found (Machado *et al.*, 1991). In our experiment, sequence of treatment C provides complete control of browing and fairly high control of contamination.

TABLE 1: Percentage of necrosis, survival and control in browning after 45 days of culture initiation.

Treatments	Modified White's Medium				MS (half nitrates) medium			
	Necrosis	Infection	Survival	Control in browning	Necrosis	Infection	Survival	Control in browning
A	64	24	12	88	60	30	10	86
B	26	28	46	90	23	59	18	92
C	46	2	52	100	55	0	45	100
D	62	3	35	100	75	0	25	100
E	32	58	10	100	34	60	6	98
F	50	35	15	96	48	40	12	98
G	22	70	8	100	6	90	4	100
H	80	12	8	100	93	3	4	98
I	92	6	2	14	88	8	4	18

Data presented is mean of two experiments with 30 explants/treatment.

Proliferation of axillary shoots (Stage 2)

Table 2: Effect of different hormones on proliferation rate after 8 weeks of first subculture

Growth pattern	BA IBA GA_3 (2 mg l^{-1}) + (2 mg l^{-1}) + (0.1 mg l^{-1})	IAA (1 mg l^{-1})
No. of axillary microshoots	8.5	1.5
Size (LxWxH) (cm.)	1.8 x 2.1 x 0.8	2.0 x 0.8 x 0.6
General appearance	2-3 elongated green shoots with many small surrounding leaves	1-2 elongated shoots with surrounding small leaves.

Data presented is mean of two experiments with 10 micro shoots/treatment.

After 45-60 days of implantation, nodal segments turned green and grew further. These were divided into 8-10 segments (with 2-3 small buds) and subcultured. Data shown in Table 2 revealed that after 8 weeks of first subculture, the best shoot proliferation and elongation was achieved with 2.0 mg l^{-1} each of BA + IBA + 0.1 mg l^{-1} GA_3. Effects of IAA (1.0 mg l^{-1}) alone in the medium were less pronounced while in third combination (0.1 mg l^{-1} GA_3 + 0.01 mg l^{-1} NAA + 2.0 mg l^{-1} calcium pentothenate), no growth was observed and all the cultures died.

Effects of different hormonal combinations on proliferation and elongation have been reviewed by Skirvin (1984). In our study, combination of BA + IBA + GA_3 was very effective in providing 2-3 elongated shoots for rooting purpose, simultaneously providing around 8-9 small axillary shoots for further proliferation and elongation.

Induction of roots (Stage 3)

2-3 cm long green shoots with 5-6 small leaves and axillary microshoots with 2-3 buds (obtained in Stage 2) were transferred to rooting culture medium. Roots appeared (42%) within 2 weeks of subculturing in the medium supplemented with 1 mg l^{-1} each of IAA + IBA (Table 3) while with 1 mg l^{-1} each of IAA + IBA + NAA, per cent rooting was less.

TABLE 3: Effect of different hormones on rooting after 2 weeks of second subculture

Growth pattern	IAA + IBA (1 mg l^{-1} each)	IAA + IBA + NAA (1 mg l^{-1} each)
Rooting (%)	42.5	30.6
No. of roots	3.6	1.2
Length of roots (cm.)	0.8	0.7

Data presented is mean of two experiments with 10 plantlets/ treatments.

TABLE 4: Effect of different hormones on proliferation and elongation of axillary microshoots after 4 weeks of second subculture

Growth pattern	IBA + Kinetin (1 mg l^{-1} each)	NAA + BA (1 mg l^{-1} each)	IAA + IBA (1 mg l^{-1} each)	IAA + IBA + NAA (1 mg l^{-1} each)
No. of axillary shoots	5.8	4.6	3.8	3.8
Length of shoots (cm.)	0.24	1.2	3.2	2.1
No. of leaves	Very small leaves	4-5	12-15	10-12
General appearance	Little callusing and many axillary buds developing on callus	Callusing and many axillary buds developing on callus	Little callusing and many axillary buds developing on callus	Profuse callusing and less no. of axillary buds on callus.

Data presented is mean of two experiments with 10 micro shoots/treatment

Different combinations of auxins have been attempted to induce rooting in stone fruits (Skirvin *et al.*, 1982) and other woody species (Reddy and Srivasuki, 1992). In our study, IAA + IBA were more effective. The *in vitro* rooted plantlets were available within 2 months of subculturing for hardening and field transfer. Axillary microshoots proliferated further in both the rooting culture media and data on proliferation and elongation are presented in Table 4.

Druart (1980) has been able to develop rooted plants in several *Prunus* species. His system too involved shifting of axillary buds from proliferation to rooting medium. Rooted plantlets were later returned to proliferation medium to induce a shoot formation and elongation.

Culture of axillary microshoots (Stage 4)

Axillary microshoots obtained in Stage 2 were cut into segments with 2-3 buds and cultured on 4 different hormonal combinations. Data presented in Table 4 indicated after 4 weeks of subculturing, more number of axillary buds with little callusing at the base were observed in medium supplemented with 1 mg l^{-1} each of IBA + Kinetin. However, elongation of shoots was very less. Maximum elongation in microshoots with 12-15 small leaves was observed with IAA + IBA each at 1 mg l^{-1}.

Plantlets were acclimatized in fumigated soil: Sand: leaf mould mixture under diffused light and high humidity before field transfer.

CONCLUSION

A protocol is developed through which high frequency *in vitro* propagation of plum can be achieved by using explants from mature trees. Following this protocol, it is possible to obtain *in vitro* rooted plantlets for hardening and field transfer and, simultaneously, proliferation of axillary buds for further elongation of shoots and rooting is also achieved.

REFERENCES

Dalal, M.A., Sharma, B.B. and Rao, M.S. 1992. Studies on stock plant treatment and initiation culture mode in control of oxidative browning in *in vitro* cultures of grapevine. Scientia Hort. 51, 35-41.

Druart, P. 1980. Plantlet regeneration from root callus of different *Prunus* species. Scientia Hort., 12, 339-342.

Farmer, R.E. and Basemann, K.D. 1974. Rooting cuttings from physiologically mature black cherry. *Silvae Genet.*, 23: 99-134.

Machado, M.L.C., Machado, A.C., Hanzer, V., Kalthoff, Weib, H., Mattanovich, D., Regner, F. and Katinger, H. 1991. A new, efficient method using 8-hydroxy-quinolinol-sulfate for the initiation and establishment of tissue cultures of apple from adult material. Plant cell., Tissue and Organ Culture, 27: 155-160.

Murashige, T. and Skoog, F. 1962. A revised medium for rapid growth and bioassay with tobacco tissue cultures. Physiol. Plant., 15: 473-497.

Nekrosova, T.V. 1964. The culture of isolated buds of fruit trees. Sov. Plant Physiol., 11: 107-113.

Reddy, K.K. and Srivasuki, K.P. 1992. Biotechnological approach for tree improvement in Red Sanders. The Indian Forester, 118: 15-20.

Robitaille, H.A. and Yu, K.S. 1980. Rapid multiplication of peach clones from sprouted nodal cuttings. Hort. Science, 15: 579-580.

Skirvin, R.M., Chu, M.C. and Rukan-Kerus, H. 1982. An improved medium for *in vitro* rooting of Harbrite peach. Fruit Var. J., 36: 15-18.

Skirvin, R.M. 1984. Stone Fruits. In: Hand Book of Plant Cell Culture (Eds. Ammirato, P.V., Evans, D.A., Sharp, W.R. and Yamado, Y.), 3, Macmillan Publishing Company, New York, pp. 402-452.

White, P.R. 1963. The cultivation of Animal and Plant Cells, 2nd Ed. The Ronald Press, New York.

Plantlet Regeneration From Seed Culture of *Arundina graminifolia* (D.Don) Hochr. and Establishment of Plantlets Under Natural Conditions

C. M. Sarma and Suchismita Roy

INTRODUCTION

Orchids, which are under the family Orchidaceae, produce flowers of an exquisite beauty with a variety of colourful patterns. The cut flowers of orchids remain fresh for a long duration, make pretty corsages and add to the decorative floral arrangements. The growing interest on orchid exports has led to a condition of near extinction of many orchid species of North Eastern India from their natural habitats. Moreover, the tiny seeds of orchids are non-endospermic and require the association of appropriate fungi for supplying them the proper nourishment for germination and subsequent growth. This limitation allows only 3-4% of seeds to germinate in nature.

India, with 1250 species and NE India alone with 650 species of orchids have a vast scope to develop it into an important industry. However, to develop this trade in India and to overcome the problem of germination of seeds, the need for perfecting the *in vitro* micropropagation technique is urgent. For this purpose, *in vitro* seed culture is the best technique as because only one capsule of orchid contains 1,300-4,000,000 number of seeds (Mehra, 1990).

The present investigation was undertaken to study the effect of some plant growth regulators in standard culture media on germination and post-germination development of *Arundina graminifolia*. The procedure adopted for obtaining seedlings through *in vitro* germination of seeds and also subsequent stages of seedling growth have been described and discussed. The technique of transplanting seedlings from *in vitro* to *in vivo* condition has also been given.

MATERIALS AND METHODS

About four months old capsule of *A. graminifolia* was harvested and the surface was sterilized with absolute alcohol for 10 seconds and rinsed with double distilled sterile water 3 times. The seeds were exposed aseptically by cutting open the capsule and inoculated on MS (Murashige and Skoog, 1962) medium (Fig. I) followed by incubation at 25 ± 1°C. The culture flasks were daily exposed to 16 hr photoperiod of 2,500-3,000 lux intensity. The MS medium was supplemented with different types of combinations of growth regulators in different concentrations. The observations were made at regular intervals of one week and the results were tabulated (Table 1).

TABLE 1: Germination response of *A. graminifolia* in different media (Mean of 3 replications)

	Medium	Germination	Time taken for plbs development (in weeks)	Remarks
1.	MS + IAA (0.2 mg/l) + KN (0.4 mg/l)	60	8	Poor response, light green coloured plb formation
2.	MS + IAA (0.2 mg/l) + KN (0.4 mg/l) + CW (15%)	85	3-4	Very good response, good plbs formation, plbs deep green, vegetative apex developed after 40 days.
3.	MS + IBA (1 mg/l) + KN (5 mg/l)	70	8-9	Moderate response, light green plbs, vegetative apex formed after 70 days.
4.	MS + NAA (4 mg/l) + KN (4 mg/l)	65	8-10	Moderate response, green plbs, vegetative apex formed after 85 days.

When the plbs developed vegetative apex, they were transferred to fresh MS and KnC Knudson (1946) media with different proportions of plant growth regulators e.g. KN, IBA, BA, NAA and IAA to compare their effects on the growth of *A. graminifolia* plbs and plantlets respectively. Sucrose (3%) and 0.8% agar were given in all initial and subculture media. pH was adjusted at 5.8 before autoclaving at 120°C for 15 minutes. The seedlings were regularly subcultured to the same subsequent media and their subsequent developments were followed at 30 days intervals (Table 2). The seedlings were directly transplanted to earthen pots where they attained a length of 7-8 cm with well defined shoot and root system.

The plantlets were deflasked, washed well in running tap water to remove agar and transplanted directed into community pots. The potting medium used in this experiment consisted of locally available, inexpensive materials, e.g. charcoal chips, brick pieces, gravel, dried fern roots, dried leaves and roots of water hyacinth etc. The pots were filled with manured soil and compost mixture of different types viz.

1. Manured soil, leaf mould, charcoal (2:1:1)
2. Manured soil, charcoal (3:1)
3. Manured soil, dried leaves and roots of water hyacinth, charcoal (2:1:1)
4. Manured soil, dried rice straw, charcoal (2:1:1)
5. Manured soil, dried *Salvinia* leaves and roots, charcoal (2:1:1)

The whole earthen pots with the prepared potting media were autoclaved. Seedlings, 4 in number were planted in each pot and were kept under diffused sunlight. They were regularly watered and sprayed with Arnon and Hoagland,s nutrient solution (Hoagland and Arnon, 1938).

TABLE 2: Observations on the growth of plbs in various media (Mean of 3 replications)

	Medium No.		Growth of plbs after	
		30 days	60 days	90 days
1.	MS + KN (1 mg/l) + IBA (0.1 mg/l) + CH (100 mg/l)	Plbs swollen upto 3 mm dia., many of them proliferated to 4-5 more plbs. Shoot system grown upto 2-2.5 cm long with 2 leaves. No root.	Shoot system attained 5 cm length with 3-4 green leaves small, 1-2 in number, 0.5 cm long roots developed.	Plantlets grown upto 6 cm in length, with 5-6 green leaves. Roots 3 cm long.
2.	MS + IBA (5 mg/l) + BA (0.1 mg/l)	Plbs swollen upto 3 mm dia. some of them proliferated to 2 or 3 more plbs.	Shoot system attained 2 cm length, some developed 1-2 roots upto 1 cm long. Almost all the plantlets show multiple shooting	Plbs, proliferated more in number, multiple shoots increased, shoot system attained 3 cm length. Roots were upto 2 cm long.
3.	KnC + KN (1 mg/l) + IBA (0.1 mg/l)	Plbs slightly swollen. Some of them developed into 0.5 cm long plantlets	Shoot system was 1.5 cm long with 2 green leaves. Roots formed in some cases.	Plantlets were 3-3.5 cm long. Roots were upto 1 cm long.
4.	KnC + KN (0.1 mg/l) + IAA (0.1 mg/l)	Plbs attained 2 mm dia. Plantlets grown upto 1 cm with 2 green leaves. About 3-5 mm long. Roots 1-2 were formed in 90% of cases.	Plantlets attained 1.5 cm length, roots 4-5 in number 3 cm long.	Shoot system attained 3 cm length with 3-4 green leaves. Roots 7-8 in number, about 5 cm long.

RESULTS AND DISCUSSIONS

Germination of seeds

The onset and frequency of germination of seeds varied with the plant growth regulators supplied to the MS medium (Table 1). When the MS medium was supplied with IAA (0.2 mg/l), and KN (0.2 mg/l), poor plb formation was observed. But on addition of 15% coconut water with the same range of PGRs, percentage of germination increased upto 85. Such stimulatory effect of CW was also reported by some other workers (Chung *et al.*, 1985, Devi *et al.*, 1990). When MS medium was supplemented with IBA (1 mg/l) and KN (5 mg/l), 70% germination was achieved which was almost same when KN (4 mg/l) with NAA (4 mg/l) was supplied in the medium. So, KN (0.4 mg/l) with IAA (0.2 mg/l) and CW (15%) was best for seed germination of *A. graminifolia*.

Subsequent growth of the plbs

Depending on the quality of growth regulators in the medium, the plbs behaved differently. The different types of PGRs are responsible for the growth of different parts of *A. graminifolia* (Table 2). MS medium supplemented with KN (1 mg/l), IBA (0.1 mg/l) and CH (100 mg/l) induced high degree of proliferation of plbs.

FIG. 1. Plantlet Regeneration From Seed Culture of *Arundina Graminifolia* (I) Inoculation of seeds of *Arundina graminifolia*. (II) Multiple plantlets formed in MS medium supplemented with IBA (5 mg/l) and BA (0.1 mg/l), (III) Plantlets with well developed shoot and root system before transplantation to earthen pots, (IV) Growth of plantlets in manured soil and leaf mould (Potting medium no. 1).

Shoot system developed in this medium very much faster than the other combinations of PGRs. This stimulatory effect of CH was also reported by some other workers (Johri and Bajaj, 1964; Bonga and Chakraborty, 1968). When KnC medium was supplemented with the same PGRs minus CH, no satisfactory growth of seedlings was recorded, whereas, KN and IAA (each 0.1 mg/l) in the KnC medium induced root initiation in 90% of the cases and the growth of the roots was also very much faster. IBA (5 mg/l) and BA (0.1 mg/l) in MS medium caused proliferation of plbs and formation of multiple shoots (Figs. II, III) but after that no satisfactory growth of seedlings was recorded.

Thus, by using the proper plant growth regulators and the combinations of auxins and cytokinins, rapid propagation of healthy plantlets of orchids in *in vitro* is possible.

Establishment of the plantlets

The length of the plantlets was measured after 6 months of transplantation in the potting media and it was found that the growth of *A. graminifolia* plantlets was best in the medium no. 1 in which manured soil and leaf mould were used (Fig. III). 95% plantlets survived in this potting medium and thus, the plantlets became established in *in vitro* condition.

ACKNOWLEDGMENT

The authors are grateful to the CSIR (India) for giving financial assistance. Grateful thanks are also due to the authorities of Gauhati University for providing laboratory facilities.

REFERENCES

Bonga, J.M. and Chakraborty, C. 1968. *In vitro* culture of a dwarf mistletoe, *Arceuthohium pusillum*. Peck. Can. J. Bot. 46: 161-164.

Chung, J.D., Chun, C.K. and Choi, S.O. 1985. Asymbiotic germination of Cymbidium ensifolium II. Effect of several supplements to the medium, pH values and/or dark culture periods on rhizome growth and organogenesis from the rhizome. J. Kor. Soc. Hort. Sci. 2: 186-192.

Devi, J., Nath, M., Devi, M. and Deka, P.C. 1990. Effect of different media on germination and growth of some North-East Indian species of *Dendrobium*. J. Orchid Soc. India 4: (1,2): 45-49.

Hoagland, D.R. and Arnon, D.I. 1938. The water culture methods for growing plants without soil. Calif. Agr. Expt. Sta. Cir. 347 Berkeley, Calif.

Johri, B.M. and Bajaj, Y.P.S. 1964. Growth of the embryos of Amyema, *Amylotheca* and *Scurrula* on synthetic media. Nature, Lond. 204: 1220-1221.

Knudson, L. 1946. A new nutrient solution for germination of orchid seeds. Am. Orchid Soc. Bull. 15: 214-217.

Mehra, P.N. 1990. Perspectives and prospects of orchid research in India. J. Orchid Soc. India 4: (1,2) 01-07.

Murashige, T. and Skoog, F. 1962. A revised medium for rapid growth and bioassay with tobacco tissue culture. Physiol. Pl. 15: 473-497.

Direct Embryo Formation and Regeneration of Plants From Embryogenic Protoplasts of Ginseng (*Panax ginseng* C.A. Mayer)[1]

Sarita Arya, Inder Dev Arya, Jang, R. Liu and Tage Eriksson

INTRODUCTION

Ginseng (*Panax ginseng*) is a medicinally important plant used as a vitalizing and stimulating agent. Some of its pharmacologically active substances are triterpene saponins and ginsenosides (Shibata *et al.*, 1985). The root has been used as drug since ancient times and is expensive due to long and complicated cultivation. A three year juvenile period has hampered conventional breeding programmes for rapid multiplication and genetic improvement of ginseng.

The introduction of foreign genetic material using protoplast fusion and direct gene transfer requires production of viable protoplasts and a protocol for rapid regeneration of plants. Embryogenesis from protoplasts may occur indirectly via callus proliferation and differentiation into embryos, or directly from the cells which develop from protoplasts into embryos. Although a lot of work has been done on protoplasts in different plant species, direct embryogenesis from protoplasts is still limited to few plant species. Despite a variety of approaches, only non-protoplasts derived somatic embryo formation and regeneration of plants is so far reported in ginseng (Butenko *et al.* 1968; Chang & Hsing 1980a, b; Lee *et al.*, 1990; Wang 1990; Arya *et al.*, 1993b) Recently, a protocol for isolation of protoplasts from somatic embryos and plant regeneration has been reported (Arya *et al.*, 1991).

Embryogenic cultures are proved better sources for protoplast isolation of cereals (Shillito *et al.*, 1989; Wei and Xu, 1990; Vasil *et al.*, 1990), conifers (Gupta and Durzan, 1987; Attree *et al.*, 1989; Arya *et al.*, 1993a) and in several other plant species (Grambow *et al.*, 1972; Sim *et al.*, 1988). Calli from embryogenic protoplasts have been used for the production of transgenic plants (Datta *et al.*, 1990) we report here the regeneration of plants from embryos developed from the individual protoplasts and protoplast derived calli, using the embryogenic callus.

---

[1] Abbreviations: 2,4-D: -2,4-dichlorophenoxyacetic acid, Kn:- kinetin, GA_3:- gibberellic acid, BA:- benzylaminopurine.

MATERIALS AND METHODS

Isolation and culture of protoplasts

Two week old callus were used for the isolation of protoplasts. The callus was preplasmolysed for 1h in a solution containing 0.3M sorbitol and 0.5M $CaCl_2$ and were incubated in enzyme solution in a petridish. The enzyme mixture consisted of 2% cellulysin, 1% macerase, 1% pectinase, 0.5% potassium destrionsulphate and 12% mannitol in half strength MS medium at pH 5.6. The enzyme solution used was filter sterilized. The cultures were kept on gyratory shaker for 5-6 h in the dark at 25°C, initially 1h at 50 rpm and then at 40 rpm. The protoplast-enzyme mixture was filtered through 100 and 50 mm, respectively. The protoplasts were washed in an equal volume of CPW 16 salts (Banks & Evans, 1976) and were centrifuged at 70 g for 5 min in a swing bucket rotor. The pellet was washed three times with the W5 medium (Menczel *et al.*, 1981) and once with the culture medium. Protoplast yields were determined with a hemocytometer. MS (Murashige and Skoog, 1962), K3 (Nagy and Maliga, 1976) and M8p (Glimelius *et al.*, 1986) media were tested for the culture of protoplasts. The protoplasts were cultured at a density of 1-1.5 x 10^5/ml in 2.5 ml of liquid or Seaplacque (0.6%) agarose M8P medium. The culture dishes were sealed with parafilm and incubated in dark at 25°C. Various osmotica were investigated for their effect on protoplast culture; 0.4 M of glucose, sorbitol, mannitol or 6% (0.33M) myo-inositol were included in 1/2 M8p culture medium. The plating efficiency was determined after 10 days and is based on the percentage of cells divided once or twice from the protoplast cultures. Percentage of cells forming colonies were calculated on the basis of total number of dividing protoplasts and the cell colonies recovered after 6 weeks of culture. Colonies consisting of 20 cells or more were scored. The results are based on at least three replicates in each experiment, and each experiment was repeated at least twice. The final protocol has been repeated around ten times.

Protoplast culture medium

Optimal medium found for protoplast culture was 1/2 M8p medium supplemented with 6% myo-inositol, 1% glucose, 1% sorbitol, 1.0 mg/l 2,4-D, 0.5 mg/l BA and 0.5 mg/l Kn at pH 5.7. Half of the culture medium was replaced by fresh medium at 7 days interval. After 4-5 weeks protoplasts were diluted with 1/2 M8p medium supplemented with 1% myo-inositol, 1.0 mg/l 2,4-D, 0.5 mg/l BA and 0.5 mg/l Kn.

Somatic embryogenesis and plant regeneration

Heart shaped embryos formed from the individual protoplasts and protocalli which reached upto 1.0-1.5 mm in size were picked and transferred on MS medium with 1.0 mg/l 2,4-D, 0.01 mg/l Kn and 3.0% sucrose for further growth and development. Somatic embryos arose from the callus in 7-9 weeks of culture. Cotyledonary somatic embryos were transferred on the regeneration medium consisting of half strength MS medium with 1.0 mg/l BA and 1.0 mg/l GA_3 where they formed plant after 4 weeks. The regeneration medium was solidified with 0.7% agarose and cultures were kept in light.

RESULTS

Source of protoplast isolation

Initial experiments were performed starting with different sources of ginseng tissue. Protoplasts were isolated from leaves of *in vitro* grown plants, and from zygotic embryo callus of two cell lines. The yields of protoplast isolated from young leaves were very low ($1\text{-}2 \times 10^6$/g leaf tissue) and protoplasts were fragile. Similar results were obtained when protoplasts were isolated from one of the cell lines which was very weakly embryogenic, the cells were oblong or oval which gave less protoplast yield. The second cell line yielded highly embryogenic callus, which was having small, dense cytoplasmic rich isodermic cells, and were compactly arranged. All subsequent experiments were performed with this highly embryogenic cell line.

Different enzyme solutions were tried to obtain maximum number of viable protoplasts. The isolation procedure described (experimental protocol) resulted in 10-15 x 10^6 protoplasts/g callus and viability ranged from 80-90% (Fig. 1).

PROTOPLAST CULTURE

Different conditions were tried for initial culturing of isolated protoplasts. When the protoplasts were cultured in liquid medium, cell division was observed only in 10 out of 1×10^4 cells. Most of the protoplasts started to die after 7-10 days. Using the same medium but embedding the protoplasts in agarose, viability and division frequency of protoplasts improved remarkably. When we evaluated agarose droplet culture and agarose plates, both the system were found equally effective. The only difference observed was that initial division frequency and colony formation was slightly better in agarose plates while development of calli was found enhanced in droplet culture system.

High division frequency was obtained when protoplasts were plated at a density of 1-1.5 x 10^5 protoplasts/ml. At a density of less than 7.5×10^4, division frequency dropped dramatically. The M8p medium was found suitable for this protoplast culture as found earlier for protoplasts isolated from somatic embryos of ginseng. The optimal concentration of 2,4-D was found 1.0 mg/l and was used in all subsequent experiments. Myo-inositol was found better osmoticum and carbon source as compared to glucose, sorbitol or mannitol. The best growth and survival of protoplasts was achieved in half strength M8p medium containing 6% myo-inositol, 1% glucose, 1% sorbitol, 1.0 mg/l 2,4-D with 0.5 mg/l BA and 0.5 mg/l Kn. The protoplasts regenerated cell wall within two days of culture. The protoplasts were densely cytoplasmic. Binucleate and multinucleate protoplasts were also observed. Protoplasts divided within 5 days and two to four celled colonies were formed at the end of first week (Fig. 2,3). Nine per cent initial division frequency was obtained (Table 1). Small compactly arranged colonies were formed within 2-3 weeks (10-15 cells). Although, initial mitotic activity was high, many cells stopped dividing after a few divisions. Around thirty per cent of dividing protoplast formed minicalli of 50-100 cells after 4 weeks of culture (Fig. 4).

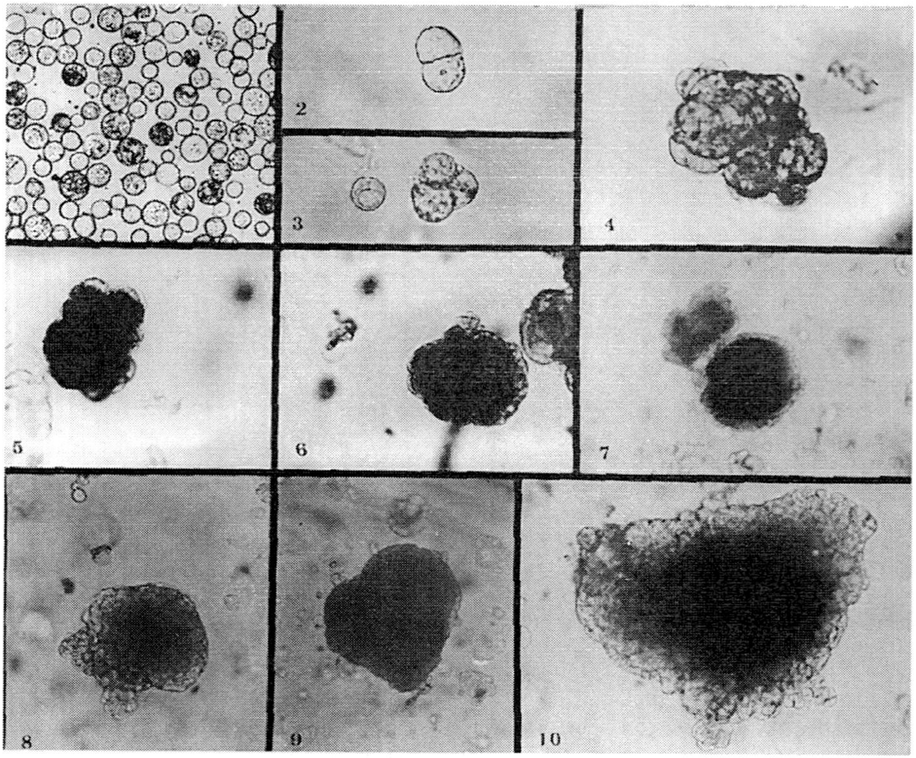

FIG. 1-14. Plant regeneration through somatic embryogenesis from protoplasts of ginseng. 1. Protoplasts isolated from embryogenic callus. 2. First division of protoplasts after 4 days of culture 3. Four celled stage. 4. Cell colony derived from protoplasts after 4 weeks of culture. 5 & 6 Compact cell clusters resulted into direct somatic embryos. 7 & 8 Globular stage embryos derived from protoplast. 9 & 10 Heart shaped somatic embryos.

Compact, yellowish cell colonies with an average diameter of 100 mm developed after 4 weeks of culture. Four to five week old colonies did not grew when placed on solid agarose medium. When colonies reached about 1 mm in size they were picked up and transferred to solid medium or agarose blocks were directly transferred on agarose solidified medium. After transfer, the colonies grew and produced somatic embryos (Fig. 11,12). Monthly subculture was necessary for continued development of callus and embryos.

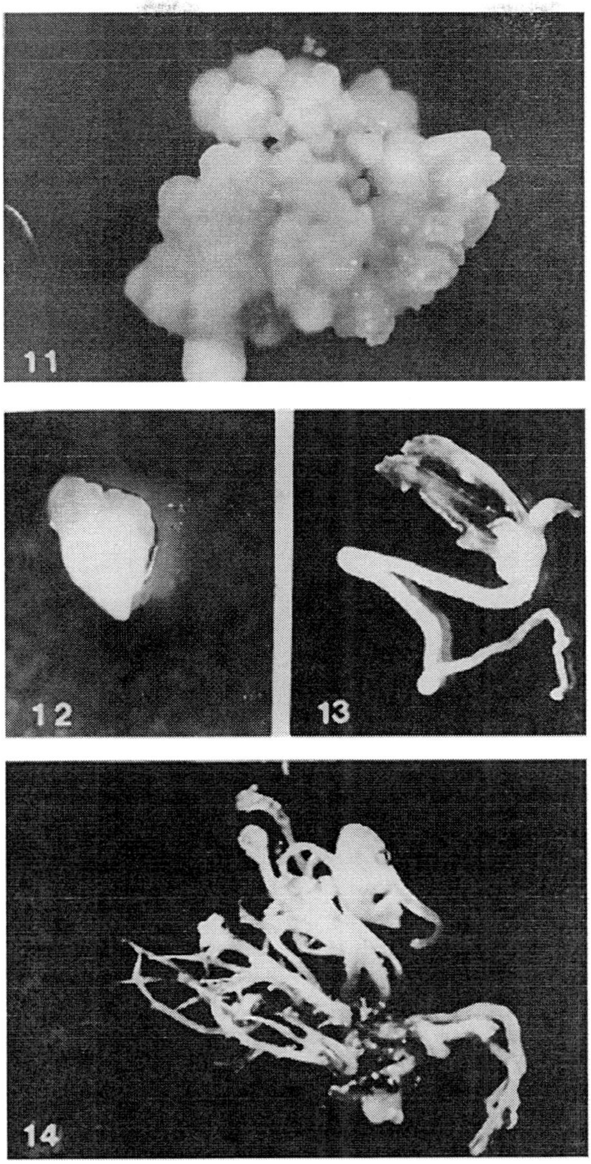

FIG. 1 CONTU... 11. Protoplast-derived embryogenic callus with different stages of somatic embryos. 12. protoplast-derived somatic embryo 13. Germination of somatic embryo. 14. Plantlet regenerated from somatic embryo.

TABLE 1: Effect of osmoticum on cell division, colony formation and direct embryogenesis. Culture medium consists of half strength M8p. 1.0 mg/l 2,4-D, 0.5 mg/l BA and 0.5 mg/l Kn.

Osmoticum used	Division Frequency	Colony Formation	Direct Embryogenesis	
			embryo formation	embryo maturation
Glucose (0.4M)	2-5	1%	0.001%	-
Mannitol (0.4M)	1	-	-	-
Sorbitol (0.4M)	2-5	1%	0.01%	25%
Myo-inositol (0.33 M)	9	30%	0.03%	25%

Direct somatic embryos were formed from the protoplast derived cells at frequency of 0.03% (Table 1). A lower frequency of direct somatic embryogenesis was found (0.01%) as against (0.03%) when myo-inositol was replaced by sorbitol (0.4M) as osmoticum. However, no direct somatic embryogenesis was observed when glucose or mannitol were used as osmoticum. During direct somatic embryo formation, the cells divided several times resulting in compact cell clusters (Fig. 5,6) which developed into globular shape embryos within 40-50 days (Figs. 7,8). These globular shape embryos grew further into heart shaped embryos in 60-80 days (Figs. 9.10). Only about 25% of the somatic embryos matured on MS medium supplemented with 1.0 mg/l 2,4-D + 0.01 mg/l Kn and 3% sucrose, rest of them initiated embryogenic callus which were redifferentiated into embryos.

When mature somatic embryos were selected and transferred on half strength MS medium supplemented with 1.0 mg/l BA and 1.0 mg/l GA_3, most of them germinated into plantlets (Figs. 13, 14). Uptil now protoplasts have been isolated more than 10 times from the embryogenic callus and in each experiment 7-13% plating efficiency was obtained. The calli obtained were highly embryogenic, which gave rise to a large number of regenerative somatic embryos.

DISCUSSION

Protoplasts isolated from embryogenic calli of ginseng, exhibited sustained growth and formed somatic embryos either directly from protoplast cell, or from protoplast derived calli. These regenerated somatic embryos formed normal plants (Fig. 14). In great variety of plants including cereals, the use of embryogenic cell suspension or calli as a source of protoplasts is essential for inducing embryogenesis and subsequent plant regeneration. Somatic embryos isolated from embryogenic callus of ginseng have also proved a good source of protoplast isolation, culture and regeneration. Selection of explant source is an important criterion for the development of direct embryos from individual protoplast as observed in ginseng. Lu *et al.* (1993) indicated that direct embryogenesis occurred most readily by the formation of small clumps of live cells. Like wise, Dijak and Brown, 1987, assumed that clumping of cells was necessary. They noted that clumps were composed mainly of dead cells and embryos developed from a few living cells. In our study, we have embedded protoplasts in agarose to avoid any cell aggregation. Furthermore, in such a system it was easier to follow the development from protoplast to an embryo. The direct embryogenesis was associated with high rates of protoplast survival and with compact cell division. Compact cell clusters developed into globular embryos which finally resulted in differentiated embryos. The similar pattern of development was found in *Daucus carrota*

where dividing protoplasts formed clusters which developed into embryoids (Grambow *et al.*, 1972). In *Citrus mitis* the cell colonies were globular in shape, latter formed mature embryos (Sim *et al.*, 1988). In *Ginkgo biloba* protoplast-derived microclones were obtained which later leads to an embryo formation (Laurain *et al.*, 1993). In *Brassica napus* mesophyll protoplast cultures, embryos arose from highly reduced callus and development appeared intermediate (Li and Kohlenbach, 1992).

Myo-inositol plays an important role in protoplast culture of ginseng. It was found superior to mannitol and glucose. It has also been found useful in protoplast culture of several conifers (Gupta and Durzan, 1987. M8p medium was the preferred medium for culturing the protoplasts of ginseng; as in the case of Brassica (Glimelius *et al.*, 1986. We could get only 1-2% cell division in MS or K3 media, furthermore, there was no colony formation. Also, plating of protoplasts in agarose was superior than to culture in liquid medium. Embedding distributed the protoplasts evenly in the culture medium. The superiority of agarose embedding has already been established for several other plant species (Shillito *et al.*, 1983; Puonti-kaerlas and Eriksson, 1988).

In many earlier studies it had been found that prolonged culturing of callus on 2,4-D supplemented medium caused loss of regeneration capacity. However, in the present case 2,4-D did not affect the callus growth with its regeneration potency.

This study showed that division and formation of plants can be achieved in regenerated cells from ginseng protoplasts using a defined medium. It further demonstrated the usefulness of the embryogenic callus as a source of totipotent protoplast of ginseng. These results might encourage efforts to transfer useful genes in ginseng either by direct delivery of DNA by electroporation or polyethlene glycol method or for use in somatic hybridization.

REFERENCES

Arya, I.D., Arya, S. and Eriksson, T. 1993a. Protoplast isolation and culture from six embryogenic cell lines of Picea abies (L) Indian. J. Exp. Biol. 31: 463-467.

Arya, S., Arya, I.D. and Eriksson, T. 1993b. Rapid multiplication of adventitious somatic embryos of Panax ginseng (C.A. Meyer). Plant Cell Tissue and organ culture. 34: 157-162.

Arya, S., Liu, J.R. and Eriksson, T. 1991. Plant regeneration from protoplasts of *Panax ginseng* (C.A. Meyer) through somatic embryogenesis. Plant Cell Rep. 10: 277-281.

Attree, S.M., Dunstan, D.I., and Fowke, L.C. 1989. Plantlet regeneration from embryogenic protoplasts of white spruce (Picea glauca). Bio/Technology 7: 1060-1062.

Banks, M.S. and Evans, P.K. 1976. A comparison of the isolation and culture of mesophyll protoplasts from several *Nicotiana* species and their hybrid. Plant Sci. lett. 7: 409-416.

Butenko, R., Brushwitzky, I.V. and Slepyan, L.I. 1968. Organogenesis and somatic embryogenesis in the tissue culture of *Panax ginseng* C.A. Meyer. Bot. Zh. 7: 906-911.

Chang, W.C. and Hsing, Y.I. 1980a. Plant regeneration through somatic embryogenesis in root derived callus of ginseng (*Panax ginseng* C.A. Meyer). Theoret. Appl. Genet. 57: 133-135.

Chang, W.C. and Hsing, Y.I. 1980b. *In vitro* flowering of embryoids derived from mature root callus of ginseng (*Panax ginseng*). Nature 284: 341-342.

Datta, S.K., Peterhans, A., Datta, K. and Potrykus, I. 1990. Genetically engineered fertile indica rice recovered from protoplasts. Bio/Technology 8: 736-740.

Dijak, M. and Brown, D.C.W. 1987. Patterns of direct and indirect embryogenesis from mesophyll protoplasts of *Medicago sativa*. Plant Cell, Tissue and Organ Cult. 9(2): 121-130.

Glimelius, K., Djupsjobacka, M. and Feldegg, H.F. 1986. Selection and enrichment of plant protoplast heterokaryons of Brassicaceae by flow sorting. Plant Sci. 45: 133-141.

Grambow, H.J., Kao, K.N., Miller, R.A. and Gamborg, O.L. 1972. Cell division and plant development from protoplasts of carrot cell suspension cultures. Planta 103: 348-355.

Gupta, P.K. and Durzan, D.J. 1987. Somatic embryos from protoplasts of loblolly pine proembryonal cells. Bio/Technology 5: 710-712.

Laurain, D., Chenieux, J.C. and Tremouillause-Guiller, J. 1993. Direct embryogenesis from female protoplasts of *Ginkgo biloba* L. a medicinal woody species. Plant Cell Rep. 12: 656-660.

Lee, H.S., Liu, J.r., Yang, S.G. and Lee, Y.H. 1990. *In vitro* flowering of plantlets regenerated from zygotic embryo derived somtic embryos of ginseng. Hort. Science 25: 1652-1654.

Li, L.C. and Kohlenbach, H.W. 1982. Somatic embryogenesis in quite a direct way in cultures of mesophyll protoplasts of *Brassica napus* L. Plant Cell Rep. 1: 209-211.

Lu, D.Y., Davey, M.R. and Cocking, E.C. 1983. A comparision of the cultural behaviour of protoplasts from leaves, cotyledons and roots of *Medicago sativa*. Plant Sci. Lett. 31: 87-99.

Menczel, L., Nagy, I., Kizz, Z.R. and Naliga, P. 1981. Teoret. Apple. Genet. 59: 191-195.

Murashige, T. and Skoog, F. 1962. A revised medium for rapid growth and bioassay with tobacco tissue cultures. Plant Physiol. 15: 437-497.

Nagy, J.I. and Maliga, P. 1976. Callus induction and plant regeneration from mesophyll protoplasts of *Nicotiana sylvestris* Z. Pflanzenphysiol. 78: 453-455.

Puonti-Kaerlas, J. and Eriksson, T. 1988. Improved protoplast culture and regeneration of shoots in pea (*Pisum sativum* L.). Plant Cell Rep. 7: 242-245.

Shibata, S., Tanaka, O., Shoji, J. and Saito, H. 1985. Chemistry and Pharmacology of Panax, p. 217-284. In: Journal of economics and medicinal plant research [Wagner, H., Hikino, H., and Farnsworth, N.R. (Eds.)] Academic Press, London.

Shillito, R.D., Carswell, G.K., Johnson, C.M., Dimaio, J.J. and Harms, C.T. 1989. Regeneration of fertile plants from protoplasts of elite inbred maize. Bio/Technology 7: 581-587.

Shillito, R.D., Paszkowski, J. and Potrykus, I. 1983. Agarose plating and a bead type culture technique enable and stimulate development of protoplast-derived colonies in a number of plant species. Plant Cell Rep. 2(5): 244-247.

Sim, G.E., Loh, C.S. and Goh, C.J. 1988. Direct somatic embryogenesis from protoplasts of *Citrus mitis* Blanco. Plant Cell Rep. 7: 418-420.

Vasil, V., Redway, F. and Vasil, I.K. 1990. Regeneration of plants from embryogenic suspension culture protoplasts of wheat (*Triticum aestivum* L.). Bio/Technology 8: 429-434.

Wang, A.S. 1990. Callus induction and plant regeneration of American ginseng. Hort. Science 25 (5): 571-572.

Wei, Z.M. and Xu, Z.. 1990. Regeneration of fertile plants from embryogenic suspension culture protoplasts of *Sorghum vulgare*. Plant Cell Rep. 9(1): 51-53.

Rapid Micropropagation of Populus × *Euramericana* Trees by Callus Culture

Veena Agrawal and Shrish C. Gupta

INTRODUCTION

Amongst forest trees, the genus *Populus* has emerged as a model system for application and understanding of the technique of genetic transformations (Fillatti *et al*, 1987; McCown, 1991; Leple *et al.*, 1992; Wilde *et al.*, 1992) as well as for producing transgenic plants that are herbicide tolerant (Riemenschneider and Haissig, 1991) and pest-resistant (McCown *et al.*, 1991). However, most of the success has been achieved only with species and hybrids of its Leuce section. On the other hand, most of the economically important species and hybrids which belong to Aigieros remain unexplored due to their recalcitrant behaviour under *in vitro* conditions leading to non-availability of their efficient micropropagation protocols (Rutledge and Douglas, 1988; Coleman and Ernst, 1989; Sellmer *et al.*, 1989). *In vitro* plantlet development from nodal and internodal explants of 25-year-old *Populus* x *euramericana* tree belonging to Aigieros section has been achieved (Agrawal and Gupta, 1991). However, several constraints, e.g. genetical and physiological differences of explants, severe microbial contamination, oxidation of phenolic substances, poor induction of rhizogenesis and seasonal variation impede *in vitro* culture significantly (Bon *et al.*, 1988; Chen and Evans, 1990; Hartney and Svensson, 1992; Gupta and Agrawal, 1992). The present communication relates a simple and efficient protocol for rapid micropropagation of *P.* x *euramericana* through callus derived from stem segments excised from old trees.

Though several publications have appeared on *in vitro* regeneration from callus in *Populus trimuloides* (Winton, 1968; Wolter, 1968; Noh and Minocha, 1986), *P. canadensis* (Berbee *et al.*, 1972) *P. euroamericana* cv. *robusta* (Chalupa, 1974), *P. alba* (Park and Son, 1988), *P. alba* x *P. grandidentata* (Son and Hall, 1990), 20 clones and hybrids of Leuce poplar (Antonetti and Pinon, 1992) but in all of them juvenile explants were used. Realising the significance of plants developed from mature tissue derived callus in tree improvement for commercial exploitation (Bychenkova and David, 1978; Lakshmi Sita *et al.*, 1979), the present work has been carried out.

MATERIAL AND METHODS

Plant material

8-10 cm long twigs of *Populus* x *euramericana* (Dode) Guinier, were cut from branches of approximately 20 to 25 year-old trees growing in the University of Delhi Campus. After removing the leaves, the twigs were washed under running tap water for half an hour and subsequently treated with 5% solution (v/v) of polysan (Polypharm Pvt. Ltd., Bombay) for 15 min. They were rewashed with tap water and dipped in 70% ethanol for 10 min. Surface sterilization was carried out using 0.5% mercuric chloride solution for 10 min. and 90% ethanol for 5 min. Finally, these were washed 3 or 4 times in sterile distilled water and inoculated under a laminar flow cabinet. 5 mm long and 3 mm thick internodal segments were cultured for raising callus.

Culture medium

Of the three basal media tried (White, 1943; Murashige and Skoog, 1962; Gamborg *et al.*, 1968), MS was the best for eliciting optimum morphogenic response. The medium contained 3% (w/v) sucrose (Central Drug House, Bombay) and was solidified with 0.8% agar (Japan Agar Company Ltd., Japan). The medium was enriched with varying ranges of different growth regulators, i.e. 2,4-D, 2iP, BA, IAA, IBA and NAA. Rooting medium was also adjuvanted with activated charcoal. The pH of the medium was adjusted to 5.8 with 1N HCl or 1N NaOH before autoclaving at 121°C and 15lb psi for 15 min. Filter sterilised glutamic acid (100 mg/l) was added to rooting medium which helped in retaining green colour of leaves and preventing defoliation. The cultures were incubated in 16h photoperiod at 25 ± 2°C under cool, white fluorescent tube light of 492 mW/cm^2 intensity and 65 ± 10% relative humidity.

In vitro hardening and transfer to soil

The tissue culture raised plantlets were taken out of the culture vessels, washed thoroughly under running tap water. They were dipped in 0.1% bavistin solution, a fungicide, (BASF Aktiengesellschaft, Germany) for half an hour to kill any microbes attached to the roots and transferred to plastic pots filled with mixture of sterile sand and garden soil (1:1). Initially, the pots were placed in plastic trays containing tap water for 8-10 days. Water was also regularly sprayed 4 or 5 times a day to keep the humidity high around the plantlets (Fig. 4A). After two months, the plantlets were transferred to earthen pots, kept them in green-house and irrigated with tap water. Finally they were transplanted in the garden soil.

OBSERVATIONS

Callus induction

The internodal segments (5 x 3 mm) excised from old trees were cultured on MS medium supplemented with 2,4-D (0.1-10 mM). Callus started appearing at cut ends of segments within a week of inoculation in all concentrations of 2,4-D. In next 15 days, rapid proliferation of callus all over the explants was achieved at 1 and 10 mM levels while at 0.1 mM it remained confined to cut ends only. After 30 days of initial inoculation, entire explants were seen to be covered with whitish-green, globular and compact callus at 1 mM 2,4-D (Fig. 1A). The callus formed at 10 mM 2,4-D was yellowish-green and less compact. Thus 1 mM 2,4-D proved best for obtaining good quality callus (Table 1).

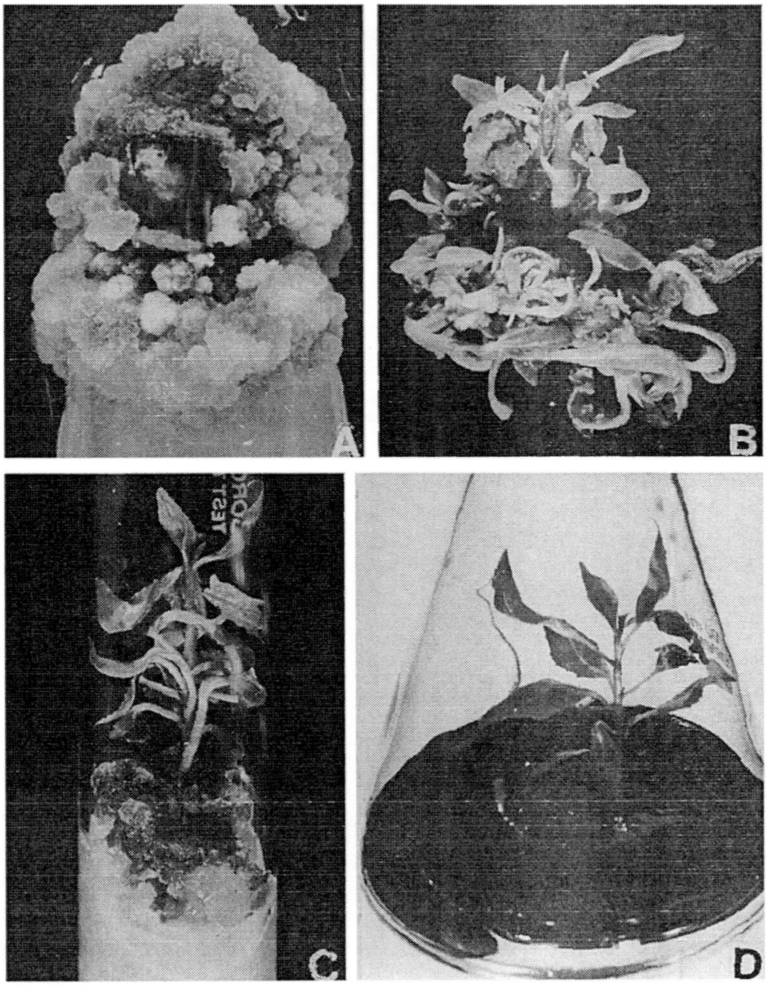

FIG.1 A-D Regulation of plantlet in stem-derived callus of Populus x euramericana old trees. (A) Whitish green, compact and globular morphogenic callus developed on a stem segment on MS + 1 μM 2,4-D after a month of inoculation. x 3. (B) Differentiation of multiple shoots in callus on transfer to MS + 1 uM BA, after 30 days of culture. x 2.5. (C) *In vitro* raised shoot developing callus instead of roots at its basal end on transfer to 1 uM IAA containing medium. x 1.24. (D) *In vitro* raised shoot organising roots in MS + 0.75% activated charcoal + 4.52 uM 2,4-D + 100 mg/l glutamic acid. x 1.

TABLE 1: Morphogenic response of *Populus* x *euramericana* stem internodal segments on 2,4-D containing MS medium, after 30 days of incubation.

2,4-D (μ M)	No. of explants	Explants forming callus (%)	Nature and amount of callus*
0	24	0	-
0.1	24	71	dark green, globular+ confined to cut ends only
1	24	79	whitish-green, compact globular+++ extending all over explants
10	24	79	yellowish-green, semi compact+++ all over explants

* Relative amount of callus: - = nil; + = poor; ++ = moderate; +++ = extensive

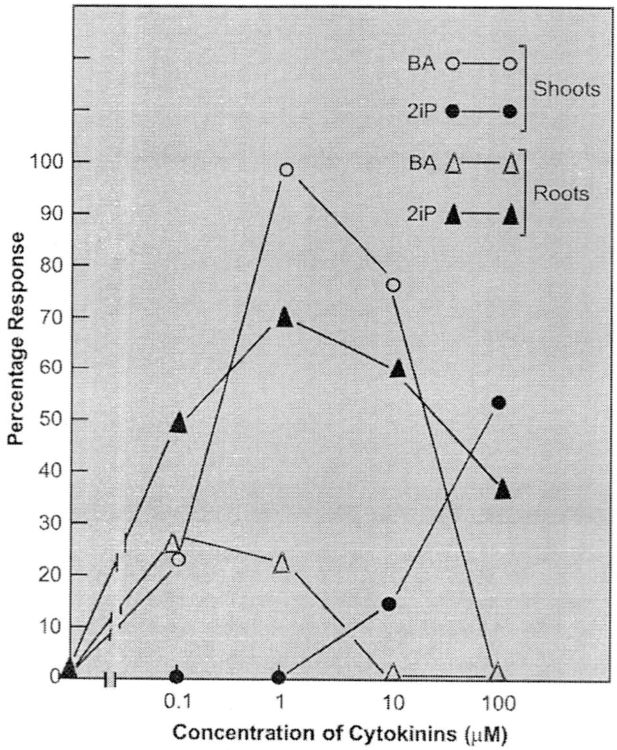

FIG. 2. Morphogenic response (shoots and roots) developed in callus cultures of *P.* x *euramericana* under the influence of cytokinins in MS medium after 30 days of incubation.

Caulogenesis

To further assess their morphogenic potential the callus pieces (1 cm diameter) were inoculated on MS medium augmented with 0.1-10 mM BA and 1 mM 2,4-D in various combinations. However, none of the combinations supported caulogenesis but the callus proliferated to three times of its size at low levels of BA (0.1 mM) and 1 mM 2,4-D. both the quality and quantity of callus declined with increase in BA levels. Such callus pieces (1 cm diameter) on further subculture to media lacking 2,4-D but containing cytokinins (BA or 2iP at 0.1 to 100 mM) individually differentiated shoot buds within 10-15 days. The shoot buds as well as callus proliferated further. A few of them elongated up to 2 to 2.5 cm in size. BA at 1 mM level proved best for inducing differentiation in (1) all cultures (100%), and (ii) maximum average number of shoots per explant (13.9; Figs. 1B, 2 & 3). In addition, roots were also organised at low levels of BA (0.1 and 1 mM) and all levels of 2iP. Maximum number of roots per culture developed at 10 mM 2iP (Figs. 2 & 3).

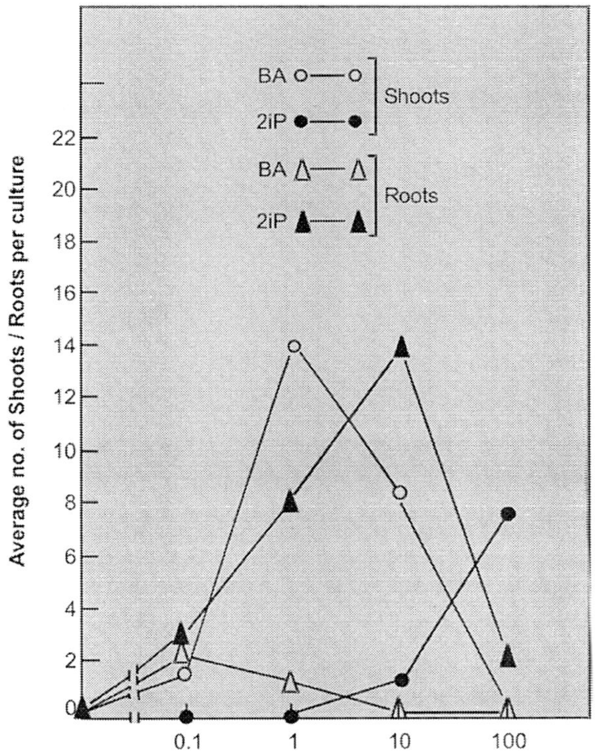

FIG. 3. Average number of shoots and roots organised per callus culture of *P.* × *euramericana* in MS + BA/2iP after 30 days of incubation.

Shoots organised at 10 mM BA were stunted and did not elongate even after subculture. Higher concentration of BA (100 mM) did not support caulogenesis and infact it dropped to zero (Fig. 2). In contrast to BA, the other cytokinin, 2iP did not favour shoot induction at lower concentrations (0.1 and 1 mM). At 10 mM 2iP only 15% of cultures differentiated shoots that too only 1 or 2 per culture (Figs 2 & #). Out of the four concentrations tried, 100 mM 2iP was the best for eliciting morphogenic response in 55% cultures (Fig. 2). The average number of shoots per explant was nearly 8 (Fig. 3). After removing the shoots, the callus masses bearing stunted shoot buds were further subcultured on the same but fresh medium. Somehow the best response was achieved at 1 mM BA. Within two weeks, some of the shoot buds further elongated which were excised and shifted to rooting medium and the callus masses bearing shoot buds were subcultured on the same medium. After third subculture, the callus cells as also the leaves of the stunted shoots turned pale yellow to brown and stopped further differentiation. Such callus masses were discarded as they seemed to lose regeneration potential.

Rhizogenesis

The *in vitro* raised shoots (more than 1 cm in length) were excised and planted on media containing various auxins which formed only callus at their basal ends (Fig. 1C) and did not support rhizogenesis.

The excised shoots, however, organised roots in medium containing activated charcoal and auxin (Fig. 1D). Of the various auxins tried, 2,4-D was the best for inducing roots. While the percentages of shoots organising roots were equal in control and 4.52 mM 2,4-D was the best for inducing roots. While the percentages of shoots organising roots were equal in control and 4.52 mM 2,4-D, the response was distinctly less in 9.04 mM 2,4-D (Table 2). The number of roots per shoot was more in 4.52 mM. Addition of activated charcoal to the rooting medium was indispensable for preventing callus formation at basal ends of shoots as well as for hastening root induction (Agrawal and Gupta, 1991). Addition of 100 mg/l glutamic acid helped in checking early leaf senescence leading to defoliation. The role of glutamic acid in chlorophyll biosynthesis is already well known (Beal, 1984; Foyer, 1984). It is not known if the earlier workers ever tried to make use of activated charcoal for improving rhizogenesis in *Populus tremuloides* (Chalupa, 1974), *P. ciliata* (Mehra and Cheema, 1980), *P. alba* (Park and Son, 1988) and *P. alba* x *P. grandidentata* (Son and Hall, 1990).

TABLE 2: Rooting response of *in vitro* raised shoots of *Populus* x *euramericana* on MS + 0.75% activated charcoal + 100 mg/l glutamic acid + 2,4-D, after 10 days of transfer.

2,4-D (μ M)	No. of shoots	Shoots organising roots (%)	Average number of roots per shoot (Mean ± S.D.)
0	27	96	1.4 ± 0.8
4.52	25	96	2.0 ± 1
9.04	25	88	1.6 ± 0.87

Soon after root induction, the young plantlets were transplanted to MS medium lacking activated charcoal and auxin but containing glutamic acid.

FIG. 4 (A-B) Tissue culture raised plantlets of *P.* x *euramericana* showing gradual transfer to field. (A) Plantlet growing in soil for two months after transfer from test tube. x 0.23. (B) Two-year-old tissue culture raised plantlet in field. x 0.07.

This step was very significant to make efficient field transfers. Nearly 80% of the plantlets are thriving in the soil for the last two years (Fig. 4B). Thus, desired number of plantlets can be obtained from stem derived callus of *Populus* trees within three months and that too at any time during an year.

DISCUSSION

The results presented here demonstrate that the number of shoots regenerated through callus phase is greatly influenced by the concentration and type of growth regulator used. Auxin in combination with cytokinin (BA) proved inhibitory for shoot induction. Of the two cytokinins, BA alone even at low concentration (1 mM) was more effective in inducing shoots. Production of roots in callus masses in the presence of only cytokinin (low levels of BA and all levels of 2iP) indicates that some of the callus cells still contain enough of auxins which they had absorbed earlier from the media (MS + 0.1 - 10 mM 2,4-D). This observation confirms the view expressed earlier by Minocha (1987). According to him, exposure of tissues to different growth regulators appears to initiate a series of events within cells that results in a highly regulated pattern of cell divisions leading to the formation of root or shoot primordia. The effect of different growth regulators are not specific in some cases. Even the growth regulators of same class, e.g. auxins or cytokinins elicit different morphogenic response in the same tissue. Recently, Ordas *et al.* (1992), while working on *Mallus* cell suspension thought that BA could selectively stimulate the synthesis of polypeptide responsible for cell growth. Perhaps for this reason, callus of hybrid poplar organizes multiple shoots in the presence of BA but in the absence of 2,4-D. The preliminary observations indicated that there was no striking morphological variation among the plants regenerated from callus (indirectly) and the nodes as well as internodes (directly).

Micropropagation via callus cultures offers numerous applications in plant tissue culture research as this ensures an incessant supply of shoots throughout the year. Even otherwise, the callus based system can be gainfully employed for protoplast isolation and regeneration (Russell and McCown, 1986), obtaining disease resistant somatic variants (Ostry and Skilling, 1988) and finally in genetic transformation studies in *Populus* species to overcome the problem of selecting transformants due to direct caulogenesis in the co-cultured leaves. The cytokinin synthesizing gene, if incorporated in this system, may prove more effective in producing numerous shoots with faster growth rate than unengineered herbaceous taxa, e.g. *Nicotiana* sp. (Smigocki and Owens, 1988) and *Solanum* sp. (Ondrej *et al.*, 1990, 1991). Smigocki & Hammerschlag (1991) have further expressed that rapid improvement of commercially important cultivars in peach could be attempted if agronomically important genes are transferred into peach cells derived from mature plants.

ACKNOWLEDGMENTS

VA is grateful to the University Grants Commission for the award of a Research Scientist `A'. Thanks are also due to Mrs. Aradhana Mittal for her generous help.

REFERENCES

Agrawal, V. and Gupta, S.C. 1991. *In vitro* plantlet development from explants of 25-year-old trees of *Populus* x *euramericana*- a hybrid poplar. Pl. Sci. 78: 99-105.

Antonetti, P. and Pinon, J. 1992. Culture de cals de peupliers de la section Leuce et regeneration de plantes entieres. Ann. Sci. For. 49: 627-636.

Beal, S.I. 1984. Biosynthesis of photosynthetic pigments. In: Baker, N.R. and Barber, J. (Eds.), Topics in Biosynthesis, Vol. 5, ?Chloroplast Biogenesis. Elsevier Scientific Pub., Amsterdam, pp. 133-205.

Berbee, F.M., Berbee, J.G. and Hildebrandt, A.C. 1972. Induction of callus and trees from stem tip cultures of a hybrid poplar. *In vitro* 7: 269.

Bon, M.C., Gendraud, M. and Franclet, A. 1988. Roles of phenolic compounds on micropropagation of juvenile and mature clones of *Sequoiadendron giganteum*: influence of activated charcoal. Sci. Hortic. 34: 283-291.

Bychenkova, E.A. and David, A. 1978. Callus formation and organogenesis in tissues of leaves of *Populus balsamifera* L. cultivated *in vitro*. Sov. Pl. Physiol. 151: 216-222.

Chalupa, V. 1974. Control of root and shoot formation and production of trees from poplar callus. Biol. Plant. 16: 316-320.

Chen, Z. and Evans, D.A. 1990. General techniques of tissue culture in perennial crops. In: Chen, Z., Evans, D.A., Sharp, W.R., Ammirato, P.V. and Sondahl, N.R. (Eds.), Handbook of Plant Cell Culture, Vol. 6. McGraw Hill Pub. Co., New York, pp. 22-61.

Chen, Z. and Sharp, W.R. 1990. Potential of biotechnology in perennial crop improvement. In: Chen, Z., Evans, D.A., Sharp, W.R., Ammirato, P.V. and Sondahl, N.R. (Eds.), Handbook of Plant Cell Culture, Vol. 6. McGraw Hill Pub. Co., New York, pp. 3-21.

Coleman, G.D. and Ernst, S.G. 1989. *In vitro* shoot regeneration of *Populus deltoides*: effect of cytokinin and genotype. Pl. Cell Rep. 8: 459-462.

El-Lakany, M.H. 1992. Rapid propagation of fast-growing tree species in developing countries: its potentials, constraints and future development. In: Baker, F.W.G. (Ed.), Rapid Propagation of Fast-growing Woody Species. C.A.B. International, U.K., pp. 102-108.

Fillatti, J.J., Sellmer, J.C., McCown, B.H., Haissig, B.E. and Comai, L. 1987. *Agrobacterium* mediated transformation and regeneration of poplar. Mol. Gen. Genet. 206: 192-199.

Foyer, C.H. 1984. Photosynthesis. John Wiley and Sons, Inc., New York, pp. 130-150.

Gamborg, O.L., Miller, R.A. and Ojima, K. 1968. Nutrient requirements of suspension cultures of soybean root cells. Expl. Cell Res. 50: 151-158.

Gupta, S.C. and Agrawal, V. 1992. Micropropagation of woody taxa and plant productivity. In: Prasad, B.N., ghimire, G.P.S. and Agrawal, V.P. (Eds.), Role of Biotechnology in Agriculture. Oxford & IBH Pub. Co., Delhi, pp. 37-52.

Hartney, V.J. and Svensson, J.G.P. 1992. The role of micropropagation for australian tree species. In: Baker, F.W.G. (Ed.), Rapid Propagation of Fast-growing Woody Species. C.A.B. International, U.K., pp. 7-28.

Lakshmi Sita, G., Raghav Ram, N.V. and Vaidyanathan, C.S. 1979. Differentiation of embryoids and plantlets from shoot callus of sandalwood. Pl. Sci. Lett. 15: 265-270.

Leple, J.C., Brasileiro, C.M., Michel, M.F., Delmotte, F. and Jouanin, L. 1992. Transgenic poplars: expression of chimeric genes using four different constructs. Pl. Cell Rep. 11: 137-147.

McCown, B.H. 1991. Genetic transformation of trees via direct gene transfer. In: Ahuja, M.R. (Ed.), Woody Plant Biotechnology. Plenum Press, New York, pp. 207-211.

McCown, B.H., McCabe, D.E., Russell, D.R., Robinson, D.J., Baton, K.A. and Raffa, K.F. 1991. Stable transformation of *Populus* and incorporation of pest-resistance by electric discharge particle acceleration. Pl. Cell Rep. 9: 590-594.

Mehra, P.N. and Cheema, G.S. 1980. Clonal multiplication *in vitro* of Himalayan poplar (*Populus cilliata*). Phytomorphology 30: 336-343.

Minocha, S.C. 1987. Plant growth regulators and morphogenesis in cell and tissue culture of forest tree. In: Bonga, J.M. and Durzan, D.J. (Eds.), Cell and Tissue Culture in Forestry, Vol.2. Martinus Nijhoff Publ. Lancaster, pp. 50-66.

Murashige, T. and Skoog, F.R. 1962. A revised medium for rapid growth and bioassays with tobacco tissue culture. Physiol. Plant. 15: 473-497.

Noh, E.W. and Minocha, S.C. 1986. High efficiency shoot regeneration from callus of quaking aspen (*Populus tremuloides* Michx.). Pl. Cell Rep. 5: 464-467.

Ondrej, M., Machackova, I., Catsky, J., Eder, J., Hrouda, M., Pospisilova, J. and Synkova, H. 1990. Potato transformation by T-DNA cytokinin synthesis gene. Biol. Plant. 32: 401-406.

Ondrej, M., Bavrina, T.V., Dudko, N., Hrouda, M., Krekule, J., Lozhnikova, V.N., Machackova, I., Seidlova, F. and Vlasak, J. 1991. Transgenic tobacco plants with T-DNA phytohormone synthesis genes. Biol. Plant. 33: 40-48.

Ordas, R.J., Ferandez, B. and Rodriguez, R. 1992. Benzyladenine controlled protein synthesis and growth in apple cell suspoension. Physiol. Plant. 84: 229-235.

Ostry, M.E. and Skilling, D.D. 1988. Somatic variation in resistance of *Populus* to *Septoria musiva*. Pl. Dis. 72: 724-727.

Park, Y.G. and Son, S.H. 1988. Regeneration of plantlets from cell suspension cultures derived callus of white poplar (*Populus alba* L.). Pl. Cell Rep. 7: 567-570.

Riemenschneider, D.E. and Haissig, B.E. 1991. Producing herbicide tolerant *Populus* using genetic transformation. In: Ahuja, M.R. (Ed.), Woody Plant Biotechnology. Plenum Press, New York, pp. 247-263.

Russel, J.A. and McCown, B.H. 1986. Culture and regeneration of *Populus* leaf protoplasts isolated from non-seedling tissue. Pl. Sci. 46: 133-142.

Rutledge, C.B. and Douglas, G.C. 1988. Culture of meristem tips and micropropagation of 12 commercial clones of poplar *in vitro*. Physiol. Plant. 72: 367-373.

Sellmer, J.C., McCown, B.H. and Haissig, B.E. 1989. Shoot culture dynamics of six *Populus* clones. Tree Physiol. 5: 219-227.

Smigocki, A.C. and Owens, L.D. 1988. Cytokinin gene fused with a strong promoter enhances shoot organogenesis and zeatin level in transformed plant cells. Proc. Natl. Acad. Sci. U.S.A. 85: 5131-5135.

Smigocki, A.C. and Hammerschlag, F.A. 1991. Regeneration of plants from peach embryo cells infected with a shooty mutant strain of Agrobacterium. J. Amer. Soc. Hort. Sci. 116: 1092-1097.

Son, S.H. and Hall, R.B. 1990. Multiple shoot regeneration from root organ cultures of *Populus alba* x *P. grandidentata*. Pl. Cell tissue Organ Cult. 20 53-57

White, P.R. 1943. Handbook of Plant Tissue Culture. Jaques Cattell, Lancaster.

Wilde, H.D., Meagher, R.B. and Merkle, S.A. 1992. Expression of foreign genes in transgenic yellow poplar plants. Pl. Physiol. 98: 114-120.

Winton, L.L. 1968. Plantlets from aspen tissue cultures. Science 160: 1234-1235.

Wolter, K.E. 1968. Root and shoot initiation in aspen callus cultures. Nature 219: 509-510.

Direct Somatic Embryogenesis From Immature Seeds of Rosewood (*Dalbergia latifolia* Roxb.)

G. Lakshmi Sita and M. Muralidhara Rao

INTRODUCTION

Somatic embryogenesis *in vitro* offers great potential in crop improvement since it allows for efficient cloning and genetic modification (Sharp *et al.*, 1980). Although shoot tip/axillary bud culture is the most commonly practiced method for the large scale production of plantlets, somatic embryogenesis is preferred as it requires fewer steps with concomitant reduction in labour, time and cost. It is evident from the increasing success of somatic embryogenesis in forest tree species, that it will gradually replace shoot tip culture method, being amenable to automation. Biotechnological approaches of tree tissue culture lag behind those of herbaceous crop species mainly because tree tissues or at least those from mature individuals, cannot be similarly manipulated in culture to yield high frequency of regeneration of plantlets from any desired starting cell, cellular aggregates or organ (Haissig *et al.*, 1987).

Recently significant progress has been made on induction of embryogenesis from immature seedling explants of coniferous trees (Pramod *et al.*, 1993) but there are very few reports of embryogenesis from tropical trees of economic value. Sandal wood is the only tropical tree wherein successful induction of somatic embryogenesis was achieved through explants derived from adult trees (Lakshmi Sita, 1979), seedling explants (Rao and Bapat, 1978) suspension cultures (Lakshmi Sita, 1980) and protoplasts (Lakshmi Sita, 1983). Somatic embryogenesis has also been reported from seedling explants of other tropical trees like rubber (Herve *et al.*, 1993), coffee (Sondahl and Sharp, 1977), mango (Richard, 1984). Leguminous trees are generally recalcitrant for *in vitro* manipulations. Somatic embryo induction has not been possible in general except for some grain legumes like soybean (Ranch *et al.*, 1989) and peanut (Hazra *et al.*, 1989) where some success has been achieved.

Rosewood (*Dalbergia latifolia*) belonging to the family Leguminosae commonly known as East Indian Rosewood or Black wood is an important timber yielding tree which is of great commercial value. Bulk of this commodity in the world is met by India and is a good source of foreign exchange to the country.

Abbreviations: MS: Murashige and Skoog medium; 2,4-D: 2,4-Dichlorophenoxy acetic acid; BAP: Benzyl amino purine; ABA: Abscissic acid.

In view of the continuous felling and lack of cultivated plantations, Rosewood has become a dearer commodity and production has substantially gone down. The gap between supply demand had widened resulting in a near crisis situation in the world market. Due to the constraints involved in the conventional propagation methods like poor seed viability, there is shortage of planting stocks. This could be overcome by *in vitro* mass production. Our laboratory has undertaken a programme on tissue culture of Rosewood with a view to develop viable tissue culture methods for mass propagation. We have earlier reported mass propagation from callus culture by organogenesis (Lakshmi Sita *et al.*, 1986) and enhanced axillary shoot multiplication (Raghava Swamy *et al.*, 1992). It is well recognised that the ultimate goal for mass propagation is by inducing somatic embryogenesis from suspension cultures. Hence we have been trying to induce somatic embryogenesis from callus cultures derived from mature trees and had very little success. We now report successful induction of direct somatic embryogenesis from immature seeds without involving callus phase.

MATERIALS AND METHODS

Plant material

Immature green pods of Rosewood were collected from 60-70 year old trees at 60, 90, 120 days after flowering. Prior to the dissection of the seeds, pods were disinfested by immersion in 90% v/v ethanol for 5-10 seconds and in mercuric chloride for 15 minutes followed by 3-4 rinses in sterile distilled water. The immature cotyledons were dissected out from green immature seeds which was aseptically removed from these pods. Cotyledons from mature seeds were also tried for the induction of somatic embryos.

Culture media

Somatic embryos were induced in two phases. In the first phase dissected out cotyledons were placed on solid Murashige and Skoog media (1962) supplemented with 20 g/L sucrose along with different combinations of 2,4-D (1.0-10.0 mg/L) with or without kinetin (0.1-1.0 mg/L). In the second phase, these cotyledons were transferred to MS media containing lower levels of 2,4-D (0.1-1.0 mg/L) supplemented with or without 100 g/l sucrose and 10% v/v coconut water. To study the influence of different sucrose concentrations, 20-100 g/l sucrose was used in the second phase along with MS basal salts supplemented with 0.5 mg/l 2,4-D and 10% coconut water. The pH of the medium was adjusted to 5.7 to 5.8 prior to autoclaving at 121°C for 15 minutes.

Maturation and Germination

Somatic embryos were induced in the second phase and these somatic embryos for further synchronous development of shoot and root, different media combinations with growth regulators like ABA (0.1, 0.2 mg/L), BAP (0.1, 0.5, 1.0 mg/L), Mannitol 1.0 g/L, MSA salts (full, half strength) alone without growth regulators were tried.

Culture conditions

The cultures were incubated at 25°2C under white fluorescent light (15mE/m/s) with a 16 hour photoperiod.

Histological procedure

In order to test the bipolar nature and also origin of the somatic embryos, the cultured explants were fixed in formalin/acetic acid/alcohol for 24 hours. After washing the material with 70% alcohol, it was dehydrated in tertiary butyl alcohol series and embedded in paraffin wax according to the procedure described by Johansen (1940). Serial sections were cut at 10 mm and mounted on DPX.

In all these studies, twenty zygotic embryos were inoculated per treatment. As the embryos could not be dissected out from the 60 day old pods, whole immature green seeds measuring 3 mm were inoculated. The study was conducted for two consecutive flowering seasons. The data on number of explants responded, number of embryos per explant was recorded twenty days after induction.

RESULTS

Direct somatic embryogenesis was induced in the second phase when immature cotyledons were transferred from high 2,4-D containing medium (MS1-MS6) to low 2,4-D medium (MS8-MS10). Globular glossy white embryos (Fig. 1A) started appearing from the surface of the cotyledons within 7-10 days after subculturing which further develop into dark green coloured embryos (Fig. 1D). There was also development of fasciated leafy structures and some were trumpet like which did not show any further development (Fig. 1C). Morphological and histological observations of somatic embryogenesis (Fig. 2A, B) indicate direct embryoid formation without callus phase. However, whether these embryos are developed from single or group of cells is not clear.

High frequency somatic embryo development was found to be dependent not only on initial induction phase but also on high sucrose concentration of 100 g/L and 10% (v/v) coconut water. Percentage of explants forming somatic embryos (92.5) was significantly high when it was transferred from MS2 to MS8 (Table 1). In the first phase when the medium was supplemented with kinetin along with 2,4-D, the response was better compared to 2,4-D alone. Moreover in the initial phase lower 2,4-D (2 mg/L) along with 0.1 mg/L kinetin was found to be superior when compared to higher levels of 2,4-D (5 mg/L) with or without kinetin.

The age of the explant was found to play an important role in the induction of somatic embryogenesis. It was seen to affect the number of somatic embryos produced per explant as well as number of days taken for induction. Days taken for the induction of somatic embryogenesis increased with the age of the immature cotyledons. Maximum number of somatic embryos (32) was observed from 90 day old cotyledons where somatic embryos started emerging within 7-10 days whereas from 120 day old explants the induction was 20-25 days after transfer. Moreover callus growth was also enhanced with the age of the explant. Callus started growing from the periphery of the cotyledons and continued to grow all over the explant making it to look as if these embryos were emerging from the callus. On the other hand 90 day old explants produced less callus and white pearl like somatic embryos arising from the surface directly on the expanded cotyledons could be clearly seen (Fig. 1A). Fully mature cotyledons, excised from the dry seeds, did not show any embryogenesis but callused profusely regardless of the media composition. Sixty day old immature seeds did not respond and only turned brown.

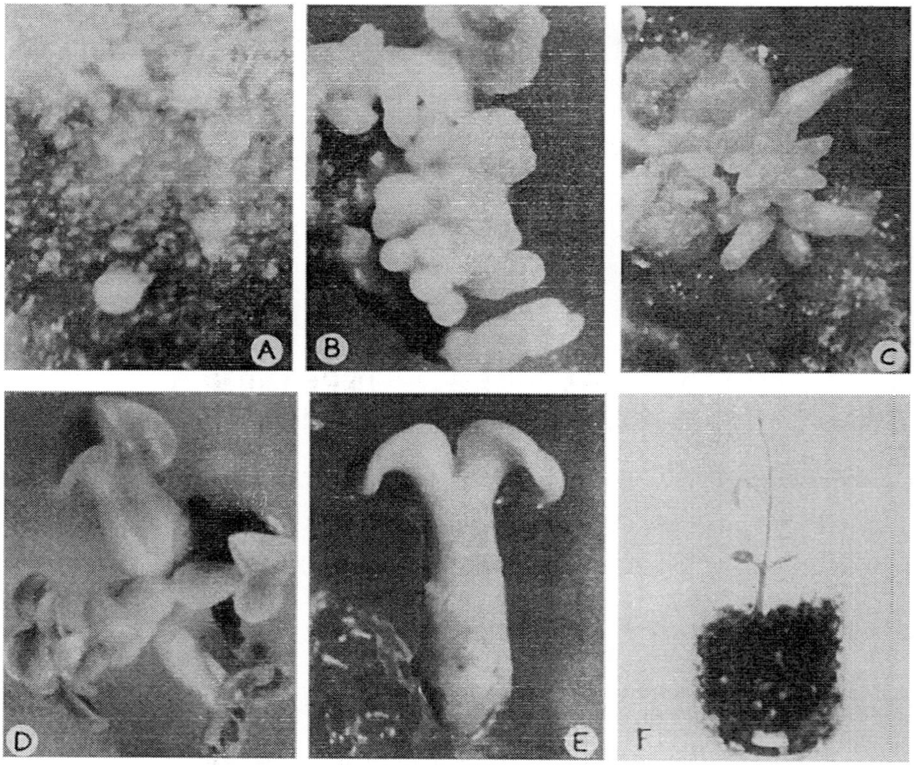

FIG.1A-F: Different developmental stages of direct somatic embryogenesis in Rosewood. (A) Globular embryos developing directly from the expanded cotyledons. (B) Cluster of somatic embryos. (C) Abnormal trumpet shaped somatic embryos (D) Cluster of fully developed embryos (E) Single somatic embryo with expanded cotyledon (F) Plantlet regenerated from somatic embryo.

TABLE 1: Effect of different media combinations and age of the explant on direct somatic embryogenesis from immature cotyledons of Rosewood

1st Phase	2nd Phase											
	MS 7			MS 8			MS 9			MS 10		
	Age of the explant											
	60	90	120	60	90	120	60	90	120	60	90	120
MS1	-	5*	C	-	10	C	-	C	C	-	10	C

1st Phase	2nd Phase											
	MS 7			MS 8			MS 9			MS 10		
	Age of the explant											
	60	90	120	60	90	120	60	90	120	60	90	120
		(4.0)**			(5.0)						(4.0)	
MS2	-	60 (18.6)	32 (16.2)	-	94.5 (32.0)	45 (22.0)	-	45 (15.6)	25 (13.4)	-	92.5 (27.0)	40 (14.8)
MS3	-	55 (15.9)	30 (13.2)	-	90 (30.0)	45 (14.3)	-	50 (14.0)	22.5 (10.2)	-	62.5 (18.0)	32.5 (12.6)
MS4	-	C	C	-	5 (6.0)	C	-	C	C	-	C	C
MS5	-	10 (6.2)	5 (3.0)	-	20 (9.0)	10 (6.0)	-	C (9.0)	C	-	15	C
MS6	-	5 (4.0)	5 (3.0)	-	15 (7.0)	5 (4.0)	-	C	C	-	15 (6.0)	C

* = Percentage of Explants forming Embryos; ** = Number of Somatic Embryos per Explant;
C = Callus
MS1: MS + 2.0 2,4-D
MS2: MS + 2.0 2,4-D + 0.1 Kin
MS3: MS + 2.0 2,4-D + 1.0 Kin
MS4: MS + 5.0 2,4-D
MS5: MS + 5.0 2,4-D + 0.1 Kin
MS6: MS + 5.0 2,4-D + 1.0 Kin
MS7: MS + 0.5 2,4-D + 10% Sucrose
MS8: MS + 0.5 2,4-D + 10% Sucrose + 10% Coconut Water
MS9: MS + 1.0 2,4-D + 10% Sucrose
MS10: MS + 1.0 2,4-D + 10% Sucrose + 10% Coconut Water

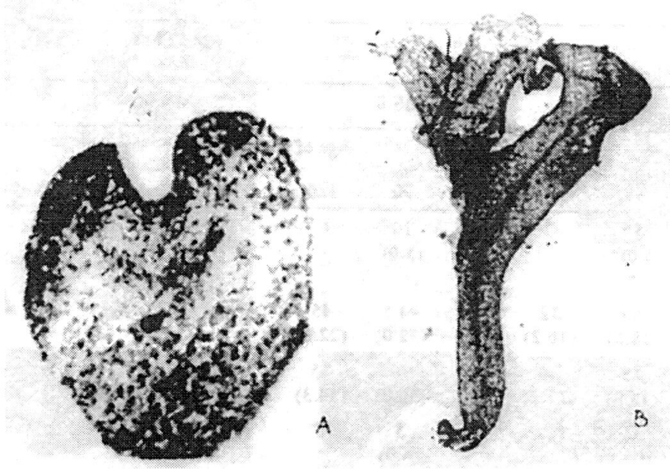

FIG. 2: Histology of somatic embryos. Longitudinal section of Heart shaped embryo (A) and fully formed embryo (B)

For further development of these embryos into complete plantlets, embryos with well developed cotyledons were transferred to growth regulator free medium with full, half strength MS major salts and other media compositions as reported earlier. But there was no further development of these embryos into complete plantlets. From these preliminary studies it suggested that preconditioning or further maturation of these somatic embryos may be necessary for the successful germination into plantlets. Out of the various media combinations tested, MS supplemented with BAP (0.5-1.0 mg/l) found to be effective for maturation giving rise to whole plantlets (Fig. 1F). Approximately 30% of the embryos of normal shape germinated within 20-30 days after transfer.

DISCUSSION

Zygotic embryogenic tissue is more amenable to somatic embryogenesis as they contain the pre embryogenic determined cells (PEDS). It has been suggested (Sharp *et al.*, 1980) that direct somatic embryogenesis in culture proceeds from cells which are already determined for embryogenic development prior to explanting requiring only growth regulators or favourable conditions to allow these PEDS to undergo cell division and expression of embryogenesis. Cotyledons derived from zygotic embryos have been reported to be suitable explants for somatic embryogenesis either directly or indirectly in other tree species like *Hevea brasiliensis* (Herve, 1993), *Azadirecta indica* (Medha Srikhande *et al.*, 1993), Walnut (Walt and gale, 1985), Black locust (Arillaga *et al.*, 1994) and crop plants like Soybean (Ranch *et al.*, 1985), Water melon (Michael and Gray, 1993) etc. The developmental stage of the explant has been reported to have a critical influence on the ability to induce somatic embryogenesis in woody legumes (Trigiano *et al.*, 1992). From our studies also it appears that size or age of the immature cotyledons of Rosewood is one of the most critical factors affecting somatic embryogenesis. Cotyledons of size 2-3mm (90 days after flowering) gave the highest frequency of embryoid formation. However, the frequency was dramatically reduced with the older cotyledons. The mature cotyledons were found to lose their potential to form somatic embryos, instead they dedifferentiated into callus tissue. This suggests that cells in younger embryos retain the embryogenic potential which is gradually lost as development proceeds (Sung, 1985).

Similar results have also been observed in Black locust, wherein it was observed that percentage of seeds forming somatic embryos was significantly affected by the age of the explant and also by the composition of the media (Arillaga *et al.*, 1994).

Somatic embryogenesis is generally believed to be triggered by auxins and in many species, 2,4-D is the most commonly used for the induction of somatic embryos. Tissues are generally transferred from high auxin containing medium to low auxin containing medium (Sharp *et al.*, 1980). Similarly in Rosewood we found that 2,4-D at lower concentrations was more effective for induction phase. Moreover kinetin at 0.1 mg/l along with 2,4-D was found to be favourable. Somatic embryos are induced when the explant is transferred to low 2,4-D (0.5-1.0 mg/L) levels containing high sucrose (10%) and coconut water (10%). Addition of coconut water was found to be favourable in the second phase of induction.

Higher sucrose levels found to favour somatic embryogenesis in several species like *Trifolium*, *Medicago sativa* (Maheswaran and Williams, 1984, 186), Lycopersicon (Young *et al.*, 1987), Soybean (Lazzeri *et al.*, 1987) *Helianthus annuus* (Finer, 1987), ground nut (Hazra *et al.*, 1991), Capsicum (Harini and Lakshmi Sita, 1993. Our results clearly indicate that sucrose upto 10% (w/v) favoured direct somatic embryogenesis in Rosewood and frequency of somatic embryogenesis decreased with the sucrose concentration. There was no embryogenesis at 2% level (Fig. 3). Although it has not been experimentally proven, sucrose most likely acts both as a nutrient as well as osmotic agent to influence somatic embryo development (Xu *et al.*, 1990).

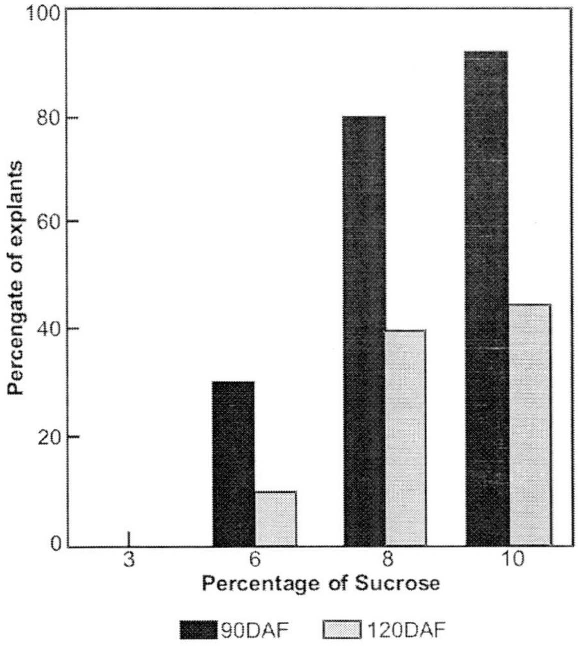

FIG. 3. Effect of different sucrose levels on direct somatic embryogenesis from immature cotyledons of Rosewood.

Most critical step in somatic embryogenesis is the maturation process, specifically the ability to germinate and produce plantlets. In other species importance of ABA (Verhagen, 1989) and Mannitol (Nadel *et al.*, 1989) for different and further germination of somatic embryos has been demonstrated. The use of ABA and mannitol in the differentiation media would allow better synchronization of the differentiation process and result in a higher percentage of germination. However, in Rosewood ABA or mannitol did not favour germination whereas BAP at 0.5-1.0 mg/L was found to be effective as in the case of *Pharbitis nil* (Shi-Rong Jia and Nam Hai Chua, 1992) and Water melon (Michael and Grey, 1993).

In conclusion, cotyledons from immature zygotic embryos provide the best source for somatic embryogenesis in Rosewood. Cotyledons from 90 day old pods was found to give the highest frequency of direct somatic embryogenesis. Direct regeneration from cotyledonary explants offers an ideal system for co-cultivation using agrobacterium mediated genetic transformation. Work on these lines using marker genes is in progress.

REFERENCES

Arillaga, I., Tobolski, J.J. and Merkle, S.A. 1994. Plant Cell Rep 13: 171-175.
Beat, Neuenschwander and Thomas, W. Baumann. 1992. Plant Cell Rep. 10: 608-612.
Finer, J.J. 1987. Plant Cell Rep. 6: 372-374.
Haissig, B.E., Nelson, N.D. and Kidd, G.H. 1987. Biotechnology 5: 33-59.
Harini, I. and Lakshmi Sita, G. 1993. Plant Science 89: 107-112.
Hazra, S., Sathaye, S.S. and Mascarenhas, A.F. 1989. Biotechnology 7: 949-951.
Herve Etienne, Bruno Sotta, Pascal Montoro, Emile Miginiac and Marc Phillipe Carron. 1993. Plant Science, 88: 91-96.
Johansen, D.A. 1940. Plant Microtechnique, McGraw-Hill, New York.
Lakshmi Sita, G., Shobha, J. and Vaidyanathan, C.S. 1980. Curr Sci. 49: 196-198.
Lakshmi Sita, G., Raghava Ram, N.V. and Vaidyanathan, C.S. 1979. Plant Sci. Lett. 15: 265-270.
Lakshmi Sita G., Chatoopadyaya, S. and Tejavathi, D.H. 1986. Plant Cell Rep. 5: 266-268.
Lakshmi Sita G., Rani, H.S. 1983. Experientia 45: 4-5.
Lazzari, P.A., Hildabrand, D.F. and Collins, C.B. 1985. Plant Mol. Biol. Rep. 3: 160-167.
Maheshwaran, G. and Williams, E.G. 1986. J. Plant Physiol. 124: 455-463.
Maheshwaran, G. and Williams, E.G. 1984. Ann. Bot. 54: 201-211.
Medha Srikhande, Thengane, S.R. and Mascarenhas, A.F. 1993. *In vitro* Cell Dev. Biol. 29P: 38-42.
Michael, E. Compton and Gray, d.J. 1993. Plant Cell Rep. 12: 61-65.
Murashige, T. and Skoog, F. 1962. Physiol Plant 15: 473-497.
Nadel, B.L., Altman, A. and Ziu, M. 1989. Plant Cell Tissue Organ Culture 18: 181-189.
Pramod, K. Guptha, Gerald Pullman, Roger Timmis, Mary Kreitinger, William C Carlson, Jim Grop and Elaine Welty. 1993. Biotechnology 11: 454-457.
Raghava Swamy, B.V., Himabindu, K. and Lakshmi Sita, G. 1992. Plant Cell Rep. 11: 126-131.
Ranch, J.P., Ogelsby, L. and Zielinski, A.C. 1986. In: Indra K. Vasil (ed.) Cell Culture and Somatic Cell genetics of Plants, Vol. 3, Academic Press, Inc., Florida, pp. 97-110.
Rao, P.S., Richard, E. Litz, Robert, J. Knight Jr. and Shmuel Gazit, 1984. Scientia Horticulturae 22: 233-240.
Sharp, W.R., Sondahl, M.R., Caldas, L.S. and Maraffa, S.B. 1980. Hort. Review 2: 268-310.
Shi-Rong Jia and Nam Hai Chua 1992. Plant Science 87: 215-223.
Sondahl, M.R. and Sharp, W.R. 1977. Z.P. flanzenphysiol 81: 395-405.
Sung, Z.R. 1985. In: Freeling, M. (ed.) Plant Genetics, Proceedings of the Annual ARCO Plant Cell Research Institute, UCLA symposium on Plant Biology, Alan R. Liss, Inc., Newyork, pp. 115-128.
Trigiano, R.N., Geneve, R.L., Merkle, S.A. and Preece, J.E. 1992. In: Jannick, J. (ed.) Horticultural Reviews, vol. 14, John Wiley and Sons, Inc, New York, pp. 265-332.
Trigiano, R.N., Beaty, R.M. and Graham, E.T. 1988. Plant Cell Rep. 7: 148-150.
Verhagen, S.A. and Wann, S.R. 1989. Plant Cell Tissue Org. Cult 16: 103-111.
Walt Tulecke and Gale, Mc Granahan 1985. Plant Sci. 40: 57-63.
Xu, N., Coulter, K.M. and Bewley, J.D. 1990. Planta 182: 382-390.
Young, R., Kaul, V. and Williams, E.G. 1987. Plant Science 52: 237-242.

Direct Organogenesis From Internodal Segments and Apical Meristem of Shoots of *Jatropha curcas* L.

Neena Nigam, O.P. Saxena and Vincent Braganza

INTRODUCTION

Current day awareness of the depletion of traditional energy resources has evoked an intense interest in alternative sources of energy. This has influenced trends in plant biotechnology, resulting in attempts to use cell and plant tissue for improvement and rapid propagation of plants producing oils with fuel application possibility. *Jatropha curcas* (L.) is one such plant that has oil that has oil may be able to substitute engine oil (Diwakar, 1993; Ogbobe, 1993).

It belongs to the Euphorbiaceae family and is one of the promising drought tolerant perennial species. The plant is very hardy and can tolerate high temperature and salinity. The latex of the plant and the seeds are used for medicinal purposes (Kossasi, 1989). The latex has also been used in local Ayurvedic pharmacopoeia according to a report from the Botany laboratory, BARC-East Beltsville, Maryland, USA. The active principle in its latex Jatrophone, has shown anti-cancer properties. The green tender leaves are useful as feed for tasar silkworms, thus the plant holds promise for sustaining small scale silk industries in rural India. The Seeds contain semi-drying oil an efficient substitute for diesel engines. (Bhasubutra and Sutiponpeibun, 1982; Raina, 1987).

The present study established that multiple shoots can be obtained directly from internodal segments of the plant. However attempts using apical meristem as explants did not give encouraging results.

MATERIALS AND METHODS

The explants were obtained from plants grown in our botanical garden. Internodal segments from the 4th to the 8th node from the apex and apical meristem were washed thoroughly with tap water and surface sterilized in 20% sodium hypochlorite with 3-4 drops of teepol for 15 min. followed by $HgCl_{2(0.1)}$%, 2-3 min.) and rinsed 4-5 times with sterile double distilled water. These explants were then transferred to solid MS medium (Murashige and Skoog, 1962).

Salts of MS medium and vitamins of B5 medium (Gamborg *et al.*, 1968) supplemented with various combinations and concentrations of different growth regulators namely 1-napthaleneacetic acid [NAA], 2,4-Dichlorophenoxy acetic acid [2,4-D], 6-Benzylamino purine [BAP], Kinetin, 6(Y,Y dimethylallylamino)-purine[2-ip], gibberellic acid [GA3] and sucrose were used for regeneration studies. The pH was adjusted to 5.8 before autoclaving at 151bs. for 15 mins.

All the cultures were maintained at 26-28°C with 24 hours light, and 32% relative humidity.

RESULTS AND DISCUSSIONS

The outcome of the various experiments conducted are summarized in table 1. Two shoots with leaves from one nodal segment were visible on the tenth day since inoculation (Fig. 1C). These shoots, when isolated and transferred singly on same medium grew and produced 8*-10 shoots with leaves, per node (Fig. 1D). The medium used in these experiments was supplemented with 0.2 mg/l BAP, 0.1 mg/l NAA and 0.035 mg/l GA_3.

TABLE 1: Response of apical meristem and internodal segments of *Jatropha curcas* (L.) cultured on MS* basal medium supplemented with growth regulators

S.No.	PGRS and Conc. in mg/l			Day of Initiation	No. of Subculture	Response of Explants	
	BAP	NAA	GA_3			1. Apical Meristem	2. Internodal Segment
1.	-	-	-	12th, 36th	-	+	+
2.	0.10	-	-	11th	-	++	+
3.	-	0.20	-	-	-	-	-
4.	-	-	0.035	8th	-	-	+
5.	0.10	0.20	-	-	-	-	+
6.	0.10	-	0.035	-	-	-	-
7.	-	0.20	0.035	-	-	-	-
8.	0.20	0.10	0.035	17th, 8th	-	+	+
9.	0.20	0.10	0.035	8th	1	-	+++
10.	0.10	0.15	0.035	11th	-	-	++++
11.	0.10	0.15	0.035	6th	1	-	++++
12.	0.05	0.20	0.035	18th	-	-	+
13.	0.20	0.15	0.035	20th	-	-	++
14.	0.15	0.05	0.035	20th	-	++	-
15.	0.25	0.10	0.035	26th	-	-	+

MS*: Supplemented with 10 mg/l Thiamine, 1 mg/l Pyridoxine, 1 mg/l Nicotinic acid
Number of shoots has been denoted as: + (1-2) shoots, ++ (2-4) shoots, +++ (4-6) shoots, ++++ (6-8) shoots; Temperature = 26-28°C, Photoperiod = 24 hrs., Humidity = : 32%

In further trials on reducing the concentration of BAP from 0.2 mg/l to 0.1 mg/l and increasing the concentration of NAA from 0.1 mg/l to 0.15 mg/l, 7-9 shoot buds from the internodal end (Fig. 1E) were observed and leaves appeared on the 13th day. On first subculture the same explant produced 7-9 shoots again in identical medium (Fig. 1F) taking the total number of shoots to 16-18. In both the cases callus was there at the base (at the other cut end of the nodal region). It was green friable and granular in nature.

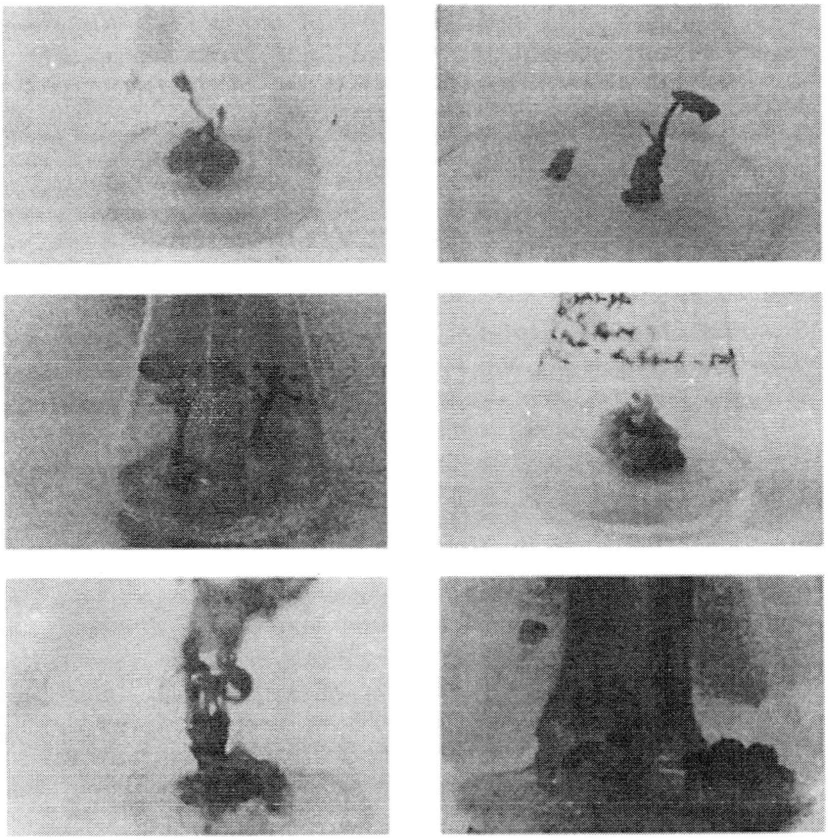

FIG. 1. Direct organogenesis from internodal regions of *Jatropha curcas* L. a. Multiple shoots from internode. b. Multiple shoots in first subculture of same explant (*). c. Shoots from internodal region. d. Multiple shoots from the first subculture of internodal region of (c). e. Internodal region (#). f. Apical meristem (#). *Same explant. #Raised on basal MS medium.

Increasing the concentration of auxin resulted in callus growth increase and reduction in the concentration of cytokinin brought about an increase in the number of shoots. The response of the apical meristem region was not significant.

With an increase in the concentration of cytokinin and a decrease in the concentration of auxin the number of shoots were reduced. Cytokinin alone (BAP 0.1 mg/l) did not respond well. The internodal region and apical meristems kept on growing and gave rise to one leaf from each node. Here the internodal region was very small and shoot grew faster.

Basal medium alone without hormones produced 2-3 shoot buds from the internodal region while from the apical meristem it gave 1 shoot with leaf on the 12th day and one more on the 30th day. On subculturing them on the same medium they failed to grow.

A few more concentrations and combinations of auxins and cytokinins were also tried but failed to give success. The results reported above were confirmed by 2-3 replicates. 16 to 18 shoots were obtained from a single explant (7-9 shoots per subculture).

Srivastava (1971, 1973) has obtained root and shoot from endosperm calli. However much detail is not available. We ourselves were unable to get differentiation from callus of cotyledons hypocotyl and epicotyl. Flick *et al.* (1983) have reported organogenesis in Euphorbia pulcherrima using 0.5 uM NAA and 10-50 uM 2-iP. No such findings are available for *Jatropha curcas* (L.). This work is therefore significant and has potential for rapid propagation of the plant.

CONCLUSION

Our experiments show that MS basal medium supplemented with higher concentration of thiamine is suitable for organogenesis in Jatropha. Moreover low concentrations of cytokinin (BAP 0.1 mg/l) and slightly higher concentrations of auxin (NAA 0.15 mg/l) along with GA_3 (0.035 mg/l) are found to be more effective in inducing multiple shoots from internodal regions in *Jatropha curcas* (L.).

ACKNOWLEDGMENT

This project was supported by grants from the Foundation for the Health & Human Sciences, administered by The Ahmedabad St. Xavier's College Society, Ahmedabad 380 009.

REFERENCES

Bhasubutra, R. and Sutiponpeibum, S. 1982. *Jatropha curcas* oil as a substitute for diesel engine oil, Renewable energy. Review Journal: 4: 56-70.

Diwarkar, V.J.M. 1993. Energy from the humble castor. Indian Express (17-11-93).

Flick, C.E., Evans, D.A. and Sharp, W.R. 1983. In hand book of plant cell culture, volume 1, Evans, D.a., Sharp, W.R., Ammirato, P.B. and Yamada, Y. (eds.) Macmillan Publishing Company, New York p. 13.

Gamborg, O.L., Miller, R.A. and Ojima, 1968. J. Exp. Cell. Res. 50: 151.

Kossasi, S., Thart, H., Van Duk and Laradier, R. 1989. J. Ethnopharmacol. 27(1/2): 1-90.

Murashige, T. and Skoog, F. 1962. A revised medium for rapid growth and bioassays with tobacco tissue cultures. Physiologia Pl. 15: 473-497.

Ogbobe, O.E. and Akan, O.V. 1993. Physico-chemical properties of the seed and seed oil of Jatropha gossipifolia. Plant foods for human nutrition V: 43(1): 197-200 (11 ref. Eng.).

Raina, A.K. and Gaikwad, B.R. 1987. Chemobotany of Jatropha sps. in India and further characterization of Curcas Oil JOTAI Oct-Dec. 81-85.

Srivastava, P.S. 1971. *In vitro* induction of triploid roots and shoots from mature endosperm of Jatropha panduraefolia, Z. Pflanzenphysiol, 66: 93-96.

Srivastava, P.S. 1973. Formation of triploid plantlets in endosperm culture of Putranjiva roxburghii. Z. Pflanzenphysiol. 69: 270-273.

Bioassay for Anti-Tumor Compounds in Callus Extracts of *Phyllanthus* Sps. and *Euphorbia* Sps.

Meena Kumar, Shobhna Saxena, M. A. Anantanandthirth,
R. K. Shah and O. P. Saxena

INTRODUCTION

Traditional remedies derived from medicinal plants are non-toxic, effective, socially acceptable and economical. Wide scope of these medicine in primary health care approaches has been recognised at national and international levels. The phytochemical analysis of plants used in folklore for treatment of cancer have yielded a number of compounds with anti-tumor activity. With this view in mind two potential plant species which are used by Ayurvedic practitioners to cure cancer/tumor were taken up to establish their callus cultures for the isolation of bioactive metabolites as well as to test the bioefficacy of callus and alkaloids content. Plants under study were *Phyllanthus fracternus* and *Euphorbia orbiculata*.

MATERIALS AND METHODS

P. fracternus and *E. orbiculata* callus cultures were raised from stem region on MS medium (Murashige and Skoog, 1962) containing 2,4-D, 1.0 ppm, Kinetin, 0.5 ppm under dark at 25° ± 2°C. The callus was further proliferated on MS medium supplemented with lower concentrations of 2,4-D along with BAP, 1 mg/l. The culture of both the species were analysed for their growth rate (GI), primary metabolites using established methods and for alkaloids using the method of Constable *et al.*, 1981. The dried callus was suspended in methanol (100 ml), stirred for 5 min. and filtered. The callus was further washed with boiling methanol (3 x 100 ml) and discarded. The combined methanol extract is concentrated in vacuo and the residue partitioned between ethyl acetate and 1N HCl and discarded. The combined aqueous portion is neutralized with sodium bicarbonate, adjusted to pH 10 with 10N NaOH and then extracted with ethylacetate. The combined ethylacetate fraction is evaporated to dryness, and the residue is triturated with ethylacetate and filtered. Evaporation of the filtrate affords the indole alkaloids. For the analysis the extract was redissolved in ethyl acetate.

Thin layer chromatography was used for separation and tentative identification. The sample was spotted on silica gel G coated plates and chromatograms were eluted in a solvent system of methanol: ethylacetate (1:9). The alkaloids were detected by means of cerric ammonium sulfate spray reagent (Farnsworth *et al.*, 1964). The reagent was made by dissolving 1 g cerric ammonium sulfate in 100 ml syrupy O-phosphoric acid under gradual heating. Before use the reagent was cooled at room temperature. For identification the chromatograms were cospotted with pure samples of expected alkaloids (Constabel *et al.*, 1981).

The bio-efficacy tests using whole callus extract and isolated alkaloids extract separately was done following the potato disc bioassay (Ferrigni *et al.*, 1982). The strain of Agrobacterium tumefaciens carrying Ti plasmid was grown on medium (0.5 g sucrose, 0.8 g nutrient broth and 0.1 g yeast extract per 100 ml) for 48 hours on shaker below 30°C. For bioefficacy tests 1.5% agar was prepared by adding 100 ml water per sample, sterilized and poured in petridishes. Washed potatoes were soaked in bleach before boring out the cylinders. Discs from these cylinders were placed (5 discs) per petridish containing 1.5% agar solution. To it was added one drop (0.05 ml) inoculum (1.5 ml water + 2 ml bacteria + 0.5 ml testing sample) per disc. Each disc was wrapped with parafilm strip, kept in dark at 27°C and tumor counted after 12 to 21 days. Per cent inhibition of crown gall tumor was calculated by following equation:

$$\% \text{ inhibition} = 100\ 1 - \frac{\text{Av. no. tumors per disc of sample}}{\text{Av. no. tumors per disc of control}} \times 100$$

RESULTS AND DISCUSSION

The plants of a number of families have been shown to accumulate a class of alkaloids with anti-HIV activity, notably castanospermine from Australian Morten Bay Chestnut tree (Stafford, 1991). There is no doubt that the biosynthetic repertoire of the plant kingdom is enormous and that many of the active metabolites produced are unique.

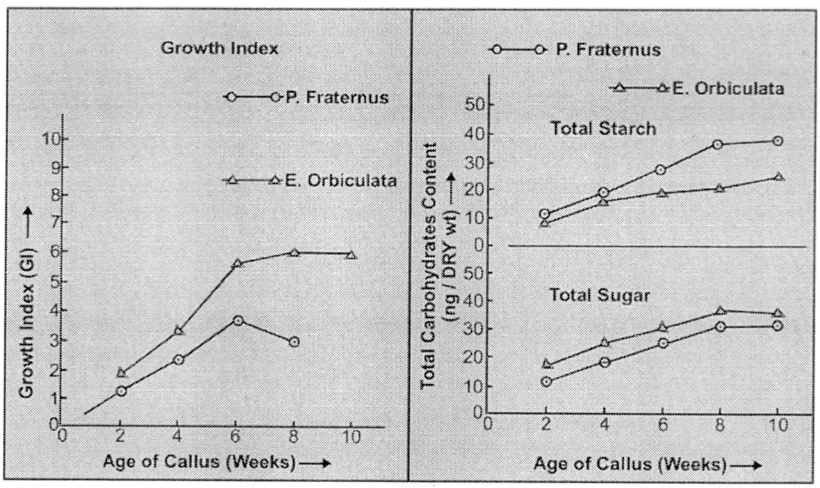

FIG. 1-2. (1) Growth Index, (2) Carbohydrate

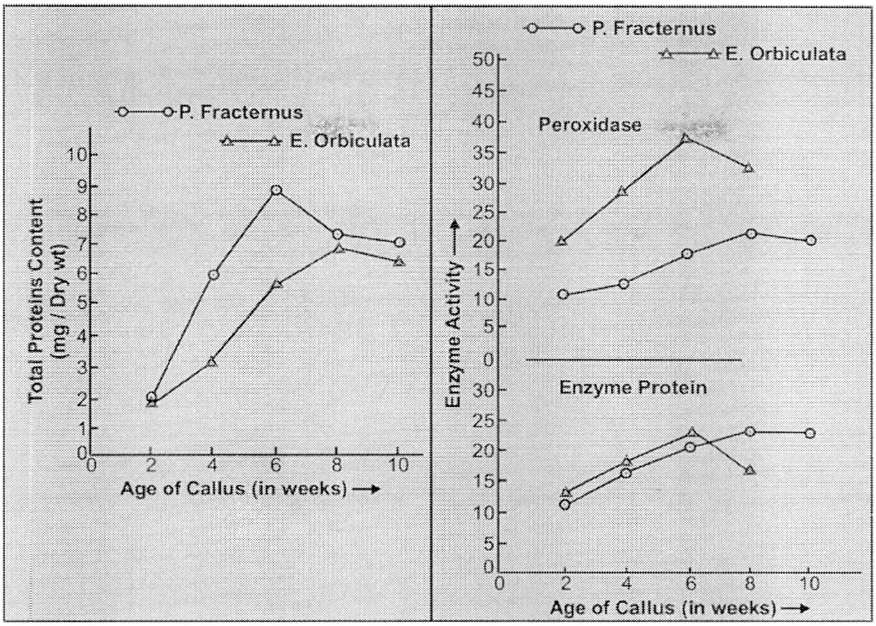

FIG. 3-4. (3) Proteins, (4) Enzymes - peroxidase, enzyme protein.

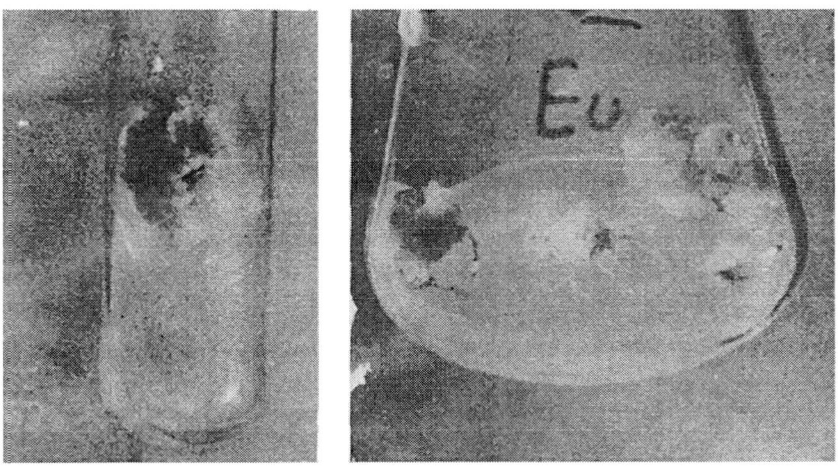

PLATE 1. (A) Calluus of P. fracternus (B) Callus of E. orbiculate L.

The callus of *P. fracternus* and *E. orbiculata* was initiated on MS medium with 2,4-D 1ppm and kinetin 0.5 ppm. But proliferation was achieved using BAP 1 mg/l with lower concentration of 2,4-D and in absence of kinetin (Plate 1). Analysis of growth at sixth week (3.70) in case of *P. fracternus* whereas it was maximum at eighth week (6.00) in *E. orbiculata* (Fig. 1).

The carbohydrates content gradually increased upto 10th week in both the plants. But the total protein content was maximum in six week old callus in both the plants (Fig. 2 and 3). Enzyme activity estimated exhibited that enzyme protein and peroxidase activity was higher at sixth week age of callus in both plants while amylase activity gradually increased upto eighth week in both the plants (Fig. 4 & 5).

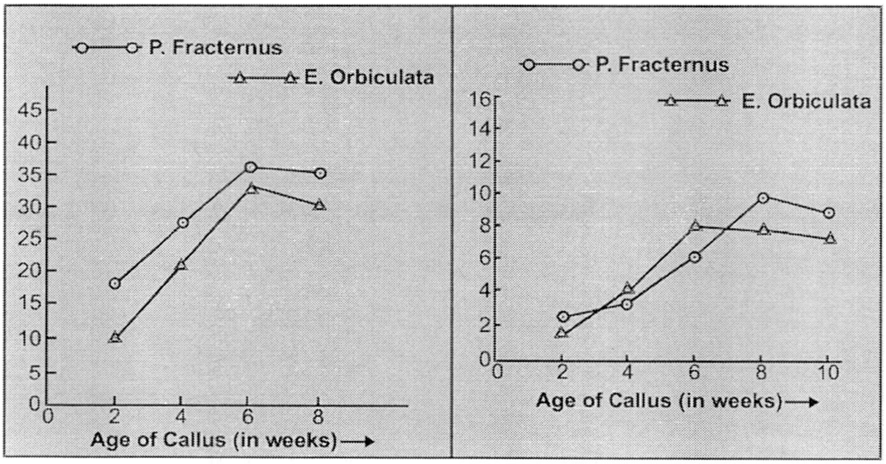

FIG. 5-6. (5) Amylase, (6) Total alkaloids contents.

Total alkaloid content estimated was also higher upto eight week but decreased in tenth week. This observation was similar for both the plants. In case of *P. fracternus* the results were supported by the data produced on *in vivo* studies of alkaloids (Jadhav *et al.*, 1993). They have reported that branches showed higher alkaloid content at 8th week stage. Earlier reports proved that in Euphorbiaceae family indole alkaloids of complicated structures are found (Manske, 1967). Hence callus was extracted using Constabel *et al.* (1981) method to get indole alkaloids. TLC examination of the chloroform extract of P. fraternus callus sample showed three spots Rf's 0.95, 0.45 and 0.25. The middle spot gave orange red colour with reagent which coincided with authentic sample of phyllalbine extracted from leaves of *P. fracternus. Euphorbia orbiculata* crude extract was mixture of five compounds from which only alkaloid which coincided with standard marker of ajmalicine (Plate 2 and Fig. 6).

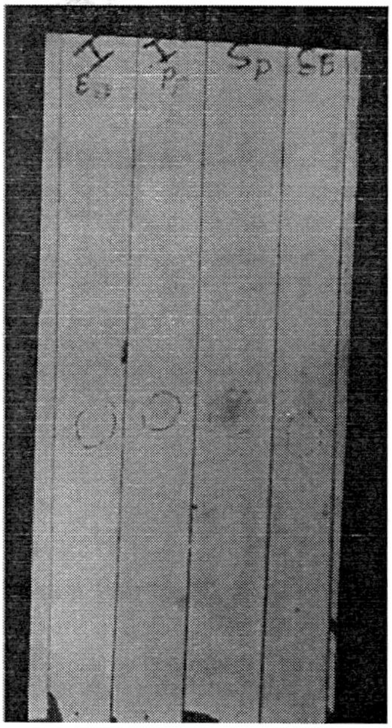

PLATE 2. TLC analysis of callus cultures of *P. fracternus* and *E. orbiculata*.

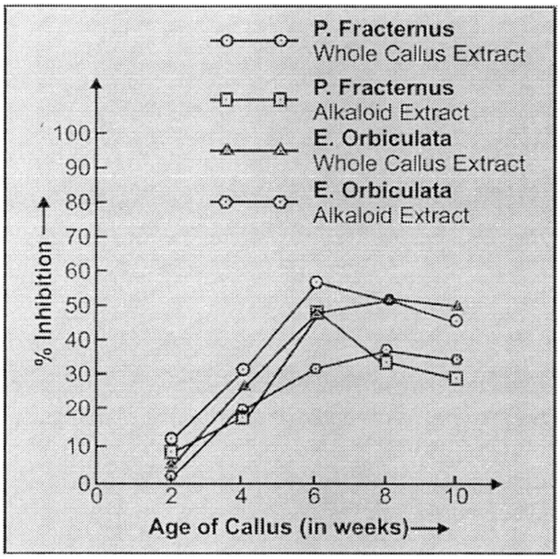

FIG. 7. Inhibition of crown gall tumors

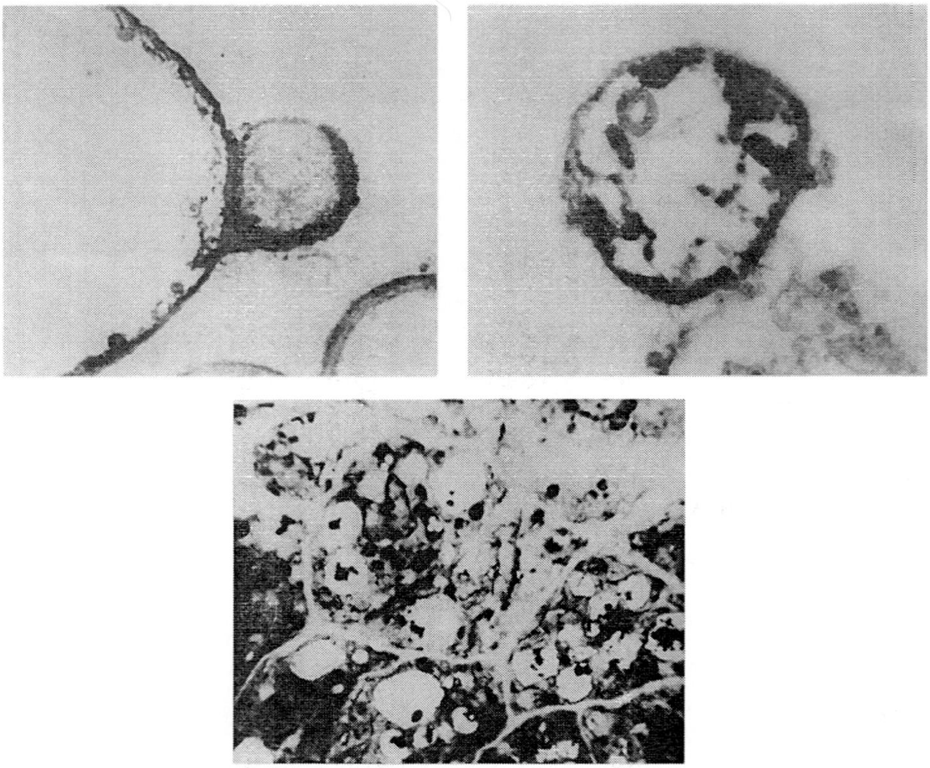

PLATE 3. TEM photographs to showTi plasmid association and per cent inhibition of tumor cells.

Bioefficacy results showed that per cent of inhibition calculated, exhibited that whole callus extract is more effective against crown gall tumor as compared to alkaloid extract of both the species under testing. The maximum inhibition (Plate 3) was by sixth week old, whole callus extract (57%) of P. fraternus followed by 52% inhibition by *E. orbiculata* whole callus extract, as well as 8th week old whole callus extract of P. fraternus. Among the alkaloid extract the maximum inhibition was 48% in *P. fracternus* six weeks old cultures followed by 37% in *E. orbiculata* 8th week culture samples (Fig. 7). The calculated kappa value for potato disc comes out to be 0.64 which represent fair to good agreement.

From the foregoing it can be concluded that these plant species possess potentiality to cure cancer/tumors. Let there be close link among Ayurvedic practitioner, botanists and chemists so that fruitful researches can be undertaken in this area of great challenge to mankind.

REFERENCES

Constabel, F., Rambold, S., Chatson, K.B., Kurz, W.G.W. and Kutney, J.P. 1981. Alkaloid production in Catharanthus roseus (L) G. Don VI. Variation in alkaloid spectra of cell lines derived from one single leaf. Plant Cell Rep. 1: 3-15.

Farnsworth, N.R., Blomster, R.N., Damratoski, D., Meer, W.A. and Camarato, L.V. 1964. Studies on Catharanthus alkaloids VI Evaluation by means of thin-layer chromatography and cerric ammonium sulfate spray reagent. J. Nat. Prod. 27: 302-314.

Ferrigni, N.R., Putnam, J.F., Anderson, B., Jacobsen, L.B., Nichols, D.E., Moore, D.S., Mclaughlin, J.L., Powell, R.G. and Smith, C.R. Jr. 1982. Modification and evaluation of the potato disc assay and antitumor screening of Euphorbiaceae seeds J. Nat. Prod. 45: 679-685.

Jadhav, R.T., Krishnamurthy, R., Joshi, N. and Parabia, M.H. 1993. Observations of the crude alkaloid content and plant age relationship in Phyllanthus fraternus Webster. National Symposium on medicinal plants, April 3-4, Surat.

Manske, R.H.F. 1967. The alkaloid chemistry and physiology. Acad. Press Inc. London, 9: 270.

Mclaughlin, J.L., Chang, C.J. and Smith, D.L. 1991. Bench-Top Bioassays for the discovery of bioactive natural products: an update. Studies in Natural Products Chemistry, 9: 383-406.

Murashige, T. and Skoog, F. 1962. A revised medium for rapid growth and bioassays with tobacco tissue cultures. Physiol Plant 15: 473-479.

Stafford, A. 1991. Natural products and metabolites from plants and plant tissue cultures in: Plant cell and tissue culture (eds. A Stafford and G. Warren) Open University Press Milton Keynes, 125-161.

Micropropagation of *Butea monosperma* (Lam) Taub.

Kavitha R. Kulkarni, Smitha Hegde and L. D' Souza

INTRODUCTION

Butea monosperma Lam (*Butea frondosa* Koen) is a deciduous tree belonging to the family Leguminosae (Papilionaceae). In summer the gorgeous canopy of scarlet flowers look like flames giving the tree its characteristics name the flame of the forest. *Butea* is a frost hardy, drought resistant tree. Due to heavy deforestation, the trees are facing extinction especially along the west coast of India. *Butea* would be very useful for afforestation of eroded, degraded lands and has a great potential as an ornamental and shade tree in parks and gardens. Leguminous trees besides being able to meet the demands for fodder, fuel and timber hold some promise as a means of maintaining soil nitrogen without the use of fertilizers (Mishra and Prasad, 1980). International bodies like FAO, NAS, ICRAF are actively promoting the planting of woody legumes (Pokhriyal *et al.*, 1990).

MATERIAL AND METHODS

Mature trees of *Butea frondosa*, from Thumbe forest about 15 Km away from Mangalore were selected for the experiments. Ripe pods and young terminal portions of the branches bearing 4-5 nodes were used for the studies. The twigs were washed under running tap water for 15 minutes, soaked in 1% ITEOL-3 for 10 minutes and treated with a solution of 0.1% $HgCl_2$ and 0.1% sodium lauryl sulphate for 5 minutes and finally washed with distilled water 3 times. Nodal segments 1 cm in length were used as explants. The nodes were first cultured on salt free medium with 6 g/l agar, 5 g/l sucrose and 60 mg/l ascorbic acid to control browning. The seeds were germinated *in vitro*. The cotyledonary nodes, cotyledons and hypocotyl segments of 10 day old seedlings were used as explants. The explants were cultured on MS medium (Murashige and Skoog, 1962), half and full strength woody plant medium (Lloyd and McCown, 1980) with various growth regulators. The cultures were incubated in a culture room having a temperature of 28 ± 2°C and relative humidity of 75% ± 5%. A 16 hour photo period with a photon flux density of 10-15 m $E/m^{-2}/S^{-1}$ was provided by cool day light fluorescent tubes.

RESULTS

Multiple shoot induction

Multiple shoots were initiated from axillary buds of mature trees of *Butea* on WP medium with BAP. These were limited in number, only few shoots could be got from an axillary bud (Table 1). The shoot tips turned brown and became necrotic.

TABLE 1: Number of shoots from axillary buds of mature trees of *Butea* on WP medium supplemented with BAP

BAP mg/l	Response of explants (%)	Mean no. of shoots[1]
2.5	91.44	9.72 ± 0.21
5	84.92	4.02± 0.01
10	87.42	5.91 ± 0.91

[1]Each value represents mean ± SE of 3 experiments each having 10 replications.

A large number of multiple buds could be obtained from cotyledonary nodes on half strength WP medium supplemented with BAP. The tips of the shoots turned brown and became necrotic. Browning of shoot tips could be controlled by incorporation of fructose in the medium (Table 2).

TABLE 2: Number and nature of shoots obtained from cotyledonary nodes of *Butea* on half strength WP medium supplemented with BAP, with and without fructose.

BAP (mg/l)	Fructose (g/l)	Response of explants (%)	Mean no. of shoots[1]	Remarks
10	-	94.83	22.12 ± 0.73	Necrosis of shoot tips
10	10	74.83	22.12 ± 0.24	No necrosis of tips

[1]Each value represents mean ± SE of 3 experiments each having 10 replications.

Callus and Morphogenesis

Cotyledon segments gave rise to white friable callus on MS medium with 5 mg/l kinetin. The callus gave rise to shoots some of which produced roots. Green nodular callus was formed from the cotyledon segments on MS medium with 5 mg/l BAP. The nodular callus gave rise to embryoids which germinated when subcultured on the same medium giving rise to plantlets with weak roots (Table 3).

Tuber formation

In nature root tubers are produced in *Butea* which give rise to new plants. Experiments were conducted to induce such tubers *in vitro*. The hypocotyl of ten day old seedlings was cut off just below the cotyledon and the seedlings were cultured on MS medium with Kin and phloroglucinol. The cut surface of the hypocotyl produced a mass of cells which developed into globular tuber like structures. These tubers when subcultured on WP medium supplemented with BAP, phloroglucinol, gave rise to shoots. Secondary tubers were formed from the primary tubers.

TABLE 3: Nature and morphogenetic response of callus from cotyledon section of *Butea* on MS medium with kinetin and BAP

Cytokinin	Concentration (mg/l)	Nature of callus	Morphogenetic response
Kinetin	5	White friable	Shoots and roots
BAP	5	Green nodular	Plantlets

Rooting

Shoots from axillary buds as well from tubers do not root easily. Roots could be induced in a few shoots by planting them in moist sand supplemented with $MnSO_4$ and IBA. Rooted plants have been transferred to pots in the green house and also to the field.

ACKNOWLEDGMENTS

The project was funded by grants received from the DST. Laboratory facilities provided by the Mangalore Jesuit Educational Society and the technical assistance given by J. Alexander are gratefully acknowledged.

REFERENCES

Lloyd, G. and McCown, B.H. 1980. Commercially feasible micropropagation of mountain laurel, Kalmia latifolia by use of shoot tip cultures. Proc. Intl. Plant Prop. Soc. 30: 421-427.

Mishra, J. and Prasad, U.N. 1980. Agricultural and silvicultural studies on raising of oil seeds like *Sesamum indicum* L., *Arachis hypogea* L. and *Glycine max*. Merril as cash crop in conjunction with *Dalbergia siso* and *Tectona grandis* L. at Ranchi. Indian Forester, 106: 675-695.

Murashige, T. and Skoog, F. 1962. A revised medium for rapid growth and bioassay with tobacco tissue cultures. Physiol Plant 15: 473-497.

Pokhriyal, T.C., Bhandary, H.C.S., Negi, D.S., Chaykiyal, S.P. and Gupta, B.B. 1990. Identification of some fast growing leguminous tree species for nitrogen fixation studies. Indian Forester 6: 504-510.

Rapid Propagation of *Eucalyptus tereticornis* Sm. Through Tissue Culture

R.I .S. Gill, S. S. Gill and S. S. Gosal

INTRODUCTION

Eucalyptus offers an ideal industrial feed stock for pulp and paper due to its short rotation cycle, by virtue of which it provides a competitive advantage against the increasing prices of superior quality, long duration timbers, such as *Tectona grandis, Dalbergia sissoo, Shorea robusta* etc. *Eucalyptus* timber is also suitable for various types of construction (for example door and window frames). To meet the increasing demand for timber in the future, fast growing *Eucalyptus* plantations are expected to make a great contribution (McComb and Bennett, 1986).

For the production of uniform plant material *Eucalyptus*, being cross-pollinated, needs to be multiplied vegetatively (Gill *et al.*, 1994). This is necessary in view of the heterogeneity observed in plantations raised from seed. In recent years, tissue culture technology has been successfully used in some *Eucalyptus* species viz., *E. citriodora* (Gupta *et al.*, 1981), *E. resinifera* and *E. maculata* (McComb and Wroth, 1986), *E. sideroxylon* (Burger, 1987), *Eucalyptus* species and hybrids (Machado and Assis, 1990), *E. camaldulensis* (Periasamy and Srinivasan, 1990) and *E. grandis* (Teixetra and De Silva, 1990) for the production of clonally uniform plantlets from the desirable germplasm. However, the success of exploiting this technique commercially has not been achieved in all the *Eucalyptus* species of commercial value. This communication deals with the establishment of shoot cultures of *E. tereticornis* by taking explants from mature plus trees and the study of various factors affecting mass multiplication of vegetative parts through tissue culture.

MATERIALS AND METHODS

The mature trees of *Eucalyptus tereticornis* (6-8 years) were cut during February from the forest area of Department of Forestry & Natural Resources, Punjab Agricultural University, Ludhiana. The coppice shoots developed within two months. The twigs (10-15 cm) comprising 5-8 nodes were excised from the stump. The leaves were pinched off and the twigs having nodal buds were first washed thoroughly in distilled water to remove the dust particles and then cleaned with a swab of cotton dipped in alcohol (absolute). The twigs were cut into stem segments (1.5 cm). The segments consisting of leaf petiole and nodal bud were surface sterilized (0.1% Mercuric chloride, 8 minutes) and washed thrice with sterile water. The material was left in water at the final washing. The nodal segments were cultured on different media modifications based on Murashige and Skoog (1962) medium (Table 1). The cultures were incubated at 25 ± 2°C under 16 h light and 8 h dark conditions.

TABLE 1: *In vitro* shoot bud regeneration (%) from nodal segment explant of *Eucalyptus* cultured on different media compositions.

Kind of growth regulator (mg/l) in MS media	Weeks after culturing	Mean	
	3WAC	6WAC	
Basal (MS)	31.66 (19/60) 34.13	35.0 (21/60) 36.22	33.33 (20/60) 35.17
BAP (1.0)	36.66 (22/60) 37.24	40.0 (24/60) 39.19	38.33 (23/60) 38.21
BAP (0.5)	16.66 (10/60) 24.03	23.33 (14/60) 28.76	20.0 (12/60) 26.40
BAP (1.0) + IBA (0.5)	43.33 (26/60) 41.14	46.66 (28/60) 43.05	45.0 (27/60) 42.10
BAP (1.0) + IBA (0.2)	25.0 (15/60) 29.91	28.33 (17/60) 32.12	26.66 (16/60) 31.02

L.S.D (P = 0.05) Media = 3.61
Weeks after culturing = 2.28
Interaction (Media x WAC) = NS

Figures in parenthesis indicate the number of responsive cultures (numerator)/ total number of cultures (denominator)., Figures given in italics are Arc Sine transformed values.

Tissue culture of *Eucalyptus* undergoes 4 main stages- initiation, multiplication, rooting and establishment. The shoot bud initiation takes 3-4 months for establishment of cultures. The multiplication of shoot buds was carried out in high cytokinin medium. The multiplication rate increased upto 3 to 4 times, when the cultures were transferred to low cytokinin medium during subcultured period of 4 weeks. For shoot elongation, the subdivided shoot but clumps were transferred on half-strength MS medium having activated charcoal (0.2%). The elongated shoots (3-4 cm) were transferred to rooting medium. Rooting occurred within 10 to 15 days on both solid as well as liquid medium. The rooted plantlets thus, obtained were hardened and established in open conditions.

TABLE 2: *In vitro* rooting (%) of tissue culture derived shoots of *Eucalyptus* in liquid media.

Kind of auxin (mg/l) in 1/2 MS media	Days after culturing		Mean
	10 DAC	20 DAC	
IBA (0.5)	61.66 (37/60) 51.73	85.0 (51/60) 67.38	73.33 (44/60) 59.55

Kind of auxin (mg/l) in 1/2 MS media	Days after culturing		Mean
	10 DAC	20 DAC	
IBA (1.0)	43.33 (26/60) 41.15	63.33 (38/60) 52.77	53.33 (32/60) 46.96
IBA (1.0) + NAA (0.2)	20.0 (12/60) 26.25	20.0 (12/60) 26.25	20.0 (12/60) 26.25
IBA (0.5) + NAA (0.2)	23.33 (14/60) 28.77	26.66 (16/60) 30.93	25.0 (15/60) 29.85

L.S.D. (P = 0.05) Media = 5.68
Days after culturing = 4.01
Interaction (Media x DAC) = 8.03

Figures in parenthesis indicate the number of responding cultures (numerator)/ total number of explants cultured (denominator). Figures given in italics are Arc Sine transformed values.

The data given in Table (1 and 2) is means of three replicate batches of 20 cultures each. All the results were analysed statistically using ANOVA and means were tested for significance using least significance difference (LSD).

RESULTS

The per cent shoot bud regeneration from nodal segments of *E. tereticornis* varied more with media composition and less with the duration of incubation during culture (Table 1). The mean per cent shoot bud regeneration from the nodal segment was significantly better on basal MS medium than the medium containing BAP (0.5) or a combination of BAP (1.0) and IBA (0.2). The explants cultured on BAP (1.0) showed significantly better per cent shoot bud regeneration than basal MS and the medium supplemented with either BAP (1.0) + IBA (0.2) or BAP (0.5). The maximum mean per cent shoot bud regeneration (45.0) was obtained when the explants were cultured on MS medium supplemented with BAP (1.0) + IBA (0.5) and minimum (20.0) on BAP (0.5).

The duration, for which the explants were incubated, exhibited less pronounced influence on the response of cultures than the composition of media. However, the present shoot bud regeneration was significantly better after 6 weeks of culturing than 3 weeks. The multiple shoot buds were subcultured either on multiplication medium for mass establishment of bud cultures (by bud proliferation) or on shoot regeneration medium for obtaining shoots. For multiplication, shoot bud clumps were subcultured on MS medium supplemented with BAP (0.5), however, for shoot regeneration, these bud clumps were subcultured on half-strength MS medium supplemented with IBA (1.0) and activated charcoal (0.2%). The shoots, with well developed leaves, were obtained from the shoot bud clumps within 4 weeks. The shoots (3-5 cm) thus obtained were either cut into nodal segments or recultured on MS medium supplemented with BAP (1.0) and IBA (0.5) for multiple bud formation or rooted *in vitro*.

The *in vitro* developed shoots cultured on rooting medium exhibited varied mean per cent rooting with incubation duration of cultures and media composition (Table 2). The rooting of elongated shoots was achieved on both solid and liquid medium. However, the roots developed on solid medium were thin coil type. The survival percentage of such plantlets in open conditions was less (38.5) as compared to plantlets developed on liquid rooting medium (82.0). The difference w.r.t. mean per cent rooting and incubation duration was non-significant among media having a combination of IBA and NAA. The maximum rooting percentage (85.0) was observed 20 DAC in shoots cultured on half-strength MS medium supplemented with IBA (0.5) and sucrose (2.0%). Rooting percentage 20 DAC was significantly better than 10 DAC in both the media having only IBA (Table 2).

The complete plantlets thus developed were partially hardened on filter paper in liquid media. Further, the liquid media was replaced by distilled autoclaved water and the tight cotton plug with loose cotton swab. After keeping the plantlets in water for two days, they were transferred to pots. The pots were kept covered with moist muslin cloth for 15 days and then the cover was completely removed gradually.

DISCUSSION

Micropropagation of *E. tereticornis* is well established, however, in most of the cases the explant was taken from seedling (Ilahi and Jamal, 1987; Kumar and Ayyapan, 1987) rather than from mature trees (Das and Mitra, 1990). Although it is easy to develop protocol from seedlings but the plants of identified desirable characters can't be produced. Thus the method for vegetative propagation of *E. tereticornis* reported in this paper through axillary branching of nodal segments is ideal for rapid propagation of selected genotypes. The plantlets thus obtained directly from the multiple buds help in retaining clonal identity as reported by Broertjes and Keen (1980). This method of shoot proliferation from mature trees would allow selection and multiplication of superior genotypes under field conditions.

REFERENCES

Broertjes, C. and Keen, A. 1980. Adventitious shoots: do they develop from one cell? *Euphytica*, 29: 73-87.

Burger, D.W. 1987. *In vitro* micropropagation of *Eucalyptus sideroxylon*. Hort. Science 22 (3): 496-497.

Das, T. and Mitra, G.C. 1990. Micropropagation of *Eucalyptus tereticornis* Smith. Plant Cell, Tissue and Organ Cult., 22(2): 95-103.

Gill, R.I.S., Gill, S.S. and Gosal, S.S. 1994. Vegetative propagation of *Eucalyptus tereticornis* Sm. through tissue culture. Plant Tissue Cult. 4 (1): 59-67.

Gupta, P.K., Mascarenhas, A.F. and Jagannathan, G. 1981. Tissue culture of forest trees- Clonal propagation of mature trees of E. citriodora Hook by tissue culture. Plant Sci. Lett. 20: 195-201.

Ilahi, Ihsan and Shabana, Jamal. 1987. Mass propagation of *Eucalyptus tereticornis* Smith. Pakistan J. Bot. 19 (1): 67-74.

Kumar, R.R. and Ayyapan, P. 1987. Rapid clonal multiplication of *Eucalyptus camaldulensis* Dehn. through *in vitro* culture. Planters Chronicle 81 (7): 225-227.

Machado, M.A. and Assis, T.F. 1990. Large scale micropropagation of Eucalyptus species and hybrids. Abstracts VIIth International Congress on Plant Tissue and Cell Culture (IAPTC), Amsterdam, A-3-123.

McComb, J.A. and Wroth, M. 1986. Vegetative propagation of *Eucalyptus resinifera* and *E. maculata* using coppice cutting and micropropagation. Australian Forest Research, 16 (3): 231-242.

Periasamy, K. and Srinivasan, S. 1990. Shoot bud formation from leaf explants of *Eucalyptus camaldulensis* in modified MS medium. Abstracts VIIth International Congress on Plant Tissue and Cell Culture (IAPTC), Amsterdam, A-3-154.

Teixetra, S.L. and De Silva, L.L. 1990. *In vitro* propagation of adult *Eucalyptus grandis* Hill Ex. Maiden from epicormic shoots. Abstracts VIIth International Congress on Plant Tissue and Cell Culture (IAPTC), Amsterdam, A-3-218.

Regeneration of Plantlets Through Tissue Culture Methods in Dicotyledonous Fruit Trees - A Brief Review

L. K. Pareek and Shashi Mathur

INTRODUCTION

Compared to herbaceous plant species, a few woody trees have been successfully, propagated *in vitro*. According to Abbott (1977), woody trees were assumed to be intractable in culture. The first report of organogenesis in tree tissue culture was made by Gautheret (1940). Fully organized plantlets of tree sps. propagated *in vitro* and capable of transfer to soil were first obtained in *Populus tremuloides* (Winton, 1968). Since then micropropagation techniques have been successfully used to regenerate plantlets of a large number of tree species. Tissue culture of fruit and forest trees and its importance in clonal improvement has also been covered in several reviews (Durzan and Campbell, 1974; Winton and Huhtinen, 1976; Reinert and Bajaj, 1977; Sommer and Brown, 1979; Rao *et al.*, 1980; Mott and Zimmermann, 1981; Rachie and Lyman, 1981; Sommer and Caldas, 1981; Jones, 1983; Broome and Zimmermann, 1984; Sommer and Wetzstein, 1984; Hammerschlag, 1986a; Litz, 1985; Litz *et al.*, 1985; Zimmermann, 1985; Bajaj, 1986; Boxus and Druart, 1986; Hammerschlag, 1986b; Litz *et al.*, 1986; Zimmermann, 1986; Bonga, 1987; Boulay, 1987, 1987; Fiorino and Loreti, 1987; Litz, 1987.

The nature of problems associated with tissue culture of trees have been reviewed by Bonga (1982).

The difficulties encountered in developing *in vitro* systems for woody plant regeneration are.

a) establishment of explants in culture due to the oxidation of polyphenols from explants (Durand-Cresswell *et al.*, 1982).
b) absence of juvenility in the explants of mature trees. (Bonga, 1982).
c) induction of roots in culture.

Some of the procedures used to face the problem of polyphenol oxidation are:

1. addition of antioxidants such as polyvinylpyrrolidone, ascorbic acid, citric acid, cysteine HCl, activated charcoal, reduced glutathione to the culture medium. (Ichihashi and Kako, 1977; McComb and Newton, 1981; Stevenson and Harris, 1980).
2. incubating the cultures in reduced light or darkness (Adams *et al.*, 1979; McComb and Newton, 1981; Monaco *et al.*, 1977; Wang and Huang, 1974).
3. frequently transferring explants on to a fresh medium whenever browning of the medium is observed (Broome and Zimmermann, 1978; Litz, 1984c; Lloyd and McCown, 1980).
4. reducing the concentration of nitrogen containing compounds (Anderson, 1975; Chevre *et al.*, 1983).

More then 60 different media have been used for tissues of various woody species. Most commonly used are that of Murashige and Skoog (1962); Miller (1965); Woody plant medium of Lloyd and McCown, 1980; Hoagland's medium (Hoagland and Arnon, 1950); Linsmaier and Skoog (1965); Lepoivre medium (Oruoirina nd Lepoivre, 1973); White's medium (White, 1963).

MS, LS and Miller formulations contain high level of medium salts including high ammonium where as Hoagland's, White's contain low level of salts and no ammonium. Compared with MS, woody plant medium is low in ammonium, nitrate, potassium, chloride and high in sulphate. Skirvin (1980), Griffis *et al.*, (1983) found beneficial to reduce the strength of MS medium by half. Increasing the concentration of reduced nitrogen in the medium is reported better for organogenesis (Kohlenbach, 1977; Lee, 1984; Thomas and Street, 1972). Several fruit trees have a high manganese requirement (Quoirin and Lepoivre, 1977; Hammerschlag, 1980). A variety of nitrogenous compounds such as asparagine (Nitsch, 1965), casein hydrolysate (Skirvin *et al.*, 1980), malt extract (Toledo *et al.*, 1980), yeast extract (Watanabe *et al.*, 1981), glutamine (Litz, 1984c) have been found beneficial in fruit tree tissue cultures.

Though regeneration *in vitro* has been achieved in some fruit trees using embryonic or seedling tissues as explants for callus induction and organogenesis, it has been difficult to regenerate plantlets using tissues of mature trees due to the loss of regenerative potential in such tissues (Bonga, 1982). Some trees which have been regenerated using mature explants are *Citrus sinensis* (Barlass and Skene, 1982) from stem via organogenesis, *Carica stipulata* (Litz and Conover, 1980) from peduncle via embryogenesis. *Coffea arabica* (Martinez and Madrigal, 1986) from leaf explants via embryogenesis, but these appear to be exceptions. Fruit mesocarp tissues have also been used for callus induction in avacado (Schroeder, 1961) and sapodilla (Bapat and Narayanaswamy, 1977), plantlet regeneration from such callus has never been reported.

Since the ability to propagate depends on the physiological juvenility of the explant, there are various reports of micropropagation of fruit trees using shoot tip of mature tree in *Malus* (Welander, 1986), *Prunus* (Chornu and Chaiz, 1981; Rugini and Verma, 1982), *Carica pentagona* (Pandey and Rajeevan, 1982), nodal explants of *Eucalyptus* (Lakshmi-Sita and Shobha Rani, 1985; McComb and Benett, 1982).

Regeneration in tree species has been achieved using juvenile tissues within mature tree as inflorescence of *Howeia forsteriana* and *Chamaedorea costaricana* (Reynolds and Murashige, 1979).

Regeneration has been achieved in fruit tree species in which there is naturally occurring polyembryony. Adventitious embryos are produced in vivo from the inner integument and nucellus in ovules of several fruit trees. Since the original report by Rangaswamy (1958) that embryogenic callus could be induced from the nucellus excised from fertilized polyembryonic *citrus* ovules, a few tropical fruit trees have been regenerated *in vitro* from the nucellus or adventive embryos or nucellar callus or whole fertilized or unfertilized ovules. Rangan *et al.* (1968, 1969) first reported the direct induction of somatic embryos from nucellus of monoembryonic *citrus*. He found the *response of monoembryonic Citrus* similar to that of polyembryonic species. There are various other reports of somatic embryogenesis in nucellar explants from fertilized ovules of monoembryonic *Citrus* (Bitters *et al.*, 1970; Navarro and Juarez, 1977). Somatic embryogenesis from nucellar explants of abortive ovules of seedless novel orange (Bitters *et al.*, 1970), unfertilized ovules of Shamouti and Valencia sweet orange (Kochba *et al.*, 1972) have been reported. There are various other reports of somatic embryogenesis from fertilized ovules of Shamouti orange (Spiegel-Roy and Kochba, 1973) unfertilized ovules of *Citrus* species (Kobayashi *et al.*, 1985; Vardi *et al.*, 1982). The *in vitro* response of *Citrus* nucellar callus hjas been the subject of several reviews (Button and Kochba, 1977; Spiegel-Roy and Kochba, 1980; Rangan, 1984, 1984; Spiegel-Roy and Vardi, 1984; Litz, 1985).

Somatic embryogenesis has been reported from nucellar callus in *Vitis vinifera* (Mullins and Srinivasan, 1976), *Malus domestica* (Eiccholtz, *et al.*, 1979), *Mangifera indica* (Litz *et al.*, 1984), *Eriobotrya japonica* (Litz unpublished). Litz and Schaffer (1986) have reported the *in vitro* response of nucellus of mango to be cultivar dependent, and on the type i.e. ovule of polyembryonic or monoembryonic mango and stage of ovule development.

Litz (1984b, a) has reported somatic embryogenesis from adventive embryos of *Eugenia* species and *Myrciaria cauliflora*.

Regeneration of plantlets from cell suspension has been achieved with Santalum (Lakshmi Sita *et al.*, 1980, Bapat and Rao, 1984); *Mangifera indica* (Litz, 1984c); *Eucalyptus* (Muralidharan *et al.*, 1989, Teulieres *et al.*, 1989), *Coffea arabica* (Sondahl and Sharp, 1977), Theobroma cacao (Kong and Rao, 1982); *Prunus* (Durzan, 1984), *Citrus* (Kochba *et al.*, 1982), Eugenia jambos (Phillip and Schraudolf, 1981). Embryo cultures in fruit tree species were first used by Tukey (1933) to overcome seed abortion in *Cherry*. Since then culture of excised embryos has been employed as a useful tool for direct regeneration in trees where seeds are dormant or recalcitrant or where embryo aborts due to degeneration of endosperm as in *Prunus domestica* (Bini and Bellini, 1973), *Prunus avium* (Zdrujkovskaja-Richter, 1981); *Prunus persica* (Jordan, 1975; toledo *et al.*, 1980; Dhanju *et al.*, 1983); *Santalum album* (Rao, 1965; Rao and Rangaswamy, 1971); a disease resistant interspecific hybrid of *Carica* (Rojkind *et al.*, 1982); *Persea americana* (Skene and Burlass, 1983); Makapuno coconut cultivar (Krikorian, 1982; Reynolds, 1982).

Since the first report or production of haploid plants by Guha and Maheshwari, 1964, haploid tissue cultures have been used for obtaining homozygous diploid plants in many species. Haploids are also useful for recessive mutant screening and the study of genetic linkages (Winton and Stettler, 1974; Sondahl and Sharp, 1979). Plantlets of tree species have been obtained either through direct embryogenesis of pollen (Radcjevic, 1978) or through organogenesis in callus of androgenetic origin (Wang *et al.*, 1975; Chen *et al.*, 1978, 1979). Embryogenesis has been reported in pollen cultures of *Citrus aurantifolia* (Chaturverdi *et al.*, 1982); *Cocos nucifera* (Than- Tuyen and de Guzman, 1983), *Malus* (Kubicki *et al.*, 1975); Organogenesis in *Prunus cerasifera* (Seirlis *et al.*, 1979).

Triploid plants have been reported in tree species from cultured endosperm. Some examples of plantlets regenerated from tree species are from endosperm of *Santalum album* (Lakshmi-Sita *et al.*, 1980); *Putranjiva roxburghii* (Srivastava, 1982). Regeneration of new polyploid trees from polyploid cells in culture has a potential application in genetic improvement of trees (Winton, 1971; Mehra and Mehra, 1974; Lee and Rao, 1980).

Although tree protoplasts have been isolated from the juvenile parts and callus cultures in *Prunus persica* (Mouras *et al.*, 1978); *Carica papaya* (Litz and conover, 1979; Litz, 1984d); *Eucalyptus gunnii* and *E. pauciflora* (Venketeswaran and Gandhi, 1980), embryoids and plantlets have been obtained only in apple (Kouider *et al.*, 1984c) and *Citrus* (Vardi *et al.*, 1982), *Manihot esculenta* (Shahin and Shepard, 1980) respectively. Somatic hybrids have been obtained from *Citrus sinensis* and *C. unshiu* (Kobayashi *et al.*, 1988). Callus formation has been reported from protoplasts of *Malus domestica* (Doughty and Power, 1988), *Pyrus communis* (Ochatt and Power, 1988), *Santalum album* (Lakshmi-Sita and Shobha Rani, 1983).

Nag and Street (1973) first reported plantlet regeneration of *Carrot* after freezing callus cultures in liquid nitrogen. A few tree species that have been regenerated from cryopreserved and subsequently thawed tissues are *Citrus* spp. (Bajaj, 1984), *Elaeis guineensis* (Grout *et al.*, 1983), *Malus domestica* (Katano *et al.*, 1983), Manihot (Kartha *et al.*, 1982), *Phoenix dactylifera* (Tisserat *et al.*, 1981), *Pyrus communis* (Sakai and Nishiyama, 1978), *Ulmus americana* (Ulrich *et al.*, 1984).

Plantlets regenerated *in vitro* from shoot apical meristems are disease free (Walkey, 1978). Tree species that have been freed of diseases by *in vitro* culture of shoot tips from young seedlings or from older plants include *Citrus* (Bitters *et al.*, 1972), *Vitis vinifera* (Barlass *et al.*, 1982), *Prunus* sps. (Boxus and Quoirin, 1977), apple (Hansena nd Lane, 1983), apricot (Pena Iglesias and Ayoso (1978), peach (Navarro *et al.*, 1983), Papaya (Sharma and Skidmore, 1988).

Regeneration of plantlets via adventitious shoot bud formation has been reported in several tree species. A brief review of regeneration of plantlets via somatic embryogenesis or organogenesis or shoot proliferation using preexisting meristematic tissue of dicotyledonous fruit tree species has been given in the Table-1.

TABLE 1.

Species	Growth Substance								Reference
	Explant	Medium	Callus	Proliferation of shoots	Regeneration Organogenesis	Embryogenesis	Formation of Plant	Plants to Soil	
Carica papaya	Petiole (S)	White	NAA, 2ip	-	-	NAA, BA	No	No	De Bruijne et al., 1974.
C. papaya	Shoot-tip (M)	MS	-	NAA, BA	-	-	Yes	Yes	Litz & Conover, 1977, 1978; P&ey & Rajeevan, 1982.
C. papaya	Stem (S)	MS	NAA, Kn	-	IAA, Kn	IAA, Kn	No	No	Yie & Liaw, 1977.
C. papaya	Shoot-tip (S)	MS	-	IAA, BA	-	-	Yes	Yes	Yie & Liaw, 1977.
C. papaya	Stem (S)	MS	NAA, Kn	-	IAA, Kn	-	Yes	Yes	Arora & Singh, 1978.
C. papaya	Anthers	MS + Glu.	-	-	NAA, BA	-	Yes	Yes	Litz & Conover. 1978.
C. papaya	Ovule	White	CW	-	-	None	Yes, Few	No	Litz & Conover 1982.
C. papaya	Cotyledon (S)	MS	NAA, BA	-	NAA, BA	-	Yes	Yes	Litz et al., 1983.
C. papaya	Anther	MS	NAA, BA	-	-	None	Yes	No	Tsay & Su, 1985.
C. papaya	Lateral Bud	MS	-	BA	-	-	Yes	Yes	Rajeevan & P&ey, 1986.
C. papaya	Shoot apex	MS	-	BA, NAA	-	-	Yes	NS	Sharma & Skidmore, 1988.
C. pentagona	Shoot-tip	MS + CH	-	BA, AS	-	-	Yes	Yes	Cohen &

Species	Growth Substance								Reference
	Explant	Medium	Callus	Proliferation of shoots	Regeneration Organogenesis	Embryogenesis	Formation of Plant	Plants to Soil	
	(M)								Cooper, 1982.
C. stipulata	Peduncle (M)	MS	NAA, BA	-	-	-NAA, BA, AC	Yes	Yes	Litz & Conover, 1980.
C. papaya x C. cauliflora	Ft. Ovule	White	CW	-	-	CW	Yes	No	Litz & Conover, 1981.
Citrus microcarpa	Ft. Ovule	White + CH	None	-	-	None	Yes	NS	Rangaswamy, 1958.
C. grandis	Nucellus of Ft. Ovule	MS+ Orange juice	-	-	-	AS, NAA	Yes	Yes	Rangan et al., 1968, 1969.
C. limon	Nucellus of Ft. Ovule	MS+ Orange Juice	-	-	-	AS, NAA	Yes	Yes	Rangan et al., 1968, 1969.
C. limon	Nucellus of Ft. Ovule	MS + ME	-	-	-	None	Yes	NS	Bitters et al., 1972.
C. limon	Stem (S,M)	MS	-	-	Kn, NAA	-	Yes	NS	Bouzid, 1975.
C. limon	Root (S, M)	MS	-	-	BA, 2,4-D	-	Yes	NS	Sauton et al., 1982.
C. microcarpa	Ft. Ovule	White + CH	None	-	-	None	Yes	NS	Rangaswamy, 1958.
C. sinensis	Nucellus of Ft. Ovule	MS + Orange juice	-	-	-	AS, NAA	Yes	Yes	Rangan et al., 1968, 1969.
C. sinensis	Nucellus of Ft. Ovule	MS + ME	-	-	-	None	Yes	NS	Bitters et al., 1972.
C. sinensis	Nucellus of unft. Ovule	MT + ME	-	-	-	IAA, Kn	Yes	Yes	Kochba et al., 1972.
C. sinensis	Nucellus	MT + ME	-	-	-	Adenine	NS	NS	Kochba & Spiegel-Roy, 1973.
C. sinensis	Stem(S) Leaf (S)	MS + ME	Kn, NAA, 2,4-D	-	BA, NAA	-	Yes	Yes	Chaturvedi & Mitra, 1974.
C. sinensis	Stem (S,M)	MS	-	-	BA, IBA	-	Yes	NS	Bouzid, 1975.
C. sinensis	Protoplast	MT	None	-	-	None	NS	NS	Vardi et al., 1975.
C. sinensis x Poncirus trifoliata	Shoot-tip (S,M)	MS	-	BA	-	-	Yes	Yes	Barlass & Skene, 1982.
C. sinensis x P. trifoliata	Root (S)	MS	-	-	BA, 2,4-D	-	Yes	NS	Sauton et al., 1982.
C. limettoides	Stem (S)	MS + ME	2,4-D	-	BA, Kn, NAA	-	Yes	No	Raj Bhansali & Arya, 1979.
C. limon	Nucellus of Ft. Ovule	MS + Orange Juice	-	-	-	AS, NAA	Yes	Yes	Rangan et al., 1968, 1969.
C. limon	Nucellus of Ft. Ovule	MS + ME	-	-	-	None	Yes	NS	Bitters et al., 1972.
C. limon	Stem (S, M)	MS			BA, IBA	-	Yes	NS	Bouzid, 1975.
C. limon	Stem (S, M)	MS	-	-	BA, 2,4-D	-	Yes	NS	Sauton et al., 1982.
C. madurensis	Stem	MS	BA, 2,4-D	-	None	-	Yes	NS	Grinblat, 1972.
C. madurensis	Shoot-tip (s)	White	-	A/CW	-	-	NS	NS	Rangaswamy, 1975.
C. paradisi	Nucellus of Unft. Ovule	MT + ME	-	-	-	IAA, Kn	Yes	Yes	Kochba et al., 1972.
C. reticulata	Nucellus of	MS + Orange	-	-	-	AS, NAA	Yes	Yes	Rangan et al.,

Species	Growth Substance								Reference
	Explant	Medium	Callus	Proliferation of shoots	Regeneration Organogenesis	Embryogenesis	Formation of Plant	Plants to Soil	
	Unft. Ovule	juice							1968, 1969.
C. reticulata	Nucellus of Unft. Ovule	MS + ME	-	-	-	None	Yes	NS	Bitters et al., 1972.
C. reticulata	Shoot-tip (M)	MS	-	BA	-	-	No	No	Barlass & Skene, 1982.
C. temple	Nucellus of Unft. Ovules	MS + ME	-	-	-	None	Yes	NS	Bitters et al., 1972.
Poncirus trifoliata	Shoot-tip (S)	MS	-	BA	-	-	Yes	Yes	Barlass & Skene, 1982.
P. trifoliata	Root (S)	MS	-		BA, 2,4-D	-	Yes	NS	Sauton et al., 1982.
Coffea arabica	Anther	LS	2,4-D, CW	-	-	2,4-D, CW	Yes	Yes	Sharp et al., 1973; Sondahl & Sharp, 1979.
C. arabica	Leaf	MS	Kn, 2,4-D	-	-	Kn, NAA	Yes	Yes	Sondahl & Sharp, 1977.
C. arabica	Leaf (M)	MS	2,4-D, BA	-	-	NAA, Kn	Yes	Yes	Martinez & Madrigal, 1986.
C. canephora	Shoot	MS	Kn, 2,4-D	-	-	Kn, 2,4-D	Yes	Yes	Staritsky, 1970.
C. congensis	Leaf	MS	Kn, 2,4-D	-	-	Kn, NAA	Yes	Yes	Sondahl & Sharp, 1979.
C. dewevrei	Leaf	MS	Kn, 2,4-D	-	-	Kn, NAA	Yes	Yes	Sondahl & Sharp, 1979
Eucalyptus citriodora	Lignotuber	MS	NAA,CW	-	NAA, CW	-	Yes	No	Aneja & Atal, 1969.
E. citriodora	Cotyledon (S)	MS+B_5 Vit	Zn, IAA	-	Zn, IAA	-	No	No	Lakshmi-Sita, 1979.
E. citriodorn	Seedling	MS	-	BA NAA	-	-	Yes	No	Lakshmi-Sita & Vaidyanathan, 1979.
E. citriodora	Leaves, Stem	MS	Kn, BA, NAA	-	Kn, BA	-	Yes	Yes	Gupta et al., 1981.
E. citriodora	Embryo Seeds	B_5, 5% Sucrose	-	-	-	NAA	Yes	No	Muralidharan & Mascarenhas. 1987.
E. citriodora	Embryo	B_5	-	-	-	WPM	Yes	No	Muralidharan & Mascarenhas. 1987.
E. citriodora	Embryo	B_5+Glu.	-	-	-	CH, NAA	Yes	Yes	Muralidharan etal., 1989.
E. camaldulensis	Leaf (M)	WPM	CH, NAA, BA	-	CW, BA	-	Yes	No	Muralidharan & Mascarenhas. 1987.
E. diuersicolor	Seedling	MS	-	BA, NAA	-	-	Yes	No	Muralidharan& Mascarenhas. 1987.
E. grandis	Node (J,M)	MS	-	BA, NAA	-	-	Yes	No	Rao & Venkateshwara, 1985.
E. grandis	Internode	MS	2,4-D	-	BA, NAA	-	No	No	Lakshmi-Sita, 1986.
E. Grandis	Node (M)	MS	BA, NAA	BA, NAA	-	-	Yes	No	Lakshmi-Sita, 1986.
E. grandis	Node	MS	-	BA, NAA	-	-	Yes	No	Lakshmi-Sita & Shobha Rani.

Species	Growth Substance								Reference
	Explant	Medium	Callus	Proliferation of shoots	Regeneration Organogenesis	Embryogenesis	Formation of Plant	Plants to Soil	
									1985.
E. Leichow	Seedling	N	BA, NAA	-	BA, NAA	-	Yes	Yes	Ouyang et al., 1980, 1981.
E. marginata	Cotyledon	MS min. +B_5Vit.	Zn, IAA	-	Zn, IAA	-	NS	NS	Bennett & McComb, 1982.
E. marginata	Stamen	MS min. +B_5Vit.	Zn, IAA	-	Zn, IAA	-	Yes	Yes	Bennett & McComb, 1982.
E. marginata	Seedling	MSmin.+B_5Vit.	-	BA, NAA	-	-	Yes	Yes	Benntt & McComb, 1982.
E. sideroxylon	Node (M)	MS	-	BA, NAA	-	-	Yes	Yes	Burger, 1987.
E. sideroxylon	Bud	MS	2,4-D, NAA	Kn, BA	-	-	Yes	Yes	Arya & Shekhawat, 1986.
E. virdis	Bud	MS	2,4-D, NAA	Kn, BA	-	-	Yes	Yes	Arya & Shekhawat, 1986.
10 Species of Eucalyptus	Seedling	MS½ mineral without vitamins	-	BA, NAA	-	-	Yes	Yes	Hartney, 1981; Hartney & Barker, 1980.
4 species of Eucalyptus	Node (M)	MS minerals	-	BA, NAA	-	-	Yes	Yes	Dur&-Cresswell, et al., 1982.
3 species of Eucalyptus	Terminal Bud (M)	MS + Ca. pentothenate + Biotin	-	BA, Kn	-	-	Yes	Yes	Mascarenhas et al., 1982.
3 species of Eucalyptus	Seedling	Mod. Gresshoff and Doy	-	BA, IBA	-	-	Yes	Yes	Mehra-Palta, 1982.
Eucalyptus	Internode	MS + Glu.	NAA, BA	-	-	AC	NS	NS	Boulay, 1986.
3 species of Eucalyptus	Node	MS	-	BA, NAA	-	-	Yes	NS	Lakshmi Sita & Chattopadhyay, 1987.
Eugenia grandis	Internode (S)	MS	CW, IAA, Kn	-	-	-	No	No	Lee & Rao, 1979.
E. grandis	Hypocotyl (S)	MS	NAA	-	-	-	No	No	Lee & Rao, 1979.
E. grandis	Internode (S)	MS	NAA	-	-	-	No	No	Kong & Rao, 1982.
E. jambos	Immature Adventive Embryos	Mod. MS + Glu + a.a	2,4-D	-	-	2,4-D	No	No	Litz, 1984a.
E. malaccensis	Immature Adventive Embryo	Mod. MS + Glu + a.a.	2,4-D	-	-	2,4-D	No	No	Litz, 1984a.
Malus domestica	Meristem	Kmacro MS Micro	-	BA	-	-	NS	NS	Jones, 1967.
M. domestica	Meristem	MS	-	IBA, BA, GA3	-	-	NS	NS	Quioirin, 1974.
M. domestica	Shoot-tip	MS	-	NAA, Kn	-	-	NS	NS	Coffin et al., 1976.
M. domestica	Shoot-tip	MS	-	IBA, BA, GA3	-	-	NS	NS	Jones et al., 1967.
M. domestica	Anther	MS	-	-	-	IAA, Kn	No	No	Milewska-Pawliczuk & Kubicki, 1977; Kubicki et al., 1975, 1976.
M. domestica	Shoot-tip	MS	-	BA	-	-	Yes	Yes	Lundergan & Janick, 1980.

Species	Growth Substance								Reference
	Explant	Medium	Callus	Proliferation of shoots	Regeneration Organogenesis	Embryogenesis	Formation of Plant	Plants to Soil	
M. domestica	Shoot-tip	MS	-	GA3, BA, IBA	-	-	Yes	Yes	Snir & Erez, 1980.
M. domestica	Shoot-tip	LS	-	BA, IBA	-	-	Yes	Yes	James & Thurbon, 1981.
M. domestica	Shoot-tip	MS	-	BA, IBA GA3	-	-	Yes	Yes	Joung, 1982.
M. domestica	Internode	MS + CH	BA, NAA	-	-	BA, NAA	Yes	Yes	Cheema & Sharma, 1983.
M. domestica	Leaf Hypocotyl Cotyledon	MS + B5 Vit.	BA, NAA	-	-	BA, NAA	Yes	Yes	Liu et al., 1983a, b.
M. domestica Cv. Red Delicious	Cotyledon	NN + ME	None	-	-	BA	Yes	Yes	Kouider et al., 1984a.
Cv. Jonathan	Leaf	MS + B5 Vit.	2,4-D, BA	-	-	BA	Yes	Yes	Kouider et al., 1984b.
Cv. Jonathan	Protoplast	MS	2,4-D	-	-	-	No	No	Kouider et al., 1984c.
Cv. Green Sleeves	Protoplast	Kao	None	-	-	-	No	No	Doughty & Power, 1988.
M. pumila	Cotyledon	White	NAA, Kn CW, GA3	-	2,4-D, Kn	-	Yes	NS	Mehra & Sachdeva, 1979.
M. pumila	Stem	MS + CH	BA, NAA	-	BA	-	Yes	NS	Chen et al., 1979.
M. pumila	Root, Hypocotyl, Leaf, Cotyledon	MS CW, GA3	NAA, Kn,	-	- BA	B_N + NAA	Yes	NS	Mehra & Sachdeva, 1980
M. pumila	Nucellus	LS or MT	2,4-D	-	-	BA	Yes	Yes	James et al., 1984.
M. pumila	Leaf	MS	NAA, BA	-	NAA, BA	-	Yes	Yes	James et al., 1988.
2 sps. of Malus	Shoot-tip	White	-	BA	-	-	NS	NS	Nemeth, 1981.
sps. of Malus	Shoot-tip	MS	-	BA, NAA	-	-	Yes	Yes	Lane & McDougald, 1982.
sps. of Malus	Shoot tip	MS	-	BA, NAA	-	-	Yes	Yes	Wangstreet, 1982.
Mangifera indica	Cotyledon	MS	NAA, Kn, CW	-	-	-	No	No	Rao et al., 1982.
**M. indica (P)*	Nucellus	mod. MS + Glu + a.a.	CW	-	-	CW	No	No	Litz et al., 1982.
M. indica	Nucellus	mod. MS + Glu + a.a.	2,4-D	-	-	None	No	No	Litz et al., 1983.
**M. indica (M)*	Nucellus	mod. MS + Glu + a.a.	2,4-D	-	-	2,4-D	No	No	Litz, 1984; Litz et al., 1984.
M. indica	Nucellus	MS + CH	CW, Kn	-	-	CW, Kn	No	No	Sahijram & Doreswamy 1988, 1989.
**Myrciaria cauliflora (P)*	Immature Adventive Embryo	mod. MS + Glu + a.a.	2,4-D	-	-	2,4-D	No	No	Litz, 1984b.
Prunus amygdalus	Leaf, Coty., Embryo	MS	NAA, CW	-	NAA, Kn	-	Yes	NS	Mehra & Mehra, 1974.
P. amygdalus	Anther	Miller	2,4-D, IAA, Kn	-	-	-	No	No	Michellon et al., 1974.

Species	Growth Substance								Reference
	Explant	Medium	Callus	Proliferation of shoots	Regeneration Organogenesis	Embryogenesis	Formation of Plant	Plants to Soil	
P. amygdalus	Seedling Mod.	MS	-	BA, IBA	-	-	Yes	Yes	Hisajima, 1982.
P. amygdalus	Shoot-tip	Anderson + m. ino. + thiamine	-	BA, IBA, GA3	-	-	Yes	Yes	Kyriakidou & Pontikis, 1983.
P. amygdalus	Epicotyl, Root (J)	MS	-	-	BA, NAA	-	Yes	Yes	Rugini (Unpublished)
P. domestica (*Plum*) *7cv.*	Shoot-tip or Meristem	LS	-	BA, IAA	-	-	NS	NS	Constantine, 1978.
P. domestica	Root	Lepoivre	BA, GA3, 2,4-D	-	GA3	-	Yes	Yes	Druart, 1980, 1986.
P. domestica	Anther	S	IBA, BA	-	IBA, BA	-	Yes	Yes	Sierlis et al., 1979.
P. cerasifera (*Plum*)	Shoot-tip	MS	-	BA, IBA	-	-	Yes	Yes	Hammerschlag, 1982.
P. insitia (*Plum*)	Shoot-tip	MS	-	BA, IBA, GA3	-	-	Yes	Yes	Jones & Hopgood, 1979.
P. salicina (*Plum*)	Shoot-tip	MS	-	BA, IBA, GA3	-	-	Yes	Yes	Rosati et al., 1980.
P. armeniaca (*apricot*)	Shoot-tip	MS	-	BA, NAA	-	-	Yes	Yes	Skirvin et al., 1979.
P. armeniaca	Immature Cotyledon	MS	-	-	BA, 2,4-D	-	Yes	Yes	Lane & Cossio, 1986.
P. avium (*cherry*)	Shoot-tip	MS	-	BA, NAA	-	-	Yes	Yes	Skirvin et al., 1979.
P. avium	Shoot-tip	MS	-	BA, IBA, GA3	-	-	Yes	Yes	Jones & Hopgood, 1979; Snir, 1982.
P. avium	Root Callus	MS	-	-	BA, NAA	-	Yes	Yes	James et al., 1982.
P. cerasus	Shoot-tip	MS	-	BA, NAA	-	-	Yes	Yes	Skirvin et al., 1979.
P. cerasus	Anther	Miller + inositol.	2,4-D, BA	-	BA	-	Yes	Yes	Sierlis et al., 1979.
P. cerasus	Shoot-tip	MS	-	BA, IBA, GA3	-	-	Yes	Yes	Snir, 1983.
P. persica (*peach*)	Shoot-tip	MS	-	BA, NAA	-	-	Yes	Yes	Skirvin et al., 1979.
P. persica	Imm. Coty.	MS	-	-	BA	-	Yes	Yes	Lane & Cossio, 1986.
Psidium guajava.	Anther	MS	BA	-	-	-	-	-	Babbar & Gupta, 1986.
P. guajava	Node & Shoot-tip (M)	MS + PVP	-	BA	-	-	Yes	Yes	Jaiswal & Amin, 1986.
Punica granatum	Seedling	MS + N	NAA, Kn, CW	-	-	NAA, BA, CW	Yes	Yes	Mehra, 1986.
P. granatum	Leaf	MS	BA, NAA	-	-	BA, NAA	Yes	No	Omura et al., 1987.
Pyrus communis var Sativa	Coty., Stem, Root Leaf Embryo, Petal	White	NAA, BA, CW, GA3	-	-	NAA, BA	No	No	Mehra & Jaidka, 1980.
P. communis	Protoplast	MS (NH4+ free)	2,4-D, BA	-	-	-	-	-	Ochatt & Power, 1988.
P. communis 2cv.	Leaf	Lepoivre or 1/2 MS	-	-	TDZ, NAA	-	Yes	No	Chevreau et al., 1989.

Species	Growth Substance								Reference
	Explant	Medium	Callus	Proliferation of shoots	Regeneration Organogenesis	Embryogenesis	Formation of Plant	Plants to Soil	
P. bretschnei	Leaf	Macronutrients Lepoivre.	-	-	TDZ, NAA	-	Yes	No	Chevreau et al., 1989.
Santalum album	Hypocotyl	MS	BA	-	-	BA	No	No	Rao & Bapat, 1978.
S. album	Stem	MS	2,4-D, Kn.	-	-	GA3	Yes	NS	Lakshmi Sita et al., 1979.
S. album	Endosperm	MS	2,4-D	- ·	-	IAA	Yes	Yes	Rao & Bapat, 1980.
S. album	Hypocotyl	MS	2,4-D	-	-	IAA, BA	Yes	Yes	Rao & Raghava Ram, 1983.
S. accuminata	Stem	MS	-	-	BA	-	Yes	Yes	Barlass et al., 1980.
S. lanceolatum	Stem	MT	-	-	NAA	-	Yes	Yes	Barlass et al., 1980.
Syzygium aromaticum	Axillary bud.	MS	NAA, BA	-	-	-	Yes	Yes	Mathew et al., 1987.
S. cumini	Immature embryos	MS	2,4-D, BAP	BAP	-	-	Yes	NS	Mathur et al., 1991, 1993 Pareek et al., 1991.

Abbreviations used- Coty. = Cotyledon; Imm. = Immature; m. ino = meso inositol; Ft. Ovule = Fertilized Ovule; Unft. Ovule = Unfertilized Ovule; Vit. = Vitamins; Min. = Minerals; B_N = Nitsch medium; mod. MS = MS medium modified; (J) = Juvenile; (S) = Seedling; (M) = Mature; NS = Not specified.
*(P) = Polyembryonic; (M) = Monoembryonic.

REFERENCES

Abott, A.J. 1977. Propagating temperate woody species in tissue culture. Sci. Hortic. 28: 155-200.

Adams, R.M., Koenigsberg, S.S. and Langhans, R.W. 1979. In vitro propagation of butterwort *Pinguicula moanensis* H.B.K. HortSci. 14: 701-702.

Anderson, W.C. 1975. Propagation of rhododendrons by tissue culture. Part I. Development of a culture medium for multiplication of shoots. Proc. Int. Plant Prop. Soc. 25: 129-135.

Aneja, S. and Atal, C.K. 1969. Plantlet formation in tissue cultures from lignotubers of *Eucalyptus citriodora* Hook. Curr. Sci. 38: 69.

Arora, I.K. and Singh, R.N. 1978. In vitro plant regeneration in papaya. Curr. Sci. 47: 867-868.

Arya, H.C. and Shekhawat, N.S. 1986. Clonal multiplication of tree species in the Thar Desert Through Tissue Culture. Forest Ecology and Management. 16: 201-208.

Babbar, S.B. and Gupta, S.C. 1986. Induction of androgenesis and callus formation in vitro cultured anthers of a myrtaceous fruit tree (*Psidium guajava* L.). The Botanical Magazine, Tokyo 99: 75-83.

Bajaj, Y.P.S. 1984. Induction of growth in frozen embryos of coconut and ovules of citrus. Curr. Sci. 53: 1215-1216.

Bajaj, Y.P.S. 1986. Biotechnology of tree improvement for Rapid Propagation and Biomass Energy Production. In: Biotechnology in Agriculture and Forestry I. (Ed. Y.P.S. Bajaj) pp. 1-23, Springer-Verlag, New York.

Bapat, V.A. and Narayanaswamy, S. 1977. Mesocarp and Endosperm culture of *Achras sapota* Linn. in vitro. Indian Journal of Experimental Biology 15: 294-296.

Bapat, V.A. and Rao, P.S. 1984. Regulatory features for in vitro multiplication of sandalwood tree (*Santalum album* Linn.). I. Shoot bud regeneration and somatic embryogenesis in hypocotyl cultures. Proc. Indian Acad. Sci., Plant Sci., 93: 19-27.

Barlass, M. and Skene, K.G.M. 1982. In vitro plantlet formation from Citrus species and hybrids. Sci. Hortic. 17: 333-341.

Barlass, M., Grant, W.J.R. and Skene, K.G.M. 1980. Shoot regeneration in vitro from native Australian fruit bearing trees Quandog and plum bush. Aust. J. Bot. 28: 405-409.

Barlass, M., Skene, K.G.M., Woodham, R.C. and Krake, L.R. 1982. Regeneration of virus free grapevines using in vitro apical culture. Ann. Appl. Biol. 101: 291-295.

Bennett, I.J. and M.C. Comb, J.A. 1982. Propagation of jarrah (*Eucalyptus marginata*) by organ and tissue culture. Aust. For. Res. 12: 121-127.

Bini, G. and Bellini, E. 1973. Premiers resultats de la technique de l' embryoculture appliquee a l' amelioration genetique du Prunier. In: Gerstetter, R. (ed.) Genetique et amelioration du prunier. Acta Hortic. 48: 51-58.

Bitters, W.P., Murashige, T., Rangan, T.S. and Nauer, E.M. 1970. Investigations on established virus free plants through tissue culture. Cal. Citrus Nursurymen's Soc. 9: 27-30.

Bitters, W.P., Murashige, T., Rangan, T.S. and Nauer, E. 1972. Investigation on establishing virus free citrus plants through tissue culture. In: Proc. 5th Conf. Int. Organ Citrus Virologists (Ed. Pierce, W.C.). pp. 267-271. Univ. Florida Press, Gainesville.

Bonga, J.M. 1982. Vegetative propagation in relation to juvenility, maturity, and rejuvenation. In: Tissue Culture in Forestry (Eds. J.M. Bonga, and D.J. Durzan). pp. 4-35. Nijhoff, The Hague.

Bonga, J.M. 1987. Tree Tissue Culture Applications. In: Advances in Cell Culture (Ed. Karl Maramorosch) pp. 209-237. Academic Press, Florida.

Boulay, M. 1986. In vitro propagation of Forest Tree Species. In: VI. Intn. Congr. Plant Tissue and Cell Culture. pp. 9. Minnesota, USA, Abst.

Boulay, M. 1987. In vitro Propagation of Tree Species. In: Plant Tissue and Cell Culture (Eds. C.E. Green, D.A. Somers, and D.D. Beisboer). pp. 367-382. Alan R. Liss, Inc., New York.

Bouzid, S. 1975. Quelques traits du comportement de boutures de Citrus en culture in vitro. C.R. Acad. Sci. Ser. D. 280: 1689-1692.

Boxus, P. and Quoirin, M. 1977. Comportement en pepiniere d'arbres fruitiers issus de culture in vitro. Acta Hortic. 78: 373-379.

Boxus, P. and Druart, P. 1986. Virus free trees through tissue culture. In: Biotechnology in Agriculture and Forestry 1 (Ed. Y.P.S. Bajaj). pp. 24-30. Springer-Verlag, New York.

Broome, O.C. Land Zimmermann, R.H. 1984. Culture of Shoot Meristems: Fruit Plants. In: Cell Culture and Somatic Cell Genetics of Plants. (Ed. I.K. Vasil). pp. 111-121. Academic Press, INC., New York.

Burger, D.W. 1987. In vitro micropropagation of *Eucalyptus sideroxylon*. HortSci. 22: 496-497.

Button, J. and Kochba, J. 1977. Tissue culture in the Citrus industry. In: Applied and Fundamental Aspects of Plant, Cell, Tissue and Organ Culture. (Eds. J. Reinert and Y.P.S. Bajaj). pp. 70-92. Springer-Verlag, Berlin.

Chaturvedi, H.C. and Mitra, G.C. 1974. Clonal propagation of Citrus from somatic callus cultures. HortScience, 9: 118-120.

Chaturvedi, H.C., Sharma, A.K., Sharma, M. and Prasad, R.N. 1982. Morphogenesis, micropropagation and germplasm preservation of some economic plants by tissue culture. In: Plant Tissue Culture (Ed. A. Fujiwara) pp. 687-688. Tokyo.

Cheema, G.S. and Sharma, D.P. 1983. In vitro propagation of apple. In: Plant Tissue Culture in Crop Improvement (Eds. S.K. Sen and K.L. Giles), pp. 309-317. Plenum Press, New York.

Chen, C., Chen, F., Chien, C., Wang, C., Chang, S., Hsu, H., Ou, H., Ho, Y. and Lu, T. 1978. Obtaining pollen plants of *Hevea brasiliensis* Muell. In: Proc. Symp. Plant Tissue Culture. pp. 11-21. Peking.

Chen, C., Chen, F., Chien, C., Wang, C., Chang, S., Hsu, H., Ou, H., Ho, Y. and Lu, T. 1979. A process of obtaining pollen plants of *Hevea brasiliensis* Murell. Sci. Sin. 22: 81-90.

Chevre, A.M., Gill, S.S., Mouras, A. and Saleoses, G. 1983. In vitro vegetative multiplication of chestnut. J. Hortic. Sci. 58: 23-29.

Chevreau, E., Skirivin, R.M., Abu-Qaoud, H.A., Korban, S.S. and Sullivan, J.G. 1989. Adventitious shoot regeneration from leaf tissue of three pear (*Pyrus* sp.) cultivars in vitro. Plant Cell Reports. 8: 688-691.

Cohen, D. and Cooper, P.A. 1982. Mocropropagation of babaco- A Carica hybrid from Ecuador. In: Plant Tissue Culture (Ed. A. Fujiwara), pp. 743-744. Tokyo.

Coffin, R., Taper, C.D. and Chong, C. 1976. Sorbitol and sucrose as carbon source for callus culture of some species of *Rosaceae*. Can. J. Bot. 54: 547-551.

Constantine, D. 1978. In vitro multiplication of woody species. In: Round Table Conf. 6-8 June, pp. 134. Gembloux, Belgium.

Chornu, D. and Chaix, C. 1981. Multiplication par culture in vitro de merisiers adultes (*Prunus avium*): Application a un large eventail de clones. In: IUFRO Proc. Colloq. Int. la Culture in vitro des Essences Forstieres. AFOCEL, Nangis, France, pp. 71-79.

Debruijne, E., Delanghe, E. and Rijck Van, R. 1974. Action of hormones and embryoid formation in callus cultures of *Carica papaya*. In: Int. Symp. Fytofarm Fyiat, 26: 637-645.

Doughty, S. and Power, J.B. 1988. Callus formation from leaf mesophyll protoplasts of *Malus* x *domestica* Borkh cv. Greensleeves. Plant Cell Reports. 7: 200-201.

Druart, P. 1980. Plantlet regeneration from root callus of *Prunus* species. Sci. Hortic. 12: 339-342.

Druart, P. and Gruselle, R. 1986. Plum (*Prunus domestica*). In: Biotechnology in Agriculture and Forestry I. (Ed. Y.P.S. Bajaj), pp. 130-153. Springer-Verlag, New York.

Durand-Cresswell, R., Boulay, M. and Franclet, A. 1982. Vegetative propagation of *Eucalyptus*. In: Tissue Culture in Forestry. (Eds. J.M. Bonga and D.J. Durzan), pp. 150-181. Martin Nijhoff/Dr. W. Junk, The Hague.

Durzan, D.J. 1984. Special problems: Adult vs. juvenile explants. In: Handbook of Plant Cell Culture. Vol. 2. Crop species. (Eds. W.R. Sharp, D.,A. Evans, P.V. Ammirato and Y. Yamada) pp. 471-504. Macmillan, New York.

Durzan, D.J. and Campbell, R.A. 1974. Prospects for the mass production of improved stock of forest trees by cell and tissue culture. Can. J. For. Res. 4: 151-174.

Eiccholtz, D.A., Robitaille, H.A. and Hasegawa, P.M. 1979. Adventive embryony in apple. HortSci. 14: 699-700.

Fiorino, P. and Loreti, F. 1987. Propagation of Fruit Trees by Tissue Culture in Italy. Hortscience. 22: 353-358.

Gautheret, R. 1940. Novelles recherches sur le bourgeonnement du tissu cambial d'*Ulmus compestris* cultive in vitro. C.R. Acad. Sci. Paris. 210: 744-746.

Griffis, J.L.J., Hennen, G. and Oglesby, R.P. 1983. Establishing tissue cultured plants in soil. In: Proc. Int. Plant Prop. Soc. 33: 618-621.

Grinbalt, V. 1972. Differentiation of Citrus stem in vitro. J. Am. Soc. Hortic. Sci. 97: 559-603.

Grout, B.W.W., Shelton, K. and Pritchard, H.W. 1983. Orthodox behaviour of oilpalm seed and cryopreservation of the excised embryo for genetic conservation. Ann. Bot. 52: 381-384.

Gupta, P.K., Mascarenhas, A.F. and Jagannathan, V. 1981. Tissue culture of forest trees-clonal propagation of mature trees of *Eucalyptus citriodora* Hook by tissue culture. Plant Sci. Lett. 20: 195-201.

Hammerschlag, F.A. 1980. Peach micropropagation. In: Proc. Conf. on Nursery Production of Fruit Plants Through Tissue Culture- Applications and Feasibility. (Ed. R.H. Zimmermann). pp. 48-52.

Hammerschlag, F.A. 1982. Factors influencing in vitro multiplication and rooting of the plum rootstock myrobalan (*Prunus cerasifera* Ehrh.) J. Am. Soc. Hortic. Sci. 107: 44-47.

Hammerschlag, F.A. 1986a. Temperate Fruits and Nuts. In: Tissue Culture as a Plant Production System for Horticultural Crops. (Eds. R.H. Zimmermann, R.J. Griesbach, F.A. Hammerschlag and R.H. Lawson). pp. 221-236. Martinus Nijhoff, USA.

Hammerschlag, F.A. 1986b. Peach (*Prunus persica* L. Batsch). In: Biotechnology in Agriculture and Forestry I. (Ed. Y.P.S. Bajaj), pp. 170-182. Springer-Verlag, New York.

Hansen, A.J. and Lane, D. 1983. Inhibition of apple chlorotic leafspot virus in apple meristem tissue. Plant Dis. Rep. (in press).

Hartney, V.J. 1981. Vegetative propagation of eucalypts in vitro. Coll. Int. Cult. In vitro Essences Fo. IUFRO, Fontainebleau, France, AFOCEL. pp. 175-180.

Hartney, V.J. and Barker, P.K. 1980. The vegetative propagation of eucalyptus by tissue culture. Fast growing trees. IUFRO Congr., Brazil.

Hisajima, S. 1982. Multiple shoot formation from almond embryos. Biol. Plant. 24: 235-238.

Hoagland, D.R. and Arnon, D.I. 1950. The water-culture method for growing plants without soil. Cal. Agric. Exp. Stn. Circ. 347: 1-39.

Ibrahim, R.K., Thakur, M.L. and Permanand, B. 1971. Formation of anthocyanins in callus tissue culture. Lloydia. 34: 175-182.

Inchihashi, S. and Kako, S. 1977. Studies on clonal propagation of *Cattleya* through tissue method. II Browning of Cattleya. J. Jap. Soc. Hort. Sci. 46: 325-330.

Jaiswal, V.S. and Amin, M.N. 1986. *In vitro* shoot proliferation and plantlet formation from somatic tissues of guava (*Psidium guajava* L.) In: VI Int. Cong. of Plant Tissue and Cell Culture pp. 279. Abstracts.

James, D.J. and Thurbon, I.J. 1981. Phenolic compounds and other factors controlling rhizogenesis *in vitro* in the apple rootstocks. M. 9 and M.26. Z. Pflanzenphysiol. 105: 11-20.

James, D.J., Passey, A.J. and Malhotra, S.B. 1984. Organogenesis in callus derived from stem and leaf tissues of apple and cherry root stocks. Plant Cell Tissue and Organ Culture. 3: 333-341.

James, D.J., Passey, A.J. and Rugini, E. 1988. Factors affecting high frequency plant regeneration from apple leaf tissues cultured *in vitro*. J.Plant. Physiol. 132: 148-154.

Jones, O.P. 1967. Effect of benzyladenine on isolated apple shoots. Nature. 215: 1514-1515.

Jones, O.P. and Hopgood, M.E. 1979. The successful propagation *in vitro* of two root stocks of *Prunus*: the plum rootstock Pixy (*P. insitia*) and the cherry root stock F12/1 (*P. avium*). J. Hortic. Sci. 54: 63-66.

Jordan, M. 1975. *In vitro* Cultur on *Prunus, Pyrus* and *Ribes-anthern*. Plant Med. Suppl. 59-65.

Joung, H. 1982. Studies on the shoot tip culture of M7, M26 and M106 and M106 apple rootstocks. MS Thes, Dep. Hortic., Coll. Agric., Seoul Natl. Univ. Kor.

Kao, K.N. and Michayluk, M.R. 1974. A method for high frequency intergeneric fusion of plant protoplasts. Planta 115: 355-367.

Kartha, K.K., Leung, N.L. and Morginski, L.A. 1982. *In vitro* growth responses and plant regeneration from cryopreserved meristems of cassava (*Manihot esculenta Crantz.*) Z. Pflanzenphysiol. 107: 133-140.

Katano, M., Ishihara, A. and Sakai, A. 1983. Survival of dormant apical shoot tips after immersion in liquid nitrogen. HortSci. 18: 707-708.

Kobayashi, S., Ikeda, I. and Uchimaya, Y. 1985. Conditions for high frequency embryogenesis from Orange (*Citrus sinesis Osb.*) protoplasts. Plant Cell Tissue and Organ Culture. 4: 249-259.

Constantine D. 1978. In vitro multiplication of woody species. In: Round Table Conf. 6-8 June, pp. 134. gembloux, Belgium.

Chorunu, D. and Chaix, C. 1981. Multiplication par culture in vitro de merisiers adultes (*Prunus avium*): Application a un large eventail de clones. In: IUFRO Proc. Colloq. Int. La Culture "in vitro" des Essences Forstieres. AFOCEL, Nangis, France, pp. 71-79.

Debruijne, E. and Rijck Van R. 1974. Action of hormones and embryoid formation in callus cultures of *Carica papaya*. In: Int. Symp. Fytofarm Fyiat 26: 637-645.

Doughty, S. and Power, J.B. 1988. Callus formation from leaf mesophyl protoplasts of *Malus* x *domestica* Borkh cv.Greensleeves. Plant Cell Reports. 7: 200-201.

Druart, P. 1980. Plantlet regeneration from root vallus of *Prunus* species. Sci. Hortic 12: 339-342.

Druart, P. and Gruselle, R. 1986. Plum (*Prunus domestica*). In: biotechnology in Agriculture and Forestry I. (Ed. Y.P.S. Bajaj) pp. 130-153. Springer-Verlag, New York.

Durand-Cresswell, R., Boulay, M. and Franclet, A. 1982. Vegetative Propagation of *Eucalyptus*. In: Tissue Culture in Forestry. (Eds. J.M. Bonga and D.J. Durzan) pp. 150-181. Martin Nijhoff/Dr.W. Junk, The Hague.

Durzan, D.J. 1984. Special problems: Adult vs. Juvenile explants. In: Handbook of Plant Cell Culture. Vol. 2. Crop species. (Eds. W.R. Sharp D.A. Evans, P.V. Ammirato and Y. Yamada) pp. 471-504. Macmillan, New York.

Durzan, D.J. and Compbell, R.A. 1974. Prospects for the mass production of improved stock of forest trees by cell and tissue culture. Can. J. For. Res. 4: 151-174.

Eiccholtz, D.A. Robitaille, H.A. and Hasegawa, P.M. 1979. Adventive embryony in apple. Horsci. 14: 699-700.

Fiorino, P. and Loreti, F. 1987. Propagation of Fruit Trees by Tissue Culture in Italy. Hortscience. 22: 353-358.

Gautheret, R. 1940. Novelles recherches sur le bourgeonnement du tissue cambial D' Ulmus compestris cultive in vitro. C.R. Acad. Sci. Paris. 210: 744-746.

Griffis, J.L.J., Hennen, G. and Oglesby, R.P. 1983. Establishing tissue cultured plants in soil. In: Proc. Int. Plant Prop. Soc. 33: 618-621.

Grinbalt, V. 1972. Differentiation of Citrus stem in vitro. J. Am. Soc. Hortic. Sci. 97: 559-603.

Grout, G.W.W., Shelton, K. and Pritchard, H.W. 1983. Orthodox behaviour of oilpalm seed and cryopreservation of the excised embryo for genetic conservation. Ann. Bot. 52: 381-384.

Gupta, P.K., Mascarenhas, A.F. and Jagannathan, V. 1981. Tissue culture of forest trees-clonal propagation of mature trees of *Eucalyptus citrodora* Hook by tissue culture. Plant Sci. Lett. 20: 195-201.

Hammerschlag, F.A. 1980. Peach micropropagation. In: Proc. Conf. on Nursery Production of Fruit Plants Through Tissue Culture Applications and Feasibility. (Ed. R.H. Zimmermann). pp. 48-52.

Hammerschlag, F.A. 1982. Factors influencing in vitro multiplication and rooting of the plum rootstock myrobalan (*Prunus cerasifera* Ehrh.) J. Am. Soc Hortic. Sci. 107: 44-47.

Hammerschlag, F.A. 1986a. Teperate Fruits and Nuts. In: Tissue Culture as Plant Production System for Horticultural Crops. (Eds. R.H. Zimmermann, R.J. Griesbach, F.A. hammerschlag and R.H. Lawson) pp. 221-236. Martinus Nijhoff, USA.

Kobayashi, S., Ohgawara, T., Ohgawara, E., Oiyama, I. and Ishil, S. 1988. A somatic hybrid plant obtained by protoplast fusion between navel orange (*Citrus sinesis*) and satsuma mandarin (*C. unshiu*). Plant Cell Tissue and Organ Culture. 14; 63-70.

Kochba, J., Spiegel-roy, P. and Safran, H. 1972. Adventive plants from ovules and nucelli in Citrus. Planta 106: 237-245.

Kochba, J., Ben-Hayyim, G., Spiegel-Roy, P., Saad, S. and Neumann, H. 1982. Selection of stable salt tolerant callus cell lines and embroyos in *Citrus sinensis* and c. aurontium. z. pflanzenphysiol. 106: 111-118.

Kohlenbach, H.W. 1977. Basic aspects of differentiation and plant regeneration from cell and tissue cultures. In: Plant Tissue Culture and its Biotechnological Application (Eds. W. Garg, E. Reinhard and M.H. Zenk). pp. 355-366. Springer-Verlag, New York.

Kong, L.S. and Rao, Ar. N. 1982. In vitro plantlet development in tropical trees *Calaophyllum inophyllum* and *Euggenia grandis*. In: Tissue Culture of Economically Important Plants (Ed. A. N. Rao) pp. 185-190. COSTED Singapore.

Koiuder, M., Skirvin, R.M. Korban, S.S., Widholm, J.M. and Hauptmann, R. 1984a Adventitious shoot formation from `red Delicious' apple cotyledons in vitro. J. Hortic. Sci. 59: 295-302.

Kouider, M. Hauptmann, R. Widholm, J.M., Skirvin, R.M. and Korban, S.S. 1984b. Callus formation *Malus domestica* cv. `Jonathan' protoplasts. Plant cell Rept. 3: 142-145.

Kouider, M., Korban, S.S., Skirvin, R.M. and Chu, M.C., 1984c. Influence of embryonic dominance and polarity on adventitious shoot formation from apple cotyledon in vitro. J. Am. Soc. Hortic. Sci. 109: 381-385.

Krikorian, A.D. 1982 Cloning higher plants from aseptically cultured tissue and cells. Biol. Rev. 57: 151-218.

Kubicki, B., Teleznka, J. and Milewska-Pawliczuk E. 1975. Induction of embryoid development from apple pollen grains. Acta Soc. bot. Poloniae. 44: 631-635.

Kubicki, b., Telezynska, J. and Milewska-Pawliezuk, E. 1976. Induction of embryoid development from apple polles grains. Acta. Soc. Bot. pol. 44: 631-635.

Kyriakidou, R. and Pontikis, C.A. 1983. Propagation of peach-almond hybrid GF 677 in vitro. Plant Propag. 29: 13-14.

Lakshmi-Sita, G. 1979. Morphogenesis and plant regeneration from cotyledonary cultures of *Eucalyptus*. Plant Sci. Lett. 14: 63-65.

Lakshmi Sita, G. 1986. Progress towards the in vitro clonal propagation of Eucalyptus grandis. In: Plant Tissue Culture and its agricultural application. (Eds. L.A. Withers and P. g. Alderson). pp. 159-165. London.

Lakshmi Sita, G. and Shobha Rani, B. 1983. Preliminary studies on isolation and culture of protoplasts from sandalwood. Experientia 45: 4-5.

Lakshmi Sita, G. and Vaidyanathan, C.S. 1979. Rapid multiplication of *Eucalyptus* by multiple shoot production. Curr. Sci. 48: 350-352.

Lakshmi Sita, G. and Shobha Rani, b. 1985. In vitro propagation of Eucalyptus grandis L. by tissue culture. Plant Cell Rept. 4: 63-65.

Lakshmi Sita, G. and Chattopadhyay, S. 1987. Improvement of forest trees by tissue culture. In: Plant Cell and Tissue Culture of Economically important Plants (Ed. g.M.Reddy.) pp. 195-198. Hyderabad,

Lakshmi-Sita, G., Reghava Ram, N.V. and Vaidyanathan, C.S. 1980. Differentiation of embryoids and plantlets from shoot callus of sandalwood. Plant Sci. Lett. 15: 265-270.

Lane, D.W. and Cossio, f. 1986. Adventitious shoots from cotyledons of immature Cherry and Apricot embryos. Can. J. Plant Sci. 66: 953-959.

Lane, W.D. and Mc Dougald, J.M., 1982. Shoot tissue culture of apple: comparative response of five cultativars to cytokini and auxin. Can. J. Plant Sci. 62: 689-694.

Lee, S.K. 1984. Propagation of certain tropical trees. Ph. D. Thesis. National University of Singapore.

Lee, S.K. and Rao, A.N. 1979. Tissue Culture of certain tropical trees. In: Plant Cell Cultures: Results and perspectives. Abstract.

Lee, S.K., and Rao, A.N. 1980. tissue Culture of certain tropical trees. In: Plant Cell Cultures: Results and Perspectives (Eds. F. Sala, B. Parisi, R. Cella and O. Ciferri) pp. 305-31, Elsevier/ North-Holland Biomedical Press, Amsterdam.

Linsmiaer, E.M. and Skoog, F. 1965. Organic growth factor requirements of tobacco tissue cultures. Physiol. 19: 100-127.

Litz, R.E. 1984b. In vitro somatic embryogenesis from callus of jaboticaba, Myraciaria cauliflora. HortSci. 19: 62-64.

Litz, R.E. 1984c. In vitro somatic embryogenesis from nucellar callus of monoembryonic Mangifera indica L. HortScience 19: 715-715.

Litz, R. E. 1984d. Papaya. In: Handbook of plant cell culture, Vol. II. Crop species (Eds. W.R. Sharp. D. Evans, P. Ammirato and Y. Yamada) pp. 349-368. Macmillan, New York.

Litz, R.E. 1985. Somatic embryogenesis in tropical fruit trees. In: tissue culture in Agriculture and Forestry (Ed. R.R. Henke) pp. 179-193.

Plenum Press, New York.

Litz, R.E. 1987. Application of tissue culture to tropical fruits. In: Plant Tissue and Cell Culture (Eds. C.E. Green, D.A. Somers, W.P. Hackett and D.D. Beisboer) pp. 407-418. Alan R liss. New York.

Litz, R.E. and Conover, R.A. 1977. Tissue culture propagation of papaya Proc. Fla. State Hortie. Soc. 90: 245-246.

Litz, R.E. and conover, R.A. 1980. Somatic embryogenesis in cell cultures of *Carica stipulata*. Hortsci.15: 733-735.

Litz, R.E. and Conover, R.A. 1981. In vitro polyembryony in *Carica papaya* L. ovules. Z. Pflanzenphyiol. 104: 285-288.

Litz, R.E. and conover, R.A. 1982. In vitro somatic embryogenesis and plant regeneration from *Carica papaya* L. Ovules callus. Plant Sci. Lett. 26: 153-158.

Litz, R. E., Knight, K.J. and Gazit, S. 1982. Somatic embryos from cultured ovules of *Mangifera indica* L. Pl. Cell Rept. 1: 264-266.

Litz, R.E., O' Hair, S.K. and Conover, R.A. 1983. In vitro growth of *Carica papaya* L. Cotyledons. Sci. Hortic. 19: 287-293.

Litz, R.E., Knight, K.J. and Gazit, S. 1984. In vitro somatic embryogenesis from Mangifera indica L. Callus. Sci. Hortic. 22: 233-240.

Litz, R.E. Moore, G.A. and Srinivasan,C. 1985. In vitro systems for propagation and improvement of tropical fruits and Palms. Hort. Rev. 7: 157-195.

Litz, R.E., Jarret, R.L. and Asokan, M.P. 1986. tropical and subtropical Fruits and vegetables. In: Tissue Culture as a Plant Production system for Horticultural crops. (Eds. R.H. Zimmermann, R.J. Griesbach, F.A. Hammerschlag and R.H.Lawson). pp. 237-252. Martinus USA.

Lloyd, G. and Mccown, B.H. 1980. Commercially feasible micropropagagation of mountain laurel, Kalmia latifolia, By use of shoot tip culture. Proc. Int. Plant Prop. Soc. 30: 421-427.

Lundergan, C. and Janick, J. 1980. Regulation of apple shoot proliferation and growth in vitro. Hortic. Res. 20: 19-24.

Martinez, G. and Madrigal, R. 1986. In vitro propagation of Coffea *arabica* L Cv. *Catura amarelo*. In: VI Int. Congr. on Plant Tissue and Cell Culture. pp. 282. Abstracts.

Mascarenhas, A.F., Hazara,S., Potdar, U., Kulkarni, D.K. and Gupta D.K. 1982. Rapid clonal multiplication of Mature forest trees through tissue culture. In: Plant tissue culture (Ed. A. Fujiwara) pp. 719-720. Maruzen, Tokyo.

Mathew, K.M., Francis, M.S. and Hariharan, M. 1987. Development of Callus in cloves (*Syzygium aromaticum* (L.) Merr. & Perry) J. of Plantation Crops 15: 123-125.

Mc Comb, J.A. and Newton, S. 1981. Propagation of Kangaroo paws using tissue culture. J. Hortic. Sci. 56: 181-183.

Mc Comb, J.A. and Bennett, I.J. 1982. Vegetative propagation of *Eucalyptus* using tissue culture and its application to forest improvement in Western Australia. In: Plant Tissue Culture. (Ed. A.Fujiwara) pp. 721-722. Maruzen, Tokyo.

Mehra, P.N. 1986. Morphogenesis in Pomegranate (*Punica granatum*) and Chikoo (*Archras sapota*) In: VI Intl. Congr. on Plant Tissue and Call. Culture pp. 145. Abstracts.

Mehra, A. and Mehra, P.N. 1974. Organogenesis and Plantlet formation in vitro in almond. Bot. Gaz. 135: 61-73.

Mehra, P.N. and Sachdeva,. S. 1979. Callus cultures and organogenesis in apple. Phytomorphology 29: 310-324.

Mehra, P.N. and Sachdeva, S. 1980. In vitro plantlet formation through embryogenesis in apple. In: Plant Tissue Culture Genetic Manipulation and Somatic Hybridization of Plant Cells. (Eds. P.S. Rao, M.R. Heble and M.S. Chadha) pp. 295-300.

Mehra, P.N. and Jaidka, K. 1980. Experimental induction of embrygenesis in pear. In: Plant Tissue Culture Genetic Manipulation and Somatic Hybridization of Plant Cells. (Eds. P.S. Rao, M.R. Heble and Chadha, M.S.) pp. 301-305. BARC, Bombay.

Mehra-Palta, A. 1982. Clonal propagation of Eucalyptus by tissue culture Plant Sci. Lett. 26: 1-11.

Michelon, R., Hugard, J. and Jonard, R. 1974. Sur l' isolement de colonies tissulairesde pecher (Prunus Persica Bastch, cultivars Dixired et Nectared IV) et d' Amandier (*Prunus amygalus* stokes, cultivar Ai) a partied antheres cultivees in vitro. C.R. Acad.Sci. Ser.D. 278: 1719-1722.

Milewska-Pawliczuk, E. and Kubicki, B. 1977. Inducation of androgenesis in vitro in *Malus domestica*. Acta Hortic. 78: 271-276.

Monaco, L.C., Sondahl. M.R., Carvalho, A., Crocomo, O.J. and Sharp, W.R. 1977. Applications of tissue culture in the improvement of coffee. In: Applied and Eundamental Aspects of Plant Cell, Tissue and Organ Culture (Eds. J. Reinert and Y.P.S. Bajaj) pp. 109-129. Springer, Berlin.

Mott, R.L., and zimmermann, R.H. 1981. Trees. Environ. Exp. Bot. 21: 415-420

Mouras, A., Salesses.G. and Lutz, A. 1978. Sur l' utilisation des protoplastes en cytologie: amelioration d'une methode recente en vue de l' identification des chromosomes mitotiques des genres Nicotiana et prunus. Caryologia 31: 117-127.

Mullins, M.G. and Srinivasan, C. 1976. Somatic embroyos and plantlets from an ancient clone of grapevine (Cv. Cabernet Sauvington) by apomixis in vitro. J.Exp. Bot. 27: 1022-1033.

Muralidharan, E.M. and Mascarenhas, A.F. 1987. In vitro plantlet formation by organogenesis in E. camaldulensis and by somatic embryogenesis in *Eucalyptus citriodora*. Pl. Cell Rpt. 6: 256-259.

Muralidharan, E.M., Gupta, P.K. and Mascarenhas, A.F. 1989. Plantlet production through high frequency somatic embryogenesis in long term cultures of *Eucalyptus citriodora*. Plant Cell Rept. 8: 41-43.

Navarro, L., Llacer, G., Cambra, M., Arregui, J.M. and Juarez, J. 1983. Shoot tip grafting in vitro for elimination of viruses in peach plants (*Prunus persica* Bastsch.) Acta Hortic. 130: 185-192.

Navarro, L. and Juarez, J. 1977. Tissue culture techniques used in Spain to rocover virus free *Citrus* plants. Acta Hortic. 78: 425-435.

Nemeth, G. 1981. Adventitious root induction by substituted 2-chloro-3-phenyl-propionitriles in apple root stocks cultured in vitro. Sci. Hortic. 14: 253-259.

Nitsch, J.P. 1965. Culture in vitro de tissus de fruits. III. Mesocarp et endocarpe de peche. Bull. Soc. Bot. Fr. 112: 22-25.

Ochatt, S.J. and Power, J.B. 1988. Rhizogenesis in callus from conference pear (*Pyrus communis* L.) protoplasts. Plant Cell Tissue and Organ Culture. 13: 159-164.

O mura, M., Matsuta, N., Moriguchi, T. and Kozaki, I. 1987. Adventitious shoot and plantlet formation from cultured pomegranate leaf explants. Hort Sci. 22: 133-134.

Ouyang, Q., Li, Q.Q. and Peng, H.Z. 1980. Preliminary report on the development of embryoids from Eucalyptus. Acta Phytophysiol. Sin. 6: 429-432.

Ouyang, Q., Peng, H.Z. and Li, Q.Q. 1981. Studies on the development of embryoids from *Eucalyptus* callus.Sci. Sliv. Sin. 17: 1-7.

Pandey, R.M. and Rajeevan, M.S. 1982. Propagation of papaya through tissue culture. In: Proc. XXI. Int. Hortic. Congr. Hamburg pp. 1350.

Pareek et al 1991 Take this reference from the page where only 3 referencos (our own are given.

Phillip, V.J. and Schraudolf, H. 1981. Origin and development of embryos in suspension cultures of Rose apple (*Eugenia Jambos* L.) Beitr. Boil. pflanzen 62: 9-21.

Pena-Iglesias, A. and Ayosa, P. 1978. Shoot apex (meristem) micro grafting and indexing of infected grapevine varieties at the same time. Proc. 7th Meet ICVG. Niagara Falls. pp. 333-338.

Quoirin, M. 1974. Premiers resulats obtenus dans la culture in vitro du meresteme apical de sujets porte-greffe de pommier. Bull. Rech. Agron. 9: 189-192.

Quoirin, M. and Lepoivre, P. 1977. Improved medium for in vitro culture of prunus sp. Acta Hortic. 78: 437-442.

Rachie, K.O. and Lyman, J.M. 1981. Genetic Engineering for Crop Improvement. New York.

Radojevic, L.J. 1978. In vitro induction of androgenetic plantlets in *Aesculus hippocastanum*. Protoplasma 96: 369-374.

Raj Bhansali, R. and Arya, H.C. 1979. Organogenesis In *Citrus limettoides* (Sweet lime) callus culture. Phytomorphology 29: 97-100.

Rajeevan, M.S. and Pandey, R.M. 1986. Lateral bud culture of papaya (*Carica papaya* L.) for clonal propagation. Plant Cell Tissue and Organ Culture 6: 181-188.

Rangan, T.S. 1984. Clonal Propagation: Somatic embryos of citrus. In: Cell culture and Somatic Cell Genetics of Plants Vol. I. (Ed. I.K. Vasil) pp. 68-73. Academic Press, INC, New York.

Rangan, T.S., Murashige, T. and Bitters. W.P. 1968. In vitro initiation of nucellar embryos in monoembryonic Citrus Hort Sci. 3: 226-227.

Rangan, T.S., Murashige, T. and Bitters w.p. 1969. In vitro studies on zygotic and nucellar embryogenesis in *Citrus*. In: Proc. Ist Int. Citrus Symp. Vol. I (Ed. H.D. Chapmann) pp. 2250229. Univ. California, USA.

Rangaswamy, N.S. 1958. Culture of nucellar tissue of *Citrus* in vitroExperientia 14: 111-112.

Rangaswamy, N.S. 1975. Morphogenesis in explants of *Citrus microcarpa* Bunge grown in vitro. I. J. Exp. Biol. 1 3: 213-215.

ιo, P.S. 1965. In vitro induction of embryo proliferation in *Santalum album* L. Phytomorphology 15: 175-179.

Rao. P.S. and Rangaswamy, N.S. 1971. Morphogenetic studies in tissue culture of the parasite *Santalum album* L. Biol. Plant. 13: 200-206.

Rao, P.S. and Bapat, V.A. 1978. Vegetative propagation of sandalwood plants through tissue culture. Can. J. Bot. 56: 1153-1156.

Rao, P.S. and Bapat, V.A. 1980. Morphogenetic investigations on tissue and organ culture of sandal wood tree. In: Plant Tissue Culture, Genetic Manipulation and Somatic Hybridization (Eds. P.S. Rao. M.R. Heble and M.S. Chadha) pp. 206-215. BARC, Bombay.

Rao, P.S. and Raghava Ram, N.V. 1983. Propagation of sandalwood (*Santalum album* Linn.) using tissue and organ culture technique. In, Plant Cell Culture in Crop Improvement (Eds. S.K. Sen and K.L. Giles) pp. 119-124. Plenum Press, New York.

Rao, P.S., Heble, M.R. and Chandha, M.S.1980. Plant Tissue Culture Genetic Manipulation and Somatic Hybridisation of Plant Cells. BARC, Bombay.

Rao, A.N., Sin, Y.M. Kothagoda, N. and Hutchinson, J.F. 1982. Cotyledon tissue culture of some tropical fruits In: Tissue Culture of Economically Important Plants (Ed. A.N.Rao) pp. 124-137. COSTED, Singapore.

Reinert, J. and Bajaj, Y.P.S. 1977. Anther Culture Culture: Haploid production and its significance. In: Applied and Fundamental Aspectes of Plant Cell, Tissue and Organ Culture (Eds. J. Reinert and Y.P.S. Bajaj) pp. 251-264. Springerverlag, New York.

Reynolds, J.E. 1982. Vegetative propagation of palm trees. In: Tissue Culture in Forestry (Eds. J.M. Bonga and D.J. Durzan) pp. 182-207. Martin Nijhoff, The Hague.

Reynolds, J.F., and Murashige, T. 1979. Asexual embryogenesis in callus cultures of palms. In vitro 15: 383-387.

Rosati, P., Marino, G. and Swierezwski, C. 1980. In vitro propagation of Japanese plum (*Prunus salicina Lindl* cv. *Calita*). J. Am. Soc. Hortic. Sci. 105: 126-129.

Rugini, E. and Verma, D.c. 1982. Micropropagation and Cell suspensions of a difficult to propagate almond (*Prunus amygdalus* Batsch) Cultivar. In: Plant Tissue Culture (Ed. A. Fujiwara) pp. 741-742. Maruzen, Tokyo.

Rao, K.S. and Venkateshwara, R. 1985. Tissue Culture of forest trees: Clonal multiplication of Euclyptus grandis L. Plant Science 40: 51-55.

Sahijram, L. and Doreswamy. R. 1988. Morphogenesis in embryo and nucellus cultures of Mango. In Intl. Confr. Research in Plant Science and its relevance to future. pp. 173. abstracts.

Sahijram. L. and Doreswamy, R. 1989, Somatic embryogenesis from Mango embryos. In: Plant Tissue Culture Conference. pp. 39. Abstracts.

Sakai, A. and Nishiyama, Y. 1978. Cryopreservations of winter vegetative buds of hardy fruit trees in liquid nitrogen. Hort Sci. 13: 225-227.

Sauton, A. Mouras, a. and Lutz, A. 1982. Plant regeneration from Citrus root meristems. J. Hortic. Sci. 57: 227-231.

Schroeder, C.A. 1961. Some morphological aspects of fruit tissues grown in vitro. Bot. Gaz. 122: 198-204.

Shahin, E.A. and shepard, J.F. 1980. *Cassava* mesophyll protoplasts: Isolation, proliferation and shoot formation. Plant Sci. Lett. 17: 459-465.

Sharma, N.K. and skidmore, D.I. 1988. In vitro expression of partial resistence to *Phytophthora palmivora* by shoot culture of papaya. Plant Cell Tissue and Organ Culture: 14: 187-196.

Sharp, W.R., Caldas, L.S. Crocomo, O.J., Monaco, L.C. and Caravalho, A. 1973. Production of *Coffea arabica* callus of three ploidy levels and subsequent morphogenesis. Phyton 31: 67-74.

Shashi Mathur et al 1991 & 1993 Take these two referanes from a separate sheet where only 3 reterences (our own) are given.

Sierlis G., Mouras, A. and Salesses, g. 1979. In vitro culture of anthers and organ fragments. *Prunus annals* Del L. Amelior. Plant. 29: 145-162.

Skene, K.g.M. and Barlass, M. 1983. In vitro culture of abscissed immature *Avocado* empryos.Ann. Bot. 52: 667-672.

Skirvin, R.M. 1980. Fruit Crops. In: Cloning Agricultural Plants via In Vitro techniques (Ed. B.V. conger) pp. 51-139. Florida.

Skirvin, R.M., chu, M.C. and Rukan, H. 1980. Tissue culture of peach, sweet and sour cherry and apricot shoot tips. Proc. III State Hortie. 113: 30-38.

Snir, I. 1982. In vitro micropropagation of sweet cherry cultivars. HortSci. 17: 192-193.

Snir, I. 1983. A micropropagation system for sour cherry. Sci. Hortic. 19: 85-90.

Snir, I. and Erez, A. 1980. In vitro propagation of *Malling merton* apple root stocks. HortSci. 15: 597-598.

Sommer, H.E. and Brown, C.L. 1979. Application of tissue culture to forest tree improvement. In: Plant Cell and Tissue Culture Principles and Applications (Eds. W.R. Sharp, P.O. Larsen, E.F. Paddock and V.Raghavan) pp. 461-491. Ohio State Univ. Press, Columbus.

Sommer, H.E. and Caldas, L.S. 1981. In vitro methods applied to forest trees. In: Plant tissue culture, methods and applications in agriculture (Ed. T.A. Thorpe) pp. 349-358. Academic Press, London, New York.

Sommer, H.E. and Wetzstein, H.Y. 1984. Hardwoods. In: Handbook of Plant Cell Culture. Vol. 3. Crop Species. (Eds. P.V. Ammirato, D.A. Evans, W.R. Sharp and Y.Yamada) pp. 511-540. Macmillan New York.

Sondahl, M.R. and Sharp, W.R. 1977. High frequency induction of somatic embryos in cultured leaf explants of *Coffea arabica* L.Z. Pflanzenphysiol. 81: 395-408.

Sondahl, M.R. and Sharp, W.R. 1979. Research in *Coffee* species and applications of tissue culture methods. In: Plant Cell Tissue Culture Principles and Applications (Eds. W.R. Sharp, P.O. Larsen, E.f. Paddock and V.Raghavan) pp. 527-584. Ohio. State Univ. Press, Columbus.

Spiegel-Roy, P. And Kochba, J. 1973. Stimulation of differentiation in Orange (*Citrus senesis*) ovular callus in relation to irradiation of the media. Rad. bot. 13: 97-103.

Spiegel-Roy, P. and Kochba, J. 1980. Embrygenesis in *Citrus* tissue culture. In: Advances in iochemical engineering (Ed A. Fietcher) pp. 27-48. Springer, Berlin Heidelborg New York.

Spiegel-Roy, P. and Vardi, A. 1984. Citrus. In: Handbook of Plant Cell Culture. Vo. 3. Crop Species. (Eds. P.V. Ammirato, D.A. Evans, W.R. Sharp and Y.Yamada) pp. 355-372. Macminnan, New York.

Srivastava, P.S. 1982. Endosperm Culture. In: Experiemental Embryology of vascular Plants. (Ed. B.M. Johri) pp. 175-194. Springerverlag, New York.

Staritsky, G. 1970. Embryoid formation in callus cultures of *Coffee*. Acta Bot. Neerl. 19: 509-514.

Stevenson, J.H. and Harris, R.E. 1980. In vitro Plantlet formation from shoot tip explants of fuchsia hybrida cv. Swingtime. can. J.Bot. 58: 2199.

Teulieres, C., Feuillet, C. and Boudet, A.M. 1989. Differential characteristics of cell suspension cultures initiated from Eucalypture gunnii clones differing by their frost tolerence. Plant Cell Reports 8: 407-410.

Than-Tuyen, N.t. and de Guzman, E.V. 1983. Formation of pollen embryos in cultured anthers of coconut (*Cocos nucifera* L) Plant Sci. Lett. 29: 81-88.

Thomas, E. and street, H.E. 1972. Factors influencing morphogenesis in excised roots and suspension cultures of *Atropa belladona*. Ann. Bot. 36: 239-247.

Tisserat, B., Ulrich, J.M. and Frinkle, B.J. 1981. Cryogenic Preservation and regeneration of date palm tissue. HortSci. 16: 47-78.

Toledo, W., Hugard, J. and Jonard, R. 1980. Amelioration de la technique d' obtention deplants de semis de pecher (*Prunus persica* Bastsch) a partir de graines du cultivar de maturite precoce 'spring time'. C.R. Acad, Sci. Ser. D. 290: 539-542.

Tsay, H.S. and Su, C.Y. 1985. Anther culture of papaya (*Carica papaya* L.) Plant Cell Rpt. 4: 28-30.

Ulrich, J.M. Mickler, A., Finkle, B.L. and Karnosky, D.F. 1984. Survival and regeneration of American elm callus cultures after being frozen in liquid nitrogen. Can. J. For. Res. 14: 751-753.

Vardi, A., Spiegel-Roy, P. and Galun, E., 1975. Citrus cell culture: isolation of protoplasts, Plating densities, Effect of mutagens and regeneration of embryos. Plant Sci. Lett. 4: 231-236.

Vardi, A., Spiegel-Roy, P. and Galun, E. 1982. Plant regeneration from citrus protoplasts: variability in methodological requirements among cultivars and species. Theor. Appl. Genet. 62: 171-176.

Vanketeshwaran, S. and Gandhi, V. 1980. Protoplast isolation and culture of tree genera for biomass production. In: 2nd Intl. Congr. Cell Biol. pp. 641. Berlin.

Walkey, D.G.A. 1978. In vitro methods for virus elimination. In: Frontiers of Plant tissue culture. (Ed. T. A. Thorpe) pp. 245-254. Univ Calgary Press, Calgary.

Wang, C., Chu, Z.C., Sun, C.S. 1975. the induction of populus pollen plants. Acta Bot. Sin. 17: 56-59.

Wang, P.J. and Huang, L.c. 1974. Studies on the shoot meristem culture of *Allium sativum* L. Chinese Hort. 20: 79-87.

Wangstreet, A. 1982. In vitro inoculation of tissue culture propagated Malus shoot tips with *Erwinia amylovora*. M.S. Thes., Ohio State Univ., Columbus..

Watanabe, K., Ideta, M. and Takahashi, B. 1981. Studies on tissue culture of peach. 1. Effects of combinations of growth regulating substances and of organic matter on callus formation from peach anther cultured in vitro. Bull. Coll. Agric. Vet. Med. Nihon Univ. 38: 101-107.

Welander, M. 1986. Plant regeneration in leaf segments raised in vitro frommature fruit trees. In: VI Intil Congr. on Plant Tissue and Cell Culture pp. 267. Abstracts.

Winton, L. 1968. Morphogenesis in colonal propagation of woody plants In: Frontiers of plant Tissue Culture (Ed. T.A. Thorpe). pp. 243-264. Calgray.

Winton, L. 1971. Tissue culture propagation of *European aspen*. For. Sci. 17: 348-350.

Winton, L. and Stettler, R.F. 1974. Utilization of haploidy in tree breeding. In: Haploids in higher plants: Advances and potential (Ed. K. J. Kasha) pp. 259-273. Guelph.

Winton, L. and Huhtinen, O. 1976. Tissue culture of trees. In: Modern Methods in Forest Genetics. (Ed. J.P. Miksche) pp. 243-264. New York.

Yie, S. and Liaw, S.I. 1977. Plant regeneration from shoot tips and callus of papaya. In vitro. 13: 564-567.

Zaprometov, M.N. 1978. Enzymology and regulation of synthesis of polyphenols. In: Frontiers of Plant Tissue Culture (Ed. T.A. Thorpe) pp. 335-343. Canada.

Zdrujkovskaja-Richter, A.I. 1981. Embroyo cultures and development of new forms of plants. Diss, Univ. Moscow Abstracts.

Zimmermann, R.H. 1985. Tissue culture propagation of woody plants: large scale production. In: Tissue culture in Forestry and Agriculture (Ed. r. Henke. K.W. Hughes, M.J. Constantin and A. Hollander) pp. 165-177. Plenum Press, New York.

Zimmermann, R.H. 1986. Propagation of fruit, nut and vegetable crops Overview. In: Tissue Culture as a Plant Production System for Horticultural Crops. (Eds. R.H. Zimmermann, R.J. Griesbach, F. A. Hammerchlag and R.H. Lawson) pp. 183-200. Martinus Nijhoff, USA.